Magnetic Resonances in Biological Research

Magnetic Resonances in Biological Research

Edited by

CAFIERO FRANCONI

Laboratory of Molecular Spectroscopy
University of Cagliari

GORDON AND BREACH SCIENCE PUBLISHERS

New York   London   Paris

Preface

This book contains the reports of physicists, chemists and biologists working in the field of magnetic resonance spectroscopy applied to systems of biological interest. These reports were read at the Third International Conference on Magnetic Resonances in Biological Research, held at S. Margherita of Cagliari, Italy.

Among the radiofrequency spectroscopies, magnetic resonance spectroscopies play a very important role chiefly because they involve electron and nuclear spins and their associated magnetic moments. In fact many spin magnetic interactions are present at molecular level and the results of their study have been quite rewarding so that new fields of research are now fully developed regarding studies of structure, conformation and electronic distribution of molecules, molecular and latticesymmetry and dynamics.

Also molecules of biological interest together with biological systems were subjected to analysis by magnetic resonance spectroscopies and many significant data at molecular level of interest to biochemists have already been collected. In fact in recent years parallel developments have occurred in the understanding of both molecular biology and spin magnetic interactions, so that today—given also the tremendous improvement of the basic instrumentation—it is possible to describe many biologically important problems in molecular terms which can be studied by magnetic resonance spectroscopies. Thus the latter methods of investigation have become a unique tool for taking further steps toward a better definition of major problems such as the relationships between molecular properties and biological specificity.

It appeared that 1969 was the right year in which to collect and review the main results of the research in the field, since the previous meeting had occurred in 1966 and a three year gap seemed long enough for such a rapidly growing field. However, besides a review of recent developments, it seemed also opportune to have a first hand account of the attempts made toward new directions in order to have an assessment of its future, given also the large accumulation of data relative to new problems of molecular biology occurred in this period. These feelings have been shared by many scientists working in the field so that with their enthusiasm and cooperation it was a simple task to organize a conference on this subject.

The reviews and reports read at the conference and included in this volume concern chiefly studies on the interactions of small molecules with proteins,

ESR studies of metalloproteins and in particular of hemoglobin, NMR structural and conformational studies of polypeptides and proteins. It must at once be said that the papers read at the conference confirmed the first expectations that new horizons in molecular biology are being opened by the particular view points of the magnetic resonances methods.

The success of the conference was assured first by the close cooperation of the members of the scientific committee: M. S. Blois, W. E. Blumberg, O. Jardetzky, B. G. Malmström, B. Mondovì, W. D. Phillips and L. H. Piette, and by the active participation at the conference of the most distinguished experts in the field. To all of them we give our thanks.

The conference was supported by the Società Italiana di Scienze Farmaceutiche (S. I. S. F.); the Assessorato alla Pubblica Istruzione and the Assessorato alla Industria of the Ente Regione Autonoma della Sardegna (E. R. A. S.), the Varian S. p. A. of Turin and the Cagliari Chamber of Commerce. Special thanks are due to Prof. P. Pratesi, President of the S. I. S. F., to Dott. A. Giagu De Martini and Dott. P. Soddu of the E. R. A. S. and to Dott. G. Ferretti of Varian S. p. A., for their earnestness in supporting the conference.

Thanks are also due to Prof. B. G. Marini-Bettolo Director of the National Institute of Health and to Prof. A. Boscolo, President of the Board of Cagliari University for their cooperation.

CAFIERO FRANCONI

Molecular Spectroscopy Laboratory
University of Cagliari

Opening address

P. PRATESI

There are valid reasons for believing that study of the structure of proteins and of their interactions may lead to the formulation of valid hypotheses about the dynamics of the so-called "receptors"; that is, of those particular receptive chemical structures of the biological substrates that interact with drugs and are believed to be responsible for the specificity of the drugs themselves. Indeed, research on the molecular properties which govern the interaction between molecules of biologically active compounds and receptors, carried out on a large number of substances, has led to the conclusion that shape, molecular size and distribution of the electronic charge play a fundamental part in the said interaction, determining the possibility that the molecule of a biologically active substance reaches the critical distance for interaction with the receptor.

Magnetic resonance spectroscopies are powerful means for examining these processes at the protein level. This is the main reason why the Società Italiana di Scienze Farmaceutiche has enthusiastically supported Professor Franconi's proposal to organize this international conference in Italy.

In the name of the Società Italiana di Scienze Farmaceutiche I wish, therefore, to express our thanks to all those taking part in the conference, and our hope that this meeting will stimulate the interest of chemists and physicists in the study of the physico-chemical aspects of the action of drugs. We are particularly grateful to the rector of the University of Cagliari for his support. This meeting well expresses the scientific awareness of the chemistry world at Cagliari. My best compliments to Professor Franconi and to the molecular spectroscopy laboratory people of Cagliari University for having invited the most qualified scientific workers in the field; their presence here guarantees the success of this conference.

Contents

Intermolecular forces and conformational changes in protein-ligand interactions

GEORGE NÉMETHY and NORA LAIKEN

The Rockefeller University, New York

Abstract

Some current problems in the calculation of stable conformations of proteins are outlined, with a list of the factors entering into such calculations. A new model for the thermodynamics of transfer of alcohols from a hydrocarbon solvent into water is presented. It is derived as an extension of the Némethy–Scheraga theory for water structure. A partition function is written for the alcohol molecule in water, based on a distribution over four energy levels, corresponding to three, two, one or zero hydrogen bonds formed by the hydroxyl group. The statistical weighting for each level are derived from those for water in a self-consistent manner. For the alcohol in hydrocarbon, one energy state has to be considered. The theory matches the experimental free energies of transfer within 0.2 kcal/mole, the enthalpies within 0.4 kcal/mole at 25 °C. The empirical binding and interaction constants used in the description of ligand binding, conformational changes, and allosteric models are discussed in terms of free energy changes. The free energy condition on subunit interactions for positive and negative cooperativity is derived. It is shown that an apparently hyperbolic (Michaelis–Menten) ligand saturation curve can be obtained even with strong interactions between subunits for a "simplest sequential model" with equivalent subunits.

1

I Introduction

THE CONFORMATION of macromolecules in solution is determined essentially
by the balance of noncovalent interactions which act both between segments
of the macromolecule, and between these segments and solvent molecules
surrounding them. Proteins in their native state and polypeptides under
some physical conditions take up a well-defined, compact, and more or
less rigid shape, i.e. the chain has a definite conformation. The conformation
can be described by listing the values of the torsional angles of internal
rotation around single bonds of the molecule[1]. In proteins, there are two
such angles per residue, ϕ and ψ. The third angle, ω, corresponding to the
C–N peptide bond, generally is assumed to be fixed, due to the partial
double bond character of the peptide bond (Fig. 1). The conformation of the
amino acid side chains is described by a series of torsional angles denoted
by χ.

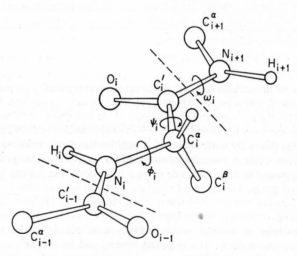

Figure 1 Two peptide units linked together in the fully extended conformation. The torsional angles of internal rotation, ϕ, ψ, and ω, defining the conformation, are shown. The
dashed lines delimit an amino acid residue[1]

Conformational changes occur frequently when protein molecules interact with each other (such as in subunit interactions) or with ligand molecules binding to them (substrates, inhibitors, activators). These conformational changes can be of great importance in biochemical functions, such
as enzyme specificity[2] or allosteric regulation.[3] The analysis of ligand bind-

ing and regulatory phenomena usually is carried out in terms of phenomeno-logical equilibrium constants.[4,5] However, the binding of ligands and con-comitant conformational changes are just as much the result of the balance of noncovalent interactions as is the establishment of stable conformations of the isolated macromolecule. A detailed understanding of binding pheno-mena, not yet available, would require a description of binding and inter-action equilibria in terms of local noncovalent interactions.[6]

II   The analysis of stable protein conformations

In recent years, great effort has been expended upon attempts to calculate stable conformations of proteins and polypeptides, as well as other macro-molecules, starting from fundamental physico-chemical principles.[7-9] The details of such computations as well as results obtained so far are discussed in several reviews.[10-11]

Most calculations are based on the fundamental hypothesis that the se-quence of the protein is sufficient to determine (given the specifications of the environment, such as temperature and solvent composition) the unique most stable conformation (or group of closely related conformations) of the macromolecule, i.e. the molecule takes up the conformation of lowest free energy. If the potential energy of the macromolecule is represented as an energy surface over a space of all the conformational parameters (torsional angles), then the most stable conformation ought to correspond to the point of lowest energy on the map (Fig. 2). In the past, most calculations were carried out by searching for a minimum of the potential energy. The existence of metastable minima, with energies above that of the most stable conforma-tion, may lead to false results. However, it must also be recognized that the most stable conformation may not correspond to that with the lowest po-tential energy.[12] Vibration around an equilibrium conformation, or equi-valently, the entropy contribution from a loosely held conformation may result in the stabilization of a conformation above the one having lowest potential energy (Fig. 3). Thus the overall free energy rather than the poten-tial energy must be minimized.[12]

The important interactions contributing to the conformational free energy are listed in Table I. The classification in the table is convenient for practical purposes of description or the subdivision of computations. In terms of fun-damental principles, the classes overlap. For example, all of the effects listed as group B in the table involve combinations of van der Waals and electro-static forces.

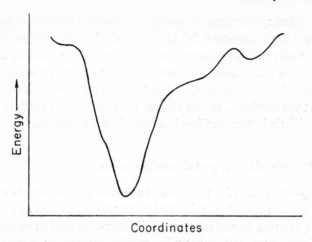

Figure 2   Schematic representation of the conformational potential energy of a protein as function of conformational parameters (atomic coordinates or torsional angles of internal rotation). While these parameters actually describe a multidimensional space, here they are symbolically represented by a single dimension on the abscissa. Theoretical calculations of stable protein conformations are directed towards the determination of the lowest point of the curve

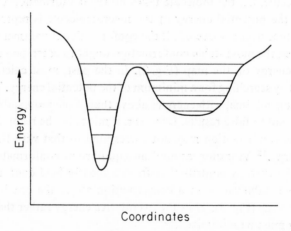

Figure 3   Comparison of a conformation of lowest potential energy but little internal freedom (narrow potential well on left) with a conformation of higher potential energy and high internal freedom (wide potential well on right). The spacing of vibrational energy levels is indicated schematically by the horizontal lines. A large vibrational partition function may stabilize the second conformation over the first one. Schematic representation as in Figure 1

Table I   Factors influencing the conformational stability of proteins

---

A. Effects intrinsic to the macromolecule

   1. Internal rotation about single bonds
   2. Distortion of covalent bond geometry
     (a) Bond angle bending
     (b) Torsion about the peptide bond
   3. Van der Waals ("nonbonded") interactions
     (a) Repulsion at short range ("steric" or "excluded volume" effects)
     (b) Attraction at close approach of atoms (London forces)

B. Effects depending strongly upon interactions with the solvent

   4. Hydrogen bonds
   5. Hydrophobic interactions in aqueous solution
   5a. Differential van der Waals interactions in nonaqueous solution
   6. Intra- and intermolecular dipole interactions
   7. Electrostatic charge interactions
   8. Preferential interaction with solvent components ("solvent binding")
   9. Specific binding of ligand molecules

In most of the computations performed so far, only the effects grouped under A, as well as *intramolecular* hydrogen bonds and dipole interactions, were considered. These are the interactions which can be described exactly as functions of the conformational parameters (torsional angles). Details of the potential functions used and of the mathematical techniques are found in the reviews cited.[10,11]

While the effects listed under B in Table I are of fundamental importance in the selection of stable conformations, their contributions are more difficult to handle. For some of the effects, theoretical models are available which correlate macroscopic thermodynamics with molecular parameters, using statistical thermodynamics. However, such an approach is not adaptable as directly to the energy calculations as are the formulations of the effects in Class A. For an exact calculation, a detailed knowledge of local liquid structure would be necessary. In the only attempt of including solvent interactions in the computations,[13] empirical functions based on averaged molecular interactions of the solvent were used.

A statistical thermodynamic model for hydrophobic interactions was developed several years ago[14], based on model theories for liquid water[15] and for the solution of hydrocarbons in water.[16] In the theory for water,[15] the presence of hydrogen bonds in the liquid was considered as the most important characteristic feature. The liquid was described in terms of a distribution of water molecules among states with various number of hydrogen bonds, assuming that the hydrogen-bonded molecules form compact clusters.*

In the application of the water model to solutions of hydrocarbons,[16] it was shown that the presence of the hydrocarbon causes a shift in the energy levels of neighboring water molecules. As a result, the equilibrium distribution of water molecules over the energy levels changes, resulting in an increase of hydrogen bonding and of ordering of the water. This is the source of the unfavorable free energy of solution. When hydrophobic interactions are established between nonpolar groups, the solution process is reversed. The source of the free energy of stabilization of such interactions is the decrease in ordering of water.

The theory of water structure was applied recently to an explanation of the thermodynamics of alcohol-water solutions.[18] The main features of the new theoretical model are summarized in Section III.

III   The thermodynamics of alcohols in aqueous and hydrocarbon solutions

In order to understand the thermodynamics of hydrogen bond formation between polar groups "buried" in the nonpolar interior of a protein molecule, it is necessary to describe the thermodynamics of transfer of a polar group from solution in a nonpolar solvent into aqueous solution. An empirical model for this process is given by the solution of aliphatic alcohols in hydrocarbons and in water, respectively.

We have derived a very simple model for the description of dilute solutions of alcohols in water,[18] based on the theories of water structure[15] and hydrocarbon solutions in water.[16] In the derivation of the model, consistency with that for water was maintained, with the introduction of the least possible number of new assumptions and parameters. Details of the derivation are published elsewhere.[18] The main physical features are summarized here.

The calculations were performed for alcohol solutions in water and in a hydrocarbon solvent at infinite dilution. Thus, any association of the solute can be disregarded. In the hydrocarbon solution, only van der Waals interactions operate between solvent molecules and the alkyl chain and hydroxyl

---

* Work is in progress (A. T. Hagler, G. Némethy, and H. A. Scheraga) on an improvement of the theory, eliminating some of the restrictive physical assumptions and avoiding some of the problems of the statistical mechanical formulation which came under criticism recently.[17]

group of the alcohol. The total interaction energy can be determined by summing over all pairwise contacts the interaction energies, using the standard formulae[10] for van der Waals interactions. A partition function is written for this single energy state, with the proper statistical weights for external and internal degrees of freedom.

In aqueous solution, the hydroxyl group can form hydrogen bonds with neighboring water molecules. A partition function for the alcohol molecule can thus be constructed in analogy with that derived earlier[15] for water.

It is assumed that G_{ROH}^{aq}, the total free energy of the alcohol in aqueous solution can be written as the sum of three contributions:

$$G_{ROH}^{aq} = G_{OH}^{aq} + G_R + \Delta G_W \tag{1}$$

Here G_{OH} is obtained from the partition function of the alcohol when its hydroxyl group interacts with water, G_R represents the interaction energy of the alkyl chain with the water molecules surrounding it, and ΔG_W is the change in the free energy of the layer of water surrounding the solute, due to structural changes occurring upon introduction of the solute molecule.[16] It should be noted that G_{OH} depends upon the nature and size of the alcohol carrying the OH group, so that eq. (1) should not be construed as representing the division of the total free energy into additive contributions by each functional group on the solute.

A hydroxyl group can form maximally three hydrogen bonds: two involving the lone pair electrons of the oxygen, and one involving the hydrogen. Thus, in analogy with the model for liquid water, one can consider the alcohol molecules as being distributed over four energy levels, E_i, where $i = 3, 2, 1$, or 0 indicates the number of hydrogen bonds formed with neighboring water molecules. In all states (except the tri-bonded one) there will be also some water neighbors to which the hydroxyl group is not hydrogen-bonded. For the completely non-hydrogen-bonded state, the average number of nearest neighbor water molecules (z_u) has been estimated as five, in analogy with the corresponding number ($z_u = 8$) for water.[15]

For the completely hydrogen-bonded state, $z_3 = 3$; z_i is assumed to change linearly for the intermediate states. (Fig. 4, water molecules indicated as b and c.) We assume that the intrinsic strength of the hydrogen bond and the van der Waals energy gained upon increasing by one the number of non-hydrogen-bonded neighbors are the same as for water,[15] namely, $\mathscr{E}_H = 3.57$ and $E_w = 2.25$ kcal/mole, respectively. Then the net energy of breaking a hydrogen bond in the liquid is $E_H = 2.07$ kcal/mole for the hydroxyl group (for water,[15] $E_H = 1.32$ kcal/mole). Thus E_H is derived from the theory for water and is not used here as an adjustable parameter. The partition function for the alcohol molecule therefore can be written as

$$Z_{OH}^{aq} = \sum_{\{x_i\}} g \prod_{i=0}^{3} [f_i^{aq} \exp(-E_i^{aq}/RT)]^{N_0 x_i^{OH}} \qquad (2)$$

In this equation, E_i^{aq} is the energy of the state with i hydrogen bonds $(E_i - E_{i+1} = 0.5 E_H)$, R is the gas constant, N_0 is Avogadro's number, T is the absolute temperature, the f_i^{aq} are statistical weighting factors, representing the contribution to the partition function from the translational and rotational degrees of freedom and of internal vibration of the alcohol molecule, and g is a combinatorial factor. The summation is taken over the possible distributions of the molecules over the four energy levels, subject to the restriction

$$\sum_{i=0}^{3} x_i^{OH} = 1 \qquad (3)$$

Most of the translational and rotational degrees of freedom of the molecule, as well as the internal rotation around the C—O bond, are restricted to vibrations due to dipole and steric interactions and hydrogen bonding. The frequencies for these vibrations were derived in a self-consistent manner from those assigned to water,[15] correcting for the differences in mass and moment of inertia, and for steric hindrance to rotation. The corrections depend on the size and mass of the group to which the hydroxyl group is attached. This is one of the sources of nonadditivity.

The partition function was evaluated in the usual manner by finding the maximum term. The free energy was then obtained as

$$G_{OH}^{aq} \approx A_{OH}^{aq} = -RT \ln Z_{OH}^{aq} \qquad (4)$$

It was assumed that the alkyl chain of the alcohol interacts with neighboring water molecules in the same manner as a hydrocarbon. Thus G_R and ΔG_w were obtained from the theory of hydrocarbon solutions.[16] ΔG_w is the free energy change for all water molecules in contact with the alkyl chain (molecules marked a and c in Fig. 4). The water molecules in contact with the hydroxyl group only (b in Fig. 4) were treated as identical with those in pure water[15] because the intrinsic strength of the hydrogen bond, $\mathscr{E}_H$, was taken to be the same for OH $\cdots$ H$_2$O and H$_2$O $\cdots$ H$_2$O hydrogen bonds.

The free energy of transfer from a hydrocarbon solution into water is then obtained as

$$\Delta G^0 = G_{ROH}^{aq} - G_{ROH}^{hc} \qquad (5)$$

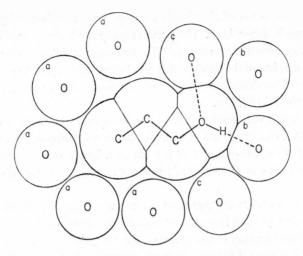

Figure 4 Schematic representation of an *n*-propanol molecule, surrounded by water molecules. The possible presence of hydrogen-bonded (dashed lines) and non-hydrogen-bonded neighbors of the hydroxyl group is indicated. For the explanation of the symbols *a*, *b*, and *c*, see the text

The computed parameters for the transfer of aliphatic alcohols from a hydrocarbon solvent into water are compared in Table II with the experimental values, compiled from several sources in recent publications.[19-26] In evaluating the extent of agreement with the experimental data, it should be

Table II Comparison of experimental and theoretical thermodynamic parameters[a] of transfer of alcohols from a hydrocarbon into water at infinite dilution (25 °C)

Alcohol	Observed			Calculated		
	ΔG^0	ΔH^0	ΔS^0	ΔG^0	ΔH^0	ΔS^0
Methanol	−2.44 to −2.37	−7.58 to −7.34	−17.5 to −16.5	−3.18	−8.03	−16.3
Ethanol	−1.68 to −1.58	−8.26 to −7.87	−22.4 to −20.4	−1.85	−7.76	−19.8
n-Propanol	−0.91 to −0.74	−8.11 to −7.58	−24.8 to −22.4	−0.83	−7.37	−21.9
n-Butanol	−0.02 to +0.11	−7.86 to −7.46	−26.8 to −24.9	+0.16	−7.04	−24.2

[a] ΔG^0 and ΔH^0 are in kcal/mole; ΔS^0 is in cal/deg · mole

noted that, while assumptions had to be made about the analogy of various parameters between water and the alcohols, no actual adjustable parameters were used in the present theory.

The errors are largest for methanol. Presumably, due to the small size of the alkyl chain, some of the simple assumptions made in the hydrocarbon theory[16] break down here. The parent theory for hydrocarbons also gave the worst agreement for methane.[16] The present calculations were not extended beyond *n*-butanol, because the treatment of longer alkyl chains requires several modifications of the computations, due to the possibility of folding of the alkyl chain.

The extent of hydrogen bonding of the hydroxyl group is higher than that of water. At 25 °C, the fraction of hydrogen bonds formed, x_{HB}, was found to be 0.62 for methanol and 0.59 for butanol. The corresponding fractions are 0.45 for pure water[15] and 0.57 for water molecules surrounding an alkyl chain.[16]

In order to calculate the thermodynamic parameters for the transfer of an OH-group on a protein from the nonpolar interior to aqueous environment (e.g. upon denaturation of the protein), the computation has to be modified to account for several factors, such as differences in mass and moment of inertia.

The contributions of the various functional groups to ΔG^0, ΔH^0, and ΔS^0 are not strictly additive. Thus it is not possible to give rigorously defined values for the "contribution of the hydroxyl group" to the thermodynamic parameters of transfer without detailed calculation. Nevertheless, the data in Table II allow a *crude* estimation of the magnitude of the contribution of the hydroxyl group if one subtracts the contributions of the alkyl chain, based on the theory for hydrocarbons[16] and for hydrophobic interactions.[14] In this manner, one obtains the approximate values $\Delta G^0 = -5.1$ to -4.7 kcal/mole, $\Delta H^0 = -7.1$ kcal/mole, $\Delta S^\circ = -7$ to -8.5 cal/deg. mole at 25 °C. The source of the favorable free energy of transfer of the hydroxyl group is the large negative enthalpy related to stable dipole interactions and to hydrogen bond formation in water. While the entropy of transfer is negative, due to the restrictions of the degrees of freedom upon hydrogen bonding, its contribution to the free energy is minor.

The model presented here should be useful in desribing the thermodynamic parameters of transfer from water to nonpolar media of other polar groups as well, including the N—H and C=O groups of the peptide units. However, several modifications will be necessary to allow for the differences in the strength and geometry of hydrogen bonding by these groups.

IV Conformational changes occurring upon ligand binding

The conformational change from A to B of a protein molecule P, due to the binding of a ligand S, corresponds to a distortion of the potential energy surface (cf. Fig. 2). Such a change implies that upon binding of the ligand, a formerly metastable conformation becomes the most stable one (Fig. 5).

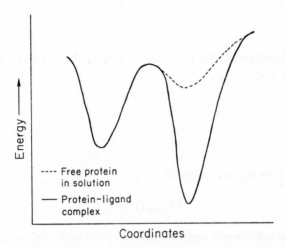

Figure 5 The distortion of the potential energy surface for protein conformations upon ligand binding accompanied by a conformational change. A schematic one-dimensional representation like that of Fig. 2 is used. Conformation A is on the left, B is on the right. On the right, the dashed line represents the potential energy surface before ligand is bound, the full line that after the binding of ligand. The difference between the potential well for A and the dashed potential well for B is $\Delta G_{conf}(P)$, the difference between the two potential wells shown by full lines is ΔG^0 (eq. 6)

The condition for a conformational change can be expressed in terms of free energies. If $G_A(P)$, $G_A(S)$, and $G_B(PS)$ represent the free energies of the protein and the ligand in their initial (non-interacting) states and that of the complex formed, respectively, then the overall free energy change for the binding process is given by

$$\Delta G^0 = G_{final}^0 - G_{initial}^0$$
$$= G_B(PS) - G_A(P) - G_A(S)$$
$$= \Delta G_{conf}(P) + \Delta G_{conf}(S) + \Delta G_{B,int}(PS) \qquad (6)$$

where $\Delta G_{conf}(P)$ and $\Delta G_{conf}(S)$ are the free energy changes for the conformational transitions of the protein and the ligand, respectively, when the two do not interact, and $\Delta G_{B,int}(PS)$ is the free energy of protein-ligand binding after both have undergone the conformational transition into state B. $\Delta G_{conf}(S)$ is included here for the sake of generality, to indicate that the ligand, too, may undergo a conformational change upon binding. This term will not be discussed here further.

The PS-complex will be stable if

$$\Delta G^0 < 0 \tag{7}$$

If the initial conformations (denoted as A) are to be the favored ones before binding, the conditions

$$\Delta G_{conf}(P) > 0 \tag{8}$$

and

$$\Delta G_{conf}(S) > 0 \tag{9}$$

must hold. This requires not only that

$$\Delta G_{B,int}(PS) < 0 \tag{10}$$

but that it is sufficiently negative to assure the satisfaction of relation (7).

The equilibrium between the two conformations of the protein is determined by the magnitude of $\Delta G_{conf}(P)$ or by ΔG^0 in the absence and the presence of ligand, respectively. If these free energy changes are sufficiently large, then the protein exists practically in only one conformation under the conditions specified. (As an arbitrary example, if the limit of detection of one conformation in the presence of the other is 1%, the free energy difference has to be at least 2.7 kcal/mole at 25 °C for the protein "to exist in one conformation only".) Thus whether conformational change upon binding can be described as the stabilization of a conformation not existing before, or as a shift in an already existing equilibrium, is not a fundamental distinction, but depends only upon the magnitude of the free energies.

The expression of the various free energies, occurring in eq. (6), in terms of elementary noncovalent interactions (Table I) would require, besides knowledge of the structure of the binding site, a description of the changes in local conformation. While no such analysis in terms of statistical thermodynamics could be performed so far, some of the principles of such calculations have been outlined earlier.[6]

V Conformational changes in multi-subunit proteins, allosteric models

In Sections V and VI, a protein E composed of n subunits will be considered. The notation E is used to differentiate this protein from the single-subunit protein discussed in Section IV. It is assumed that each of the n subunits can bind one molecule of the ligand S. If some or all of the subunits change their conformation upon binding of ligand molecules to one or more of the subunits, eq. (6) and the arguments of the preceding section still hold. However, the equivalent of $\Delta G_{conf}(P)$ now contains not only the free energy of conformational change of the subunit(s), but also changes in the free energy of interaction between neighboring subunits, caused by conformational changes. If this term varies with the extent of binding of ligand, positive or negative cooperative effects may be observed,[27] depending on whether the binding of the initial ligand molecule(s) makes it more or less favorable to bind more molecules of the ligand. In describing such cooperatice effects and allosteric interactions, usually two states, A and B, are assumed for the protein subunits,[4,5] with relative stabilities as indicated in the previous section. In the most general case, if ligand can bind to both A and B, though with unequal affinities[5] ("nonexclusive" binding), and if "mixed" conformations, such as $A_{n-i}B_iS_j$ can exist, a general binding scheme can be written as shown in Fig. 6. Special cases, in which some of the equilibria are considered as negligible, lead to simpler schemes, such as the assumption of the absence of "mixed" conformational states in the "concerted" transition model of

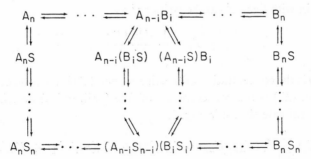

EQUIVALENT SUBUNITS
General binding equation

Figure 6 Possible forms of a protein composed of n subunits, each of which can exist in two conformations A and B. Each subunit can bind one molecule of ligand S. Binding occurs to both forms A and B. The most general scheme is shown here, corresponding to eq. (13)

Monod, Wyman, and Changeux[5], or the assumption that the conformation change $A \to B$ of each subunit must occur concurrently with the binding of a ligand to that subunit, as in the "simplest sequential model" of Koshland, Némethy and Filmer.[4] The equilibrium constants appearing in the descriptions of these models[4,5] can be correlated with free energy terms like those appearing in the preceding section, using the thermodynamic relation

$$\Delta G^0 = -RT \ln K \tag{11}$$

The nomenclature used by Koshland *et al.*[4] will be followed here, with some minor changes. The binding constant of ligand S to the subunit in conformation B is K_{BS} (K_s in ref. 4), and corresponds to $\Delta G_{B,\text{int}}(PS)$ in eq. (6). If conformation A can bind the ligand, the analogous terms are K_{AS} and $\Delta G_{A,\text{int}}(PS)$. The *intrinsic* equilibrium constant for the transformation $A \to B$ of a protein subunit was written as K_t (more specifically, K_{tAB}),[27] with free energy ΔG_{tAB}. This corresponds to $\Delta G_{\text{conf}}(P)$ only if there are no changes in inter-subunit interactions during the transformation. Otherwise $\Delta G_{\text{conf}}(P)$ must contain, in addition, terms reflecting the changing inter-subunit interactions. Taking an $A \cdots A$ subunit contact as the standard state (native conformation of the protein), the formation of an $A \cdots B$ subunit contact occurs with a free energy change ΔG_{AB}, with the corresponding equilibrium constant K_{AB}. For the formation of a $B \cdots B$ subunit contact one has ΔG_{BB} and K_{BB}. In general, for the formation of any "mixed" conformation $A_{n-i}B_i$ (or B_n as the extreme) $\Delta G_{\text{conf}}(P)$ will be the sum of $i \Delta G_{tAB}$ and of $-RT \ln \Phi_i$, which is a function of both ΔG_{AB} and ΔG_{BB}. The form of this function, written as $\Phi_i (K_{AB}, K_{BB})$ in equilibrium constant notation, depends on the nature of the arrangements of the subunits.[4] $\bar{Y}_s$, the fractional saturation of the protein with ligand[5] can be written[28] as

$$\bar{Y}_s = \frac{1}{n} \frac{d \ln (E_t)}{d(S)} \tag{12}$$

where (S) is the concentration of free ligand, and (E_t) is the total protein concentration. (E_0) is the concentration of free (unliganded) protein. For the most general case shown in Fig. 6,

$$(E_t) = (E_0) \sum_{i=0}^{n} \Phi_i (K_{AB}, K_{BB}) \cdot (K_{tAB})^i [1 + K_{AS}(S)]^{n+i} [1 + K_{BS}(S)]^i \tag{13}$$

Special cases can be derived from this general equation. For example, "exclusive binding" of the ligand to form B corresponds to setting $K_{AS} \ll 1$. The "concerted transition"[5] arises if "mixed" conformational states are assumed to be absent (with or without nonexclusive binding). This occurs if $K_{AB} \ll 1$ so that $\Phi_i = 0$ for all i except $i = 0$ and $i = n$. ($\Phi_0 = 1$ by definition.) In this case $[(K_{tAB})^n \, \Phi_n]^{-1} = L$, the "intrinsic allosteric transition constant" of Monod *et al.*[5] for the equilibrium $B_n \rightleftharpoons A_n$. The "simplest sequential model"[4] is obtained by assuming (a) $K_{AS} = 0$ and (b) that form B exists only when S is bound to it. The latter condition corresponds to the inequalities

$$\Phi_i \,(K_{AB},\, K_{BB}) \cdot (K_{tAB})^i \ll 1 \qquad (14)$$

and

$$\Phi_i \,(K_{AB},\, K_{BB}) \cdot (K_{tAB})^i \,(K_{BS})^i > 1 \qquad (15)$$

Haber and Koshland have shown[29] the distribution of protein molecules over the possible states for the various assumptions about the equilibrium constants discussed here. They also demonstrated how the various simple models represent special cases of the general binding equation. Some of the free energy relationships for subunit interactions have also been indicated schematically by Koshland.[30] The considerations presented here can be extended easily to equilibria involving more than two conformations of the protein.[29,30]

VI   The occurrence of various kinds of cooperativity

In the case of cooperative binding (more exactly, positive cooperativity), the binding of ligand promotes the binding of further molecules of the ligand. In the absence of cooperativity, the ligand saturation function $\bar{Y}_s$ is a hyperbolic function of the ligand concentration (S), corresponding to a Michaelis-Menten kinetic curve. Cooperative binding results in a sigmoidal curve shape (Fig. 7a). In terms of the previous discussion, a sigmoidal binding curve is obtained if $\Phi_i \,(K_{AB},\, K_{BB})$ is small for low values of i, but becomes larger for large values of i. Physically, this implies that the conformational change of one subunit facilitates similar changes in other subunits. (In the limiting case, all subunits must change their conformation simultaneously in the "concerted transition" model.) The ligand saturation function is steepened in its central portion.

It is also possible that $\Phi_i \,(K_{AB},\, K_{BB})$ is larger for low values of i than for high values. In this case, the binding of a few ligand molecules makes the binding of further ligand more difficult, giving rise to "negative cooperativity".[31] This results in flattening of the ligand saturation function, and possibly in the appearance of a plateau region (Fig. 7). The possibility of such binding curves was pointed out by Koshland *et al.*[4] on the basis of a

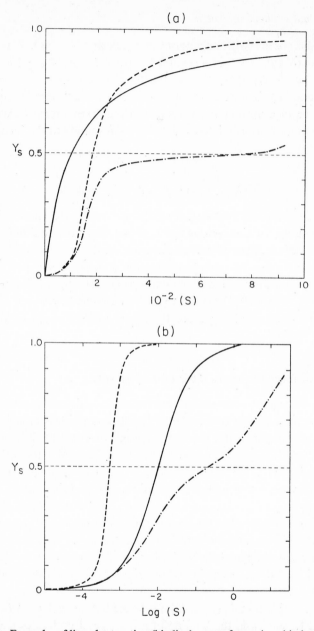

Figure 7 Examples of ligand saturation (binding) curves for various kinds of cooperativity. The fractional saturation of binding sites, $\bar{Y}_s$ is plotted against free ligand concentration (S) on (a) a linear, (b) a logarithmic scale. —— = zero cooperativity (Michaelis-Menten or hyperbolic binding, ---- = positive, -·-·- = negative cooperativity. Note that on the logarithmic scale (b), all curves have a sigmoidal character.

theoretical analysis of relationships between the various interaction constants. Recently, the actual occurrence of such binding curves has been confirmed.[31,32]

Usually, it is assumed that a hyperbolic (Michaelis-Menten type) ligand saturation curve indicates the absence of cooperative interactions, i.e. $K_{AB} = K_{BB} = 1$, between subunits. However, with sequential binding, a hyperbolic curve results for certain relationships between the subunit interaction constants K_{AB} and K_{BB}. Some examples were shown earlier (ref. 4, eqs. 40 and 41). A general derivation is given below.

If one assumes a simple sequential model, with ligand binding only to conformation B, then eq. (13) reduces to

$$(E_t) = (E_0) \sum_{i=0}^{n} \Phi_i (K_{AB}, K_{BB}) \cdot (K_{tAB})^i [K_{BS}(S)]^i \qquad (16)$$

The equation for a hyperbolic ligand saturation curve is

$$\bar{Y}_s = \frac{K_{app}(S)}{1 + K_{app}(S)} \qquad (17)$$

where K_{app} is the apparent equilibrium constant of binding S to the protein.

$$K_{app} = \frac{(ES)}{(E_0)(S)} \qquad (18)$$

By substituting eq. (16) in eq. (12) it is seen that eq. (17) can be satisfied only if eq. (16) is of the form

$$(E_t) = (E_0) [1 + K_{app}(S)]^n \qquad (19)$$

It follows that for the binding of i molecules of ligand, eq. (20) must hold.

$$\frac{(ES_i)}{(E_0)(S)^i} = \binom{n}{i} (K_{app})^i = \Phi_i (K_{AB}, K_{BB}) \cdot (K_{tAB})^i (K_{BS})^i \qquad (20)$$

This in turn requires that Φ_i be of the form

$$\Phi_i (K_{AB}, K_{BB}) = \binom{n}{i} [\Phi'_M (K_{AB}, K_{BB})]^i \qquad (21)$$

where Φ'_M is independent of i. (The subscript M indicates that we are seeking the form of Φ' corresponding to the Michaelis-Menten curve.) Equation (21) implies (a) that the subunits must be equivalent, since then $\binom{n}{i}$ is the number of ways of selecting i subunits out of a total of n, and (b) that the inter-subunit interactions change in a fashion which is independent of the extent of binding. The latter condition obviously hold for $K_{AB} = K_{BB} = 1$, i.e. for independent subunits, but it can also be satisfied by other choices of the K's.

Let us assume that each of the equivalent subunits interacts with v neighboring subunits ($v \leqslant n$), so that there is a total of $nv/2$ subunit interactions of all kinds, i.e. either $A \cdots A$, $A \cdots B$, or $B \cdots B$. Let there be now i subunits in the B form. They can be chosen in $\binom{n}{i}$ ways. For a particular choice, let there be $z(i)$ $B \cdots B$ neighbors and $y(i)$ $A \cdots B$ neighbors. The number of neighbors must obey the restriction

$$y(i) + 2z(i) = i \cdot v \tag{22}$$

Therefore, for the conversion of a particular set of i subunits from form A to form B, the total change in the free energy of subunit interactions is

$$\Delta G_i = -iRT \ln [\Phi'_M (K_{AB}, K_{BB})]$$

$$= y(i) \Delta G_{AB} + z(i) \Delta G_{BB}$$

$$= [i \cdot v - 2z(i)] \Delta G_{AB} + z(i) \Delta G_{BB}$$

$$= i \cdot v \cdot \Delta G_{AB} + z(i) [\Delta G_{BB} - 2\Delta G_{AB}]. \tag{23}$$

ΔG_{AB}, ΔG_{BB} and v are independent of i. Therefore Φ'_M will be independent of i (as required by eq. 21) only* if

$$\Delta G_{BB} = 2\Delta G_{AB} \tag{24}$$

or in terms of equilibrium constants,

$$K_{BB} = K_{AB}^2, \tag{25}$$

confirming the result obtained earlier.[4]

---

* Another possibility would be
$$z(i) = \text{constant} \cdot i$$
This solution is easily demonstrated to be false by showing ⸺ there exists no constant satisfying it: counting the number of $B \cdots B$ interactions, one finds

$$z(0) = z(1) = 0, \quad z(n-1) = v \cdot \left(\frac{n}{2} - 1\right), \quad z(n) = vn/2, \quad \text{etc.}$$

A hyperbolic curve can be obtained for any positive or negative value of ΔG_{BB}, even for large changes in the strength of subunit interactions, as long as eq. (24) is satisfied. This relationship means that the free energy is the same for the change $A \cdots A \rightarrow A \cdots B$ of neighboring subunit interactions as for the change $A \cdots B \rightarrow B \cdots B$. This implies physically that the number and kind of noncovalent interactions between the two subunits which are newly formed (or broken) must be identical during the two processes, as must be any changes in the conformational freedom of the subunits. The condition can be satisfied most easily if the contact area between *each pair of subunits* in both the $A \cdots A$ and the $B \cdots B$ interactions has a pseudo-symmetry axis (not necessarily implying symmetry *beyond* each contact area), similar to those indicated by Monod *et al.* for "isologous" association (ref. 5, fig. 9).

If the subunits are not equivalent or if the transition is not strictly sequential (e.g. in the "concerted transition"), no expressions analogous to eqs. (21) to (25) can be derived and it can be shown that there exists no combination of nonzero interaction free energies giving a hyperbolic ligand saturation curve.

For a sequential transition model with equivalent subunits eq. (24) also provides a criterion for the occurrence of various kinds of cooperativity, including hyperbolic binding as "zero" cooperativity in the phenomenological sense:

$$\left.\begin{array}{l} \text{Positive} \\ \text{Zero} \\ \text{Negative} \end{array}\right\} \text{cooperativity}: \quad \Delta G_{BB} \left\{\begin{array}{c} < \\ = \\ > \end{array}\right\} 2\Delta G_{AB} \qquad (26)$$

Equations (24) and (26) imply that the observation of a hyperbolic binding or kinetic curve in itself is not sufficient to prove that subunits bind ligands without conformation changes which alter inter-subunit interactions. A distinction would require additional, independent experimental information.

It cannot be said *a priori*, in the absence of detailed structural information, how likely is the occurrence of either one of the three kinds of cooperativity in eq. (26) for any protein system. Any analysis of the various free energies discussed in this section ($\Delta G_{AB}, \Delta G_{BB}, \Delta G_{tAB}, \Delta G_{BS}$, etc.) in terms of microscopic molecular interactions will require the prior completion of analyses indicated at the end of Section IV.

References

1   (a) The IUPAC-IUB Commission on Biochemical Nomenclature has recently adopted
     a set of abbreviation and symbols for the description of the conformation of poly-
     peptide chains. *J. Mol. Biol.* **52**, 1 (1970).
2   D. E. Koshland, Jr., *Cold Spring Harbor Symp. Quant. Biol.* **28**, 473 (1963).
3   J. Monod, J.-P. Changeux, and F. Jacob, *J. Mol. Biol.* **6**, 306 (1963).
4   D. E. Koshland, Jr., G. Némethy, and D. Filmer, *Biochemistry*, **5**, 365 (1966).
5   J. Monod, J. Wyman, and J.-P. Changeux, *J. Mol. Biol.* **12**, 88 (1965).
6   For an outlined approach to the problem, see: G. Némethy, in: *Conformation of
     Biopolymers*, G. N. Ramachandran, Ed., Academic Press, London 1967, p. 365.
7   P. de Santis, E. Giglio, A. M. Liquori, and A. Ripamonti, *J. Polymer Sci.*, **A1**, 1383
     (1963).
8   G. N. Ramachandran, C. Ramakrishnan, and V. Sasisekharan, *J. Mol. Biol.* **7**, 95
     (1963).
9   G. Némethy and H. A. Scheraga, *Biopolymers* **3**, 155 (1965).
10   H. A. Scheraga, *Adv. Phys. Org. Chem.*, **6**, 103 (1968).
11   G. N. Ramachandran and V. Sasisekharan, *Adv. Protein Chem.*, **23**, 283 (1968).
12   K. D. Gibson and H. A. Scheraga, *Physiol. Chem. and Physics*, **1**, 109 (1969).
13   K. D. Gibson and H. A. Scheraga, *Proc. Natl. Acad. Sci. U.S.*, **58**, 420 (1967).
14   G. Némethy and H. A. Scheraga, *J. Phys. Chem.*, **66**, 1773 (1962).
15   G. Némethy and H. A. Scheraga, *J. Chem. Phys.*, **36**, 3382 (1962).
16   G. Némethy and H. A. Scheraga, *J. Chem. Phys.*, **36**, 3401 (1962).
17   S. Levine and J. W. Perram, in *Hydrogen-Bonded Solvent Systems*, A. K. Covington
     and P. Jones, Eds., Taylor and Francis, Ltd., London, 1968, p. 115.
18   N. Laiken and G. Némethy, *J. Phys. Chem.*, **74**, 3501 (1970).
19   G. J. Pierotti, C. H. Deal, and E. C. Derr, *Ind. Eng. Chem.*, **51**, 95 (1959); American
     Documentation Institute, Document No. 5782 (1959).
20   H. C. Van Ness, C. A. Soczek, and N. K. Kochar, *J. Chem. Eng. Data*, **12**, 346 (1967).
21   H. C. Van Ness, C. A. Soczek, G. L. Peloquin, and R. L. Machado, *J. Chem. Eng. Data*,
     12, 217 (1967).
22   C. G. Savani, D. R. Winterhalter, and H. C. Van Ness, *J. Chem. Eng. Data*, **10**, 168
     (1965).
23   G. L. Bertrand, F. J. Millero, C. Wu, and L. G. Hepler, *J. Phys. Chem.*, **70**, 699 (1966).
24   D. M. Alexander and D. J. T. Hill, *Austr. J. Chem.*, **22**, 347 (1969).
25   E. M. Arnett, W. B. Kover, and J. V. Carter, *J. Am. Chem. Soc.*, **91**, 4028 (1969).
26   R. Aveyard and R. W. Mitchell, *Trans. Faraday Soc.*, **64**, 1757 (1968).
27   D. E. Koshland, Jr., these Proceedings, p. 21.
28   J. Wyman, *Adv. Protein Chem.*, **4**, 407 (1948).
29   J. E. Haber and D. E. Koshland, Jr., *Proc. Natl. Acad. Sci. U.S.*, **58**, 2087 (1967).
30   D. E. Koshland, Jr., *Adv. Enzyme Regulation*, **6**, 291 (1968).
31   A. Conway and D. E. Koshland, Jr., *Biochemistry*, **7**, 4011 (1968).
32   A. Levitzki and D. E. Koshland, Jr., *Proc. Natl. Acad. Sci. U.S.*, **62**, 1121 (1969).

A curve-fitting procedure for the study of the binding of small molecules to multi-subunit proteins

ATHEL CORNISH-BOWDEN, ROBERT A.COOK and
D. E. KOSHLAND, JR.

Department of Biochemistry, University of California

Introduction

PHYSICAL METHODS have become increasingly important in the elucidation of the structure of proteins, and particularly for determining the nature of the binding of small molecules to multi-subunit proteins. A classic example is the binding of oxygen to haemoglobin, which has been termed "cooperative", because the binding becomes progressively stronger as the proportion of bound ligand to protein increases. Various theories have been proposed to account for this behaviour in terms of the protein structure, among them the "symmetry" theory of Monod, Wyman and Changeux (1), and the "sequential" theory of Koshland, Némethy and Filmer (2). These theories both give an adequate fit to the binding data for haemoglobin and therefore the binding curve cannot be used to distinguish between them. Physical methods have proved appropriate to provide evidence that can distinguish in this case. Ogawa and McConnell (3) have used electron spin resonance, and Shulman and co-workers (4) have used nuclear magnetic resonance, both in the study of haemoglobin.

In the case of haemoglobin, the theoretical models which fit the data are relatively simple. However, some other proteins, such as glyceraldehyde-3-phosphate dehydrogenase (GPD) from yeast, behave in a way which cannot be reconciled with the simplest models, as we shall show, and in these cases curve-fitting procedures are needed to determine the correlation of the binding data and theoretical models. In such cases this type of curve-fitting is a necessary prelude to analysis of physical data such as NMR or electron spin resonance. We shall therefore describe this procedure and its application to yeast glyceraldehyde-3-phosphate dehydrogenase.

Since the Adair equation (equation 1) (5) for the binding of a small

$$N_s = \frac{K_1(S) + 2K_1K_2(S)^2 + 3K_1K_2K_3(S)^3 + 4K_1K_2K_3K_4(S)^4}{1 + K_1(S) + K_1K_2(S)^2 + K_1K_2K_3(S)^3 + K_1K_2K_3K_4(S)^4} \quad (1)$$

molecule to a four subunit protein is not a linear equation, and cannot be rearranged into linear form, the saturation curve for a multi-site protein cannot be analysed by the standard statistical techniques, which apply to linear equations. In order to analyse curves of this type, a curve-fitting procedure has been developed which is efficient, and which is sufficiently simple to be completely under the control of the user. The standard generalized procedures in the literature for analysing non-linear equations (6) suffer from the disadvantage that they are rather too complex to be readily manipulated without some advanced mathematical knowledge. Moreover, since it is not in general possible to estimate the standard errors of parameters determined for a non-linear equation, the general methods do not give measures of the standard errors, or any other indications of the precision of the determined values. The method which we have used, on the other hand, does give a measure of the standard errors of the parameters. The procedure and its application in a typical case are described here.

Methods

Glyceraldehyde-3-phosphate dehydrogenase was purified from Red Star brand baker's yeast following the procedure of Krebs (7). The enzyme appeared to be homogeneous when tested by cellogel electrophoresis in tris-borate buffer, pH 8.6 (8). Before equilibrium dialysis studies, crystals of the enzyme were centrifuged down and dissolved in and dialysed against 0.05 M sodium pyrophosphate buffer, pH 8.5, containing 0.001 M EDTA. Protein concentration was determined spectrophotometrically at 280 mμ using a molecular extinction coefficient determined by Krebs (7) of 1.35×10^5 (corrected for a molecular weight of 145,000).

NAD-^{14}C was prepared from nicotinamide-7-^{14}C (New England Nuclear) by enzymatic exchange as described by Colowick and Kaplan (9). The specific activity of the NAD-^{14}C thus prepared was 77,538 cpm/μmole and it was found to be free of ^{14}C-nicotinamide by cellogel electrophoresis. NAD was purchased from Boehringer und Söhne, Mannheim, Germany. Glyceralde-hyde-3-phosphate diacetal barium salt wes purchased from Sigma and converted to the free acid as described by Sigma.

Equilibrium dialysis was routinely carried out in 0.3 ml cells at 4°C or 25°C. Controls of NAD versus buffer indicated that equilibrium was reached in 12 hours. After equilibrium was reached, ligand concentration was determined on aliquots from each cell compartment. The enzyme was checked periodically for any denaturation during experiments by using a standard reaction mixture which contained 50 μmoles sodium pyrophosphate, pH 8.5, 10 μmoles sodium arsenate, 0.468 μmoles glyceraldehyde-3-phosphate and 0.650 μmoles NAD in 1.2 ml total volume.

Computations were performed on a Control Data Corporation 6400 Digital Computer, using a program written in CDC Fortran IV, copies of which are available on request.

Theoretical

The Adair equation (5) for a tetrameric protein (equation 1) can be written as equation 2, if the substitutions shown in equations 3–6 are made.

$$N_s = \frac{\psi_1(S) + 2\psi_2(S)^2 + 3\psi_3(S)^3 + 4\psi_4(S)^4}{1 + \psi_1(S) + \psi_2(S)^2 + \psi_3(S)^3 + \psi_4(S)^4} \tag{2}$$

$$K_1 = \psi_1 \tag{3}$$

$$K_2 = \psi_2/\psi_1 \tag{4}$$

$$K_3 = \psi_3/\psi_2 \tag{5}$$

$$K_b = \psi_4/\psi_3 \tag{6}$$

(We have found that if equation 1 is used, the curve fitting procedure is extremely slow, because the successive corrections in the K_i terms are highly correlated.)

The goodness of fit of the theoretical curve to the experimental points is determined using the criterion of the unweighted sum of the square deviations in N_s, i.e., the quantity ε as defined by equation 7. The curve fitting

$$\varepsilon = \sum (N_{s(\text{obs.})} - N_{s(\text{calc.})})^2 \tag{7}$$

procedure is not dependent on the definition of ε, however, so that in any particular experiment there is no reason why a different definition could not be used (e.g., one with weighting factors included). For this reason we shall describe the procedure in terms of ε, without further reference to its definition.

If three of the parameters in equation 2 are hold constant while the fourth, ψ_i, is varied, and the value of ε is calculated at each value of ψ_i, the plot ε versus ψ_i will have the general form of an inverted bell-shaped curve. It is clear that the value of ψ_i which gives a minimum value of ε for the particular values of the other three parameters can readily be estimated from a plot of this kind. However, since each measurement of ε is a laborious procedure, this would be a very time-consuming method. Instead, a much more rapid approximate method can be used, which requires only three measurements of ε. Since the portion of the plot close to the minimum resembles a parabola to a first approximation, it can be assumed that the curve is approximately described by equation 8 over at least part of the range. For this equation, it

$$\varepsilon = a\psi_i^2 + b\psi_i + c \tag{8}$$

may readily be shown that the relationship between ψ_{i1}, the value of ψ_i at the minimum of equation 8, and ψ_{io}, the value of ψ_i at some other point on the parabola, is given by equation 9, where the derivatives are measured at the

$$\psi_{i1} = \psi_{io} - \frac{\partial\varepsilon/\partial\psi_i}{\partial^2\varepsilon/\partial\psi_i^2} \tag{9}$$

starting value of ψ_i, ψ_{io}. In practice approximate values of the derivatives can be determined by calculating ε at two values of ψ_i close to ψ_{io}, and using equations 10 and 11. Since equation 8 is not an exact description of the curve,

$$\frac{\partial\varepsilon}{\partial\psi_i} \approx \frac{\Delta\varepsilon}{\Delta\psi_i} \tag{10}$$

$$\frac{\partial^2\varepsilon}{\partial\psi_i^2} \approx \frac{\Delta\,(\Delta\varepsilon/\Delta\psi_i)}{\Delta\psi_i} \tag{11}$$

it is clear that application of equation 9 will not locate the minimum of the function exactly, but in general the value of ψ_{i1} will be a much better approximation to the minimum than the starting value, and excellent results can be obtained if certain modifications are made.

A somewhat better approximation to a parabola is obtained if ε is plotted against the logarithm of ψ_i instead of ψ_i itself, and for this reason the formula shown in equation 12 generally gives a better approximation to the

$$\mathrm{Ln}\,\psi_{i1} = \mathrm{Ln}\,\psi_{io} - \frac{\partial\varepsilon/\partial\,\mathrm{Ln}\,\psi_i}{\partial^2\varepsilon/\partial\,(\mathrm{Ln}\,\psi_i)^2} \tag{12}$$

minimum than equation 9. Equation 12 has the added advantage that it prevents negative values from being assigned to ψ_i, which are physically meaningless, but which can arise in some instances.

If the starting value, ψ_{i0} is a very poor approximation to the minimum, the value of $\partial^2\varepsilon/\partial (\ln \psi_i)^2$ may be negative, in which case equation 12 will be grossly in error. The predicted value of ψ_{i1} will then be worse than the starting value, because the approximation parabola will be concave downwards, instead of upwards. In this case equation 12 cannot be used, and a fixed change in ψ_i is made, in the direction indicated by the sign of $\partial\varepsilon/\partial \ln \psi_i$. Thus, if $\partial\varepsilon/\partial \ln \psi_i$ is negative, $\ln \psi_i$ is increased by a fixed amount, and if $\partial\varepsilon/\partial \ln \psi_i$ is positive, $\ln \psi_i$ is decreased by a fixed amount. The amount of the change must be determined by trial and error, and we have found that satisfactory results are obtained with an increment of about 2.3 (Ln 10), provided that a check is made that the new value is in fact an improvement. In practice we make such a check even when equation 12 is used, and if the new value of ε is found to be worse than the starting value (which we have found to be very rare), a smaller change in the parameter is tried. Moreover, because very small positive values of $\partial^2\varepsilon/\partial (\text{Ln } \psi_i)^2$ will tend to predict grossly excessive changes in ψ_i, we set a limit of ± 2.3 on the amount of the change permitted in any one step for any parameter.

Repeated application of the iterative procedure described above can be used to minimize ε with respect to any one parameter. For minimization with respect to all four parameters in equation 2, there are two alternative methods which can be used: either the iteration formula can be applied sequentially to the four parameters until no further improvement is obtained, or the iteration formula can be used to calculate the necessary changes in the parameters, and then the four parameters are all changed simultaneously. Both of these methods work, and in the examples we have tried they give identical final results, but the sequential method produces somewhat faster convergence to the minimum, and consequently we have generally used it, with the parameters changed in the order $\psi_1, \psi_2, \psi_3, \psi_4, \psi_1,$ etc. However, in particular cases it may be that the other method is better, and it may sometimes be preferable to change the parameters in a different order. An outline of the method we have used for minimization with respect to four parameters is presented in Table 1.

The procedure described assumes that there is only one minimum in the ε profile. This is not necessarily the case, and data an be devised which would give rise to more than one minimum, in which case the minimization procedure would be expected to converge towards the nearest minimum, which

Table 1 Procedure used for fitting the Adair equation to binding data

The details of the method should not be regarded as fixed, as many modifications can be made to suit a particular problem. For example, the limit set for the change in a parameter in a single step, and the criteria for termination of the program are details which will be expected to vary considerably from problem to problem.

(1) Set starting guesses for the parameters. This can be done on the basis of a previous run on the computer, or from inspection of the experimental results, or most simply (but least efficiently) the starting guesses can be assigned by the computer on the basis of the value $S_{0.5}$, the value of (S) at half-saturation, using $\psi_1 = 4/S_{0.5}$, $\psi_2 = S_{0.5}^2$, $\psi_3 = S_{0.5}^3$, $\psi_4 = 1/S_{0.5}^4$, which correspond to a Michaelis-Menten curve.

(2) Set the iteration number to 0.

Start of calculation cycle

(3) Increase iteration number by 1.

(4) Using the current values of the parameters, calculate the value of N_s at each (S) value, compare it with the observed value, and sum the squares of the differences to get ε.

(5) Set $i = 1$, and ...

(6) Calculate the value of ε at two values of ψ close to the current value of ψ_i, and use the values of the small increments in ε to determine approximate values of $\partial\varepsilon/\partial \operatorname{Ln} \psi_i$ and $\partial^2\varepsilon/\partial (\operatorname{Ln} \psi_i)^2$ by first principles.

(7) Replace ψ_i with $\psi_i \exp (a)$, where $a = (\partial\varepsilon/\partial \operatorname{Ln} \psi_i)/[\partial^2\varepsilon/\partial (\operatorname{Ln} \psi_i)^2]$, unless a has an absolute value greater than 2.3, in which case its reduce its absolute value to 2.3, keeping the sign unchanged. If $\partial^2\varepsilon/\partial (\operatorname{Ln} \psi_i)^2 \leqslant 0$ do not use the formula, but put $a = \pm 2.3$, with a sign opposite to that of $\partial\varepsilon/\partial \operatorname{Ln} \psi_i$.

(8) Calculate ε at the new value of ψ_i, and compare the new value of ε with the old value. If $\varepsilon_{\text{new}} \leqslant \varepsilon_{\text{old}}$, proceed to step (9). Otherwise go back to step (7), but use a value of a which is half the value used previously.

(9) Set $i = 2, 3, 4$ in succession and carry out steps (6–8) for each value of i.

(10) Test for termination criteria. Those used in this work were:
(a) Every $\partial\varepsilon/\partial \operatorname{Ln} \psi_i$ term less than 0.001.
(b) Every parameter unchanged by more than 0.2% in the last iteration.
(c) ε less than 0.00001 × number of data points.
(d) Iteration limit exceeded.
If none of these is satisfied, go back to step (3).

End of calculation cycle

(11) Determine values of all the partial derivatives at the final values of the parameters as described in step (6). Assuming that $\partial\varepsilon/\partial \operatorname{Ln} \psi_i$ is small, we can write $\partial^2\varepsilon/\partial\psi_i^2 = 1/\psi_i^2 \cdot \partial^2\varepsilon/\partial (\operatorname{Ln} \psi_i)^2$, and substitute the value of $\partial^2\varepsilon/\partial\psi_i^2$ obtained into

$$\sqrt{\frac{2\varepsilon \sum (S_j)^2}{(n - p) (\partial^2\varepsilon/\partial\psi_i^2) \sum [(\bar{S}) - (S_j)]^2}} ,$$

where (S_j) are the set of substrate concentrations, $(\bar{S})$ is the mean substrate concentration, n is the number of data points, p is the number of parameters (4 in this case); to get a value for the standard error of each parameter.

(12) Print out results.

would not necessarily be the deepest. To avoid this danger, it is desirable to apply the minimization procedure to the same set of data several times with different starting guesses for the parameters. It is unlikely that this will be necessary in fact, however, because all of the sets of data which we have been able to devise which give rise to multiple minima are extremely implausible, and we believe that it is very unlikely that the problem will arise with real data.

Precision of estimates

Because the Adair equation is non-linear, the conventional methods for determining standard errors of the assigned values of the parameters are not applicable to it. However, this does not mean that a measure of the precision of the estimates is impossible. A true confidence region can be defined, which is a region around the minimum in which ε is less than some limiting value ε'. In the linear case the actual confidence level of this region, e.g., 95%, can be defined, but in the non-linear case it cannot. Moreover, the determination of the constant ε contour is not a simple procedure, since it follows no simple equation, and must therefore be constructed from a large number of measurements of ε.

A second difficulty which can arise in the linear case, and will usually be expected to arise in the non-linear case, is that the errors in the various parameters are correlated, i.e., the value of a parameter β_1 which gives a minimum value of ε varies with the value of another parameter β_2. In this case any measure of the precision of β_1 at constant β_2 is of little value as a measure of the true precision of β_1. In the case of the Adair equation, this problem is very severe if the equation is written as equation 1, but is far less so if it is written as equation 2. In other words, it is possible to define the values of the ψ_i terms much more accurately than the K_i terms. For this reason equation 2 is much to be preferred for curve-fitting purposes.

Another effect of correlation between parameters is that the minimization procedure is greatly slowed down. This is true even in the case of two parameter equations, such as equations 13 and 14, which describe, respectively, the

$$N_s = \frac{4K(S) [1 + K(S)]^3}{L + [1 + K(S)]^4} \tag{13}$$

$$N_s = \frac{4c^2 K(S) + 4c^2(c^2 + 2) K^2(S)^2 + 12c^2 K^3 (S)^3 + 4K^4(S)^4}{1 + 4c^2 K(S) + 2c^2 (c^2 + 2) K^2(S)^2 + 4c^2 K^3 (S)^3 + K^4(S)^4} \tag{14}$$

"symmetry" model of Monod, Wyman and Changeux (1) (with binding to one state only), and the "simplest sequential" model of Koshland, Némethy and Filmer (2) ("square" case). If typical cooperative data, e.g., for the binding of oxygen to haemoglobin, are fitted to these equations by the methods described, the minimization takes many iterations for equation 13, and the final values of K and L are highly correlated, but only a few iterations are needed for equation 14, and the final values of K and c are almost uncorrelated. It is clear, then, that in cases where the correlation between parameters is small, such as equations 2 and 14, the error in any parameter can validly be estimated with the other parameters held at their best fit values.

For a simple linear model, the standard error of a parameter is the increase or decrease in that parameter which raises the sum of squares, ε, by a factor ϕ, given by equation 15, in which n is the number of measurements,

$$\phi = \frac{\Sigma x_i^2}{(n - p) \Sigma (x_i - \bar{x})^2} \tag{15}$$

p is the number of parameters, and x_i is the set of observations corresponding to (S) in the specific case of a binding equation. As a working hypothesis, we shall assume that a standard error for the non-linear Adair equation can be defined similarly. In view of the fact that the plot of ε against any parameter is not a simple function of the parameter, this definition is difficult to apply exactly. If we make the same assumption used in the minimization, namely that the curve is approximated by a parabola, a simple definition of standard error can be used, which is given in equation 16. This definition cannot be

$$\text{s.e. } (\psi_j) = \sqrt{\frac{2\varepsilon \Sigma x_i^2}{(n - p) (\partial^2\varepsilon/\partial\psi_j^2) \Sigma (x_i - \bar{x})^2}} = \sqrt{\frac{2\varepsilon\phi}{(\partial^2\varepsilon/\partial\psi_j^2)}} \tag{16}$$

expected to be exactly accurate, but preliminary tests of it with synthetic data containing known errors suggest that it does predict the true error in the parameters almost as well as a more rigorous standard error. Further discussion on the validity of equation 16 will appear in a later paper.

Results

Studies of the binding of NAD to yeast GPD were carried out at 4°, 25° and 37°C. The results at 4°C were the best defined, presumably because very little denaturation of the enzyme occurred during equilibrium dialysis, and the results at the other temperature agreed qualitatively with those at 4°C. For this reason we shall only discuss the results at 4°C in detail.

A plot of N_s against $\log (NAD_{free})$ for the 4°C experiment is shown in Fig. 1, together with best-fit curves for the "symmetry" model, described by equation 13, the "square" model, described by equation 14, and for the general model, described by equation 2, with no restrictions on the values permitted for the parameters. It is clear that the first two models fit the data very poorly, but that equation 2 fits them closely, indicating that neither of the simple models can be regarded as adequate to describe the system. Two

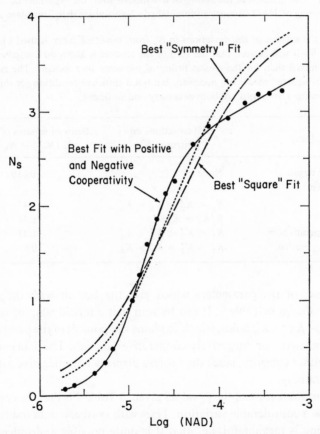

Figure 1 Comparison of experimental data for the binding of NAD to yeast glyceraldehyde-3-phosphate dehydrogenase at 4°C with best-fit curves for three models. Solid line indicates fit with the model which allowed a mixture of positive and negative cooperativity. Dotted line indicates best fit assuming the concerted model of Monod, Wyman and Changeux. Dashed line indicates best fit with the simplest sequential model using square geometry.

other models were also fitted to the data, but are not shown in Fig. 1, to avoid confusion. These both used equation 2 but with the restrictions (a) that the intrinsic binding constant* must increase progressively, i.e., $K_1' < K_2' < K_3' < K_4'$, and (b) that they must decrease progressively, i.e. $K_1' > K_2' > K_3' > K_4'$. A comparison of all five models is shown in Table 2, and it will be seen from the values of the sum of squares of errors that no simpler model approaches the general model in its ability to describe the data.

Table 2 Comparison of the ability of five models to fit the experimental data for the binding of NAD to yeast glyceraldehyde-3-phosphate dehydrogenase at 4°C

The sum of the squares of the deviations in N_s from the fitted curve is used to measure closeness of fit. It will be seen that the sum of the squares is about 40× smaller for the unrestricted model than for the closest fitting of the other four models. The restriction given for the "square" model is a necessary but not a sufficient condition for this model. In other cases the restrictions are both necessary and sufficient.

Model	Restrictions on intrinsic K's	Sums of squares of errors $\Sigma (N_{s_{obs}} - N_{s_{calc}})^2$
Positive and negative cooperative	None	0.019
Symmetry	$K_1 < K_2' = K_3' = K_4'$	1.47
"Square"	$K_1'K_4' = K_2'K_3'$	1.39
Positive cooperative	$K_1' < K_2' < K_3' < K_4'$	1.47
Negative cooperative	$K_1' > K_2' > K_3' > K_4'$	0.81

The values of the parameters which give the best fit with the general model are shown in Table 3. It can be seen that a relationship of the form $K_1' \ll K_2' > K_3' > K_4'$ holds, which explains the poor fit of the purely "positively cooperative" or "negatively cooperative" models. The data cannot be placed in either category, since the protein displays both negative and positive cooperativity.

An important feature of the ψ_i values given in Table 2 is that the calculated errors show a considerable variation. Tests with synthetic data indicate that this variation is meaningful, i.e., that it is quite possible and indeed usual for an experiment to yield much better measures of some parameters than

* The "intrinsic" association constants, K_i', are obtained from the measured constants, K_i, by dividing by the statistical factors, 4, $\frac{3}{2}$, $\frac{2}{3}$ and $\frac{1}{4}$, respectively. The K_i terms are the parameters in equation 1, and are obtained from the ψ_i terms using equations 3–6.

Table 3 Binding parameters for the binding of NAD to yeast glyceraldehyde-3-phosphate dehydrogenase at 4°C

The ψ terms are the parameters from equation 2, which were used in the minimization procedure; the K terms are the association constants for the four sites of the enzyme; and the K' terms are these association constants corrected for statistical factors.

Site	ψ	K	K'
1	$[1.84 \pm 0.47] \times 10^4$	1.84×10^4	4.60×10^3
2	$[3.82 \pm 0.31] \times 10^9$	2.08×10^5	1.39×10^5
3	$[1.89 \pm 0.09] \times 10^{14}$	4.95×10^4	7.42×10^4
4	$[1.66 \pm 0.20] \times 10^{17}$	8.78×10^2	3.51×10^3

Table 4 Investigation of the possibility that the curve-fitting procedure might converge to a "local minimum"

Starting values are the values given the computer to begin the procedure of Table 1. Final values are the values obtained by the computer at the termination point.

Essentially the same final values are obtained in every case, even though the starting values of the parameters were in error by factors as high as 3000. The conclusion is that in these cases at least there is only one combination of parameters which corresponds to a minimum in the sum of squares. The "data" used in this experiment were simulated to resemble the shape of the curve for the 4°C NAD-GPD experiment, but with much larger errors, i.e., a standard deviation of 0.124 from the "true" values, in order to provide the method with a more severe test than the real data offered. Similar experiments were carried out with the real data, with essentially identical results.

#	Starting values				Final values			
	ψ_1	ψ_2	ψ_3	ψ_4	ψ_1	ψ_2	ψ_3	ψ_4
1	100	10,000	1	0.01	0.28	34.7	191	1.17
2	0.01	1	10,000	100	0.25	34.8	190	1.19
3	100	1	1	100	0.28	34.7	191	1.17
4	0.01	10,000	10,000	0.01	0.25	34.8	190	1.19
5	100	1	10,000	0.01	0.29	34.7	191	1.17
6	0.01	10,000	1	100	0.25	34.8	190	1.19
7	1000	100	100	1	0.28	34.7	191	1.17
8	1	100,000	100	1	0.28	34.7	191	1.17
9	1	100	100,000	1	0.28	34.7	191	1.17
10	1	100	100	1000	0.28	34.7	191	1.17

of others. This is an important result because it implies that useful information about subunit interactions can be obtained from binding data even when it is not possible to assign definite values to all of the parameters.

All of the yeast GPD results were analysed several times with different starting guesses in order to be sure that the fitted curves were indeed the best that could be obtained. In all cases the same answers were obtained regardless of starting point, even when the starting guesses were very poor. In order to test the procedure more rigorously, a set of data was devised, similar in general characteristics to the GPD data, but with rather large deviations from the "true" values of N_s. Results are shown in Table 4. It will be seen that even with very large errors in the starting guesses, the procedure was able to reach the correct termination point in every case.

Discussion

The results which we have presented show that the binding of NAD to yeast GPD cannot be adequately described by the simplest models for the binding of small molecules to oligomeric proteins, but that it can be very accurately described by the general model for a four subunit protein, with no restrictions on the allowed values for the successive association constants. If any restriction is made, such as that the intrinsic constants must increase or decrease progressively, a very poor fit to the data is obtained. The assigned constants actually follow the relationship $K'_1 \ll K'_2 > K'_3 > K'_4$, which can be described as a mixture of positive and negative cooperativity. This appears to be a new concept in the study of multi-site proteins, and implies that there are at least some proteins which are by no means as simple as the simplest theories which restrict the protein to two symmetry states (1) or two conformational states of an individual subunit (2) would suggest. Since there is evidence that the four subunits of yeast GPD are identical, the explanation of the observed binding behaviour cannot lie in the amino acid sequences of the monomer units, but must be derived from the arrangement of the four subunits in the tetramer.

These results are of interest further because they show that it is possible for binding data alone to distinguish between theoretical models. It has sometimes been said that, because there are a number of models which fit some data equally well, the differences between models cannot be distinguished by binding behaviour. In the case of yeast GPD, however, the best-fit curves for the "symmetry" and "simplest sequential" models are not only quite different from one another, they are also quite different from the data, and thus the binding data alone are sufficient to reject these alternatives. A sequential mechanism involving more than two conformational states (10) would be compatible with the data and is suggested as a working hypothesis for this enzyme.

To summarize, we have shown that the binding of NAD to yeast glyceraldehyde-3-phosphate dehydrogenase cannot be reconciled with simple progressive changes in binding of multi-subunit proteins, but requires a model which permits a mixture of positive and negative cooperativity. Since more complex models of this kind require complex equations of state, i.e., equations relating structural features and binding constants, a curve-fitting procedure which can cope with several independent parameters is needed. Such a method has been developed and been used to fit the glyceraldehyde-3-phosphate dehydrogenase data. The method appears to be suitable for the fitting of complex equations of state in general.

Acknowledgments

This work was supported in part by research grants from the National Science Foundation (GB-7057) and the U.S. Public Health Service (AM-GM-09765).

References

1 J. Monod, J. Wyman and J.-P. Changeux, *J. Mol. Biol.*, **12**, 88 (1965).
2 D. E. Koshland, Jr., G. Némethy and D. Filmer, *Biochemistry*, **5**, 365 (1966).
3 S. Ogawa and H. McConnell, *Proc. Natl. Acad. Sci. U.S.*, **58**, 19 (1967).
4 R. G. Shulman, S. Ogawa, K. Wuthrich, T. Yamane, J. Peisach and W. E. Blumberg, *Science*, **165**, 251 (1969).
5 G. S. Adair, *J. Biol. Chem.*, **63**, 529 (1925).
6 N. R. Draper and H. Smith, *Applied Regression Analysis*, ch. 10, Wiley, New York, 1966.
7 E. G. Krebs, in *Methods in Enzymology*, Vol. I (S. P. Colowick and N. O. Kaplan, eds.), Academic Press, New York, 1955, p. 407.
8 S. H. Boyer, D. C. Fainer and M. A. Naughton, *Science*, **140**, 1288 (1963).
9 S. P. Colowick and N. O. Kaplan, in *Methods in Enzymology*, Vol. IV (S. P. Colowick and N. O. Kaplan, eds.), Academic Press, New York, 1957, p. 848.
10 A. Conway and D. E. Koshland, Jr., *Biochemistry*, **7**, 4011 (1968).

Nuclear magnetic resonance studies on selectively deuterated staphylococcal nuclease

O. JARDETZKY and J. L. MARKLEY

Merck Institute of Therapeutic Research

THE USEFULNESS of high resolution NMR in the solution of a wide variety of structural problems in protein and nucleic acid chemistry has been established beyond reasonable doubt. Particularly prominent among the established and potential applications are problems of the structure of binding sites for small molecules on macromolecules, conformational changes in proteins and nucleic acids and mechanisms of protein folding. A complete survey of the literature exceeds the purpose of this presentation, but is given in a review by G. C. K. Roberts and O. Jardetzky, "High Resolution NMR of Amino Acids, Peptides and Proteins", Vol. 24 of the *Advances in Protein Chemistry*, 1970, p. 447.

The principal purpose of the study we wish to report here was to demonstrate that a dramatic simplification of high resolution NMR spectra of proteins can be achieved by selective deuteration of the protein. The simplification is sufficient to allow complete assignment of all resonances in the aromatic region of the spectrum to specific amino acid residues in the protein, and to make extensive assignments in the aliphatic region. Thus it is possible to study the structure and behavior of a protein in solution in unprecedented detail. The method is uniquely well-suited for the definition of conformational changes and the delineation of small molecule binding sites.

The main findings resulting from this study to date were the following:

(1) *Staphylococcal Nuclease (Nase) exists in solution in two conformational forms (E and E') in slow equilibrium with one another. The conformational change is local and affects one His residue (H 2). The rate and position of the conformational equilibrium depend on the pH and the enzyme concentration. Calcium ion and pdTp bind to one conformational form (E') and shift the position and rate of the conformational equilibrium.*

These results follow from the existence of two NMR peaks for His residue H2 (H2a and H2b). These two peaks define two conformational forms of the enzyme: E, characterized by peak H2a; and E', characterized by peak H2b. Upper limits for the lifetime of the enzyme in either conformational form may be obtained from the NMR data. At pH 4.7 the lifetime of each of the two forms is longer than 40 msec. Values for the conformational equilibrium constant may be derived from the ratio of the areas of peaks H2b and H2a. The equilibrium constant varies from 0.5 at pH 4.7 to 3–5 at pH 7.5. Only conformational form E' is affected by pdTp or Ca^{2+} binding.

(2) *The pK values of the four His residues in several forms of Nase, Nase Foggi, Nase-V 8, and Nase-CT have been determined under a variety of conditions.*

Titration curves for the four His residues were obtained by plotting the chemical shifts of the C2-H peaks corresponding to these residues as a function of pH in D_2O. The pK values were determined from the inflection points of these curves. Since the chemical shifts from which the pK values were derived reflect the micro environments of the individual His residues these are microscopic pK values. For Nase Foggi they are 6.50 (H1), 6.55 (H2a), 5.80 (H2b), 5.75 (H3) and 5.55 (H4).

(3) *Nase and Nase-CT are conformationally very similar, but they differ with regard to both the rates and equilibrium value of the conformational equilibrium affecting His residue H 2.*

His residue H2 of Nase-CT (a derivative of Nase prepared by chymotrypsin cleavage) shows two separate peaks. The titration curves of these peaks do not exhibit exchange shifts in contrast to those of the uncleaved enzyme (Nase). Thus, the rate of the conformational equilibrium of Nase-CT is larger than peak H2b, indicating that the conformation having the "buried" His residue predominates throughout the pH region studied (pH 4.5–8.0). No other significant differences were observed between the other His peaks (H1, H2a, H3 and H4) or the aromatic envelopes of Nase-CT and Nase.

Since the chemical shifts of all these aromatic residues appear to be unchanged it is unlikely that there are major conformational differences between Nase-CT and Nase.

(4) *His peak H3 is assigned to His 124 of the Nase molecules.*

This assignment is based on a comparison of the His ring proton regions of NMR spectra of nuclease from the Foggi and V8 strains of *S. aureus*. Peak H3 is present in spectra of Nase, but is absent in spectra of Nase-V8. The assignment relies on the amino acid sequences of these two enzymes which indicate that residue 124 is His in Nase and Leu in Nase-V8 (Cusumano, Taniuchi, and Anfinsen, *J. Biol. Chem. 243*, 4769, 1968; Taniuchi, Cusumano, Anfinsen, and Cone, *J. Biol. Chem. 243*, 4775, 1968).

(5) *The Tyr residues of Nase exist in dissimilar chemical environments and have different pK values.*

The natures of the chemical environments of the seven Tyr residues of Nase are easily determined from the spectra of selectively deuterated enzyme analogs. The Tyr residues of these analogs are fully deuterated except for the C2 and C6 ring positions. Each Tyr residue is represented by a single peak in the spectrum. Only one Tyr residue has a chemical shift equivalent to free Tyr in solution in neutral pH. Two Tyr peaks lie downfield, and the remaining four lie upfield. True microscopic pK values for the Tyr residues of Nase cannot be obtained from the spectra because the enzyme denatures in the Tyr titration region. However, titration shifts occurring before the enzyme denatures may be followed. These indicate that Tyr residues Y2 and Y3 show "normal" titration behavior, whereas the other five Tyr residues are "buried", i.e., have abnormally high pK values in the native enzyme.

(6) *The binding of the inhibitor molecule pdTp to Nase both in the presence and absence of Ca^{2+} causes an additional local conformational change in the enzyme which affects Tyr peak Y4.*

This may be inferred from the anomalous binding curve for this tyrosine residue on plots of its chemical shift vs. inhibitor concentration.

(7) *Three Tyr residues of Nase are involved in binding the inhibitor, pdTp. One Tyr residue (Y5) binds near the 5' phosphate and another (Y1) near the 3' phosphate of pdTp. The other Tyr residue (Y2) probably binds to the ribose group of pdTp.*

This follows from the finding that normal binding curves may be obtained for three of the separately resolved tyrosine peaks.

(8) *The pK values of the 5'-3'-phosphate groups of pdTp bound to Nase have been determined. The values indicate that both phosphates contribute appreciable binding energy to the enzyme-inhibitor complex in the presence of Ca^{2+}. The pdTp trianion binds more strongly than the dianion; and the tetraanion binds more strongly than the trianion. The binding energy contributed by the 5' phosphate is approximately 2.3 times that contributed by the 3' phosphate.*

The pK values of the second 3'- and 5'-phosphate dissociations of pdTp are about 6.7. In the presence of Nase both pK values are lowered to around 6.5. Significant decreases in the pK values of pdTp are observed in the presence of both Nase and Ca^{2+}.

(9) *The structure of the pdTp-enzyme complex is different depending on whether Ca^{2+} ion is present or not.*

This is apparent from the differences in the changes observed in both the inhibitor and the enzyme spectrum under the two conditions.

(10) *The denaturation of Nase by extremes of pH, 8M urea, and heat appears to proceed in stages, and the resulting denatured structure depends upon the means of denaturation employed.*

This follows from the finding that the chemical shifts of different amino acid peaks change from the values found in the native protein to the values found in the fully denatured protein at different temperatures or different concentrations of the denaturing agent or, in the case of pH denaturation, at different pH values.

References

1   Markley, J.L., Putter, I., and Jardetzky, O. "High-Resolution Nuclear Magnetic Resonance Spectra of Selectively Deuterated Staphylococcal Nuclease". *Science*, **161**,1249, (1968).
2   Markley, J.L., Putter, I., and Jardetzky, O. "An analysis of the Aromatic Region of the Proton Magnetic Resonance Spectrum of Staphylococcal Nuclease". *Z.anal. Chem.*, **243**, 367 (1968).
3   Putter, I., Barreto, A., Markley, J.L., and Jardetzky, O. "Nuclear Magnetic Resonance Studies of the Structure and Binding Sites of Enzymes. X. Preparation of Selctively Deuterated Analogs of Staphylococcal Nuclease". *Proc. Nat. Acad. Sci.*, **64**, 1396 (1969).
4   Putter, I., Markley, J.L., and Jardetzky, O. "Nuclear Magnetic Resonance Studies of the Structure and Binding Sites of Enzymes. XI. Characterization of Selectivity Deuterated Analogs of Staphylococcal Nuclease." *Proc. Nat. Acad. Sci.*, **65**, 395 (1970).
5   Markley, J.L., Williams, M., and Jardetzky, O. "Nuclear Magnetic Resonance Studies of the Structure and Binding Sites of Enzymes. XII. A Conformational Equilibrium

in Staphylococcal Nuclease Involving a Histidine Residue." *Proc. Nat. Acad. Sci.*, **65**, 645 (1970).

6   Williams, M., Markley, J.L., and Jardetzky, O. "Nuclear Magnetic Resonance Studies of the Structure and Binding Sites of Enzymes. XIII. Identification of the Histidine Residues." *Proc. Nat. Acad. Sci.*, in press.

7   Markley, J.L., and Jardetzky, O. "Nuclear Magnetic Resonance Studies of the Structure and Binding Sites of Enzymes. XIV. Inhibitor Binding to Staphylococcal Nuclease." *J. Mol. Biol.*, **50**, 223 (1970).

A review of E.S.R. studies on haemoglobin and its derivatives

D. J. E. INGRAM

Physics Department, University of Keele

1 Introduction

ELECTRON SPIN RESONANCE INVESTIGATIONS on haemoglobin, myoglobin and their various derivatives, have now been undertaken for over ten years, and it therefore seemed appropriate that a general review of the way in which these studies were initiated and have progressed, might be of interest. This paper therefore starts by considering the basic reasons why electron resonance was attempted on haemoglobin and its various derivatives in the first place, and how these investigations then developed into both a detailed study of the structure of the molecule itself, as well as of the electronic state of the iron atom in its various configurations.

Some of the more recent work is also described towards the end of the article, including the measurements that are now being made at very short wavelengths, in order to obtain accurate values for the electronic splitting parameters, and also the systematic study of linewidths at these high frequencies, since their analysis suggests that there are interactions present which are still not yet fully understood.

2 The binding of the iron atom

In common with other biochemically important molecules, such as catalase and peroxidase, both haemoglobin and myoglobin have their iron atoms contained at the centre of the haem or porphyrin plane. The iron atom itself is surrounded by a square of four nitrogen atoms with the rest of the haem plane extending from these, as indicated in Fig. 1. The fifth coordination point below the iron atom then links with the polypeptide chains and the rest of the protein part of the molecule, while the sixth coordination point above

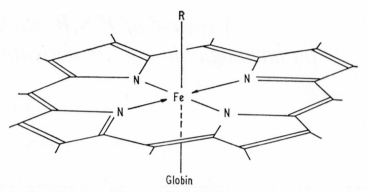

Figure 1 Central structure of haemoglobin molecule. The iron atom is in the centre of a square of nitrogens which form part of the haem, or porphyrin plane

the haem plane, is occupied by the oxygen, carbon monoxide, or other sub-stituted group. It is this particular bond which is intimately associated with the actual respiratory process, and the initial investigations by electron spin resonance were undertaken to obtain detailed information on this particular feature of the molecule. As is often the case with research, however, it transpired that some of the most interesting information actually came from quite a different feature of the molecule, and the electron resonance has not only been able to provide details on the nature of the chemical bonding, but also very significantly help in the elucidation of the structure of the molecule itself.

Before discussing the electron resonance signals observed from haemoglo-bin and its derivatives, some consideration must be given to the actual states of the central iron atom, and the energy levels associated with these. The iron atom can exist in either a ferrous or ferric state, and, to a first approx-imation, either of these can be considered as bound ionically, or covalently. There are thus four ways in which the outer electrons of the iron atom may be considered as occupying the $3d$ and $4s$ orbitals, and these are illustrated in Table 1. In this table each of the $3d$ and $4s$ orbitals is indicated as a circle, and each electron as an arrow, with its spin aligned up or down. The ferrous Fe^{++} ion has six electrons in its $3d$ shell to distribute amongst these orbitals, while the ferric Fe^{3+} has five.

Table I   Occupied electron orbitals and resultant total spin states of the ionic and covalently bound derivatives of ferrous and ferric ions

		Electron configuration				spin
		3d 4s 4p				
Ionic bonds	Fe^{++}	(⇅)(↑)(↑)(↑)(↑) ○○ ○○○				2
	Fe^{+++}	(↑)(↑)(↑)(↑)(↑) ○○ ○○○				$\frac{5}{2}$
Octahedral covalent d^2sp^3 bonds	Fe^{++}	(⇅)(⇅)(⇅)○○ ○○ ○○○				0
	Fe^{+++}	(⇅)(⇅)(↑)○○ ○○ ○○○				$\frac{1}{2}$

If ionic bonding is present, there will be no sharing of electrons with the ligand atoms, and hence all of these orbitals will be available for the 3d electrons of the iron atom itself. Following the normal Hund's rule, these will, therefore, enter the orbitals to maximise the electron spin and hence in the case of the ferric derivatives, all five electrons will enter five different 3d orbitals, with their spins aligned parallel, to give a total electron spin of $S = \frac{5}{2}$ whereas the sixth electron of the ferrous iron will have to enter an orbital which already contains one spin, as indicated in the table, giving a resultant total spin $S = 2$. If strong covalent bonding exists to the octahedron of ligand atoms on the six coordination points, there will be a sharing with the electrons of these ligand atoms and the d^2sp^3 orbital required for this will, therefore, take up the two higher 3d orbitals of the iron atom, as well as the outer 4s orbitals, as indicated in the lower half of Table 1. In this case of covalent bonding, the six electrons of the ferrous iron are forced into the three remaining 3d orbitals, and thus all of them pair to give a total spin $S = 0$, and thus diamagnetic compounds. In the case of the ferric derivatives, the five electrons enter these three orbitals, and as a result leave one unpaired spin with an $S = \frac{1}{2}$.

The difference between the ionic and covalent bonds as represented in Table 1, can be also considered in terms of the energy level diagram of Fig. 2 where it is seen that the first effect of an octahedral symmetry on the 3d orbitals is to separate these into two groups, the lower one containing the

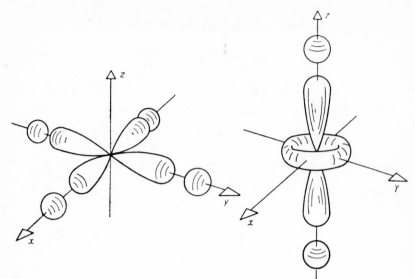

Figure 2a₁

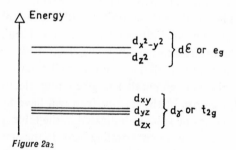

Figure 2a₂

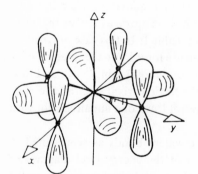

Figure2b

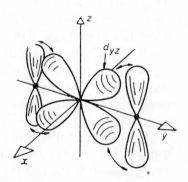

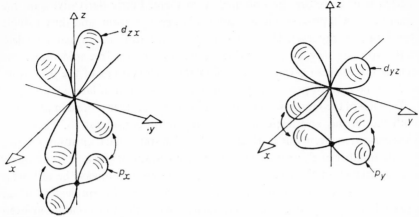

Figure 2c

Figure 2 Effect of surrounding ligands on d-orbital energies. (a) Splitting of orbitals by proximity of ligand atoms along x, y and z axes, which produce extra repulsion for the $d_{x^2-y^2}$ and d_{z^2} orbitals. (b) Additional interaction of p-orbitals on nitrogens in xy plane with d_{yz} but not d_{xy} orbitals. (c) Further interaction of p-orbital on nitrogen at 5th or 6th co-ordination point, either with d_{zx} or d_{yz} orbital

three d_{xy}, d_{yz} and d_{zx} orbitals, while the upper contains the two $d_{x^2-y^2}$ and d_{z^2} orbitals. The actual way in which electrons are fed into these orbitals will depend crucially on the splitting between these two groups and the interaction between the electrons themselves; thus, if the splitting between the two groups is rather small, as corresponds to the ionic case, the electrons will feed in, one at a time, to each orbital, before any are paired. If, however, the splitting between the two groups of orbitals is large, as corresponds to the covalent case, more energy would be required to put the electrons into a higher orbital than to pair them in the lower group, and the pairing, with resultant smaller spin values, thus takes place.

It can be seen from this analysis that both the ferrous and ferric derivatives can thus be subdivided into "high spin" and "low spin" compounds, and these are effectively a measure of the strength of the binding between the iron atom and its surrounding ligands. We will see later, however, that this classification is only a first approximation, but does serve as a very useful preliminary analysis of the binding of the iron atom.

The case of the low spin derivatives may be considered first, since these are relatively simple. Thus the ferrous iron in the covalent bonding produces a zero resultant electron spin, and hence all the compounds containing covalently bound ferrous iron will be diamagnetic, and no electron resonance spectra will, therefore, be obtained from them. Ferric derivatives, on the other hand, will possess one unpaired spin per iron atom, and thus a single degenerate energy level. This will split on the application of a magnetic field in the ordinary way, to give g-values spread around the free spin value. The anisotropy of these g-values will reflect the surroundings of the iron atom and can be used to probe their symmetry, as discussed in later sections.

The case of the high spin derivatives is, however, rather more complex, and both the ferrous and ferric cases require careful consideration. There is a vary general theorem due to Kramers[1], which states that if an atom contains an odd number of electrons, the internal electrostatic fields can never remove the degeneracy of the levels completely, and there will always be at least a two-fold degeneracy which can only be removed by the application of an external magnetic field. If, however, the paramagnetic atom contains an even number of unpaired electrons, it is possible for the internal molecular field to remove the degeneracy of all the levels completely so that even those corresponding to $M_S = \pm 1$ are split in the absence of the external magnetic field. Moreover, it is possible for this zero-field splitting to be larger than the energy corresponding to the microwave quantum used in the electron resonance spectrometer. If this is so, it will be impossible to observe electron resonance absorption from this system of energy levels, since the available quantum energy will never be able to induce transitions between the two lowest levels. This situation is, in fact, often found for the case of ionically bound ferrous derivatives, and hence very little information has, in fact, been obtained by e.s.r. on such compounds.

This comment does not apply to the ferric derivatives, however, since these do possess an odd number of unpaired electrons, and, therefore, Kramers' theorem applies to them, and each level will still be doubly degenerate in the absence of an externally applied magnetic field.

3 High spin derivatives of haemoglobin and myoglobin

The total S of $\frac{5}{2}$ which is produced by the 5 unpaired electrons associated with the high spin ferric state, may be represented vectorally as shown in Fig. 3(a), where it is seen that this total vector can take up six different orientations with respect to the axis of the internal molecular field, which will be normal to the haem plane. These six different orientations can be characterised by the quantum number $M_S = +\frac{5}{2}$ etc., and according to Kramers' theorem, the energies of the $\pm\frac{5}{2}$ levels will be equal, and hence the six orientations produce only three separate pairs of energy levels, which are,

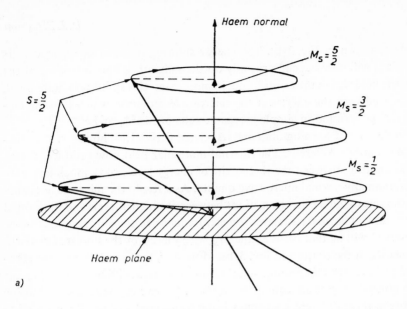

Haem normal

$M_S = \frac{5}{2}$

$M_S = \frac{3}{2}$

$M_S = \frac{1}{2}$

$S = \frac{5}{2}$

Haem plane

a)

There are similarly 3 components for
$M_S = -\frac{5}{2}, -\frac{3}{2}$ and $-\frac{1}{2}$ below the plane

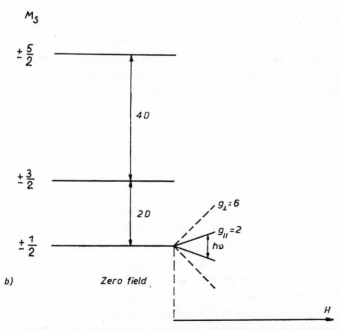

M_S

$\pm \frac{5}{2}$

$4D$

$\pm \frac{3}{2}$

$2D$

$g_\perp = 6$

$g_{//} = 2$

$h\nu$

$\pm \frac{1}{2}$

b)   Zero field

H

Figure 3 Resolved components and associated energy levels of high spin ferric state. (a) Different orientations of total $S = \frac{5}{2}$ vector. (b) Resultant energy level scheme

in fact, shown in Fig. 3(b). The energy splitting between these three pairs of levels will now depend crucially on the magnitude and symmetry of the electric field within the crystal, and, in most cases, this splitting is quite small, compared with the energy of the microwave radiation which is being employed, as is found in simple inorganic salts of Mn^{2+}, or $Fe^{3+(2)(3)}$.

In the case of haemoglobin and its related derivatives, however, the situation is entirely different. The internal molecular field now produces a very large splitting between the three pairs of energy levels, very much larger, in fact, than any normal microwave quanta. As a result, the microwave quanta are unable to induce any transitions between the higher levels and the ground state, and, at normal e.s.r. spectrometer frequencies, the only resonance to be observed will be that between the $M_S = \pm\frac{1}{2}$ levels of the lowest state itself, as shown at the bottom of Fig. 3(b). Moreover, the calculation of the effective g-values for this ground state will now be quite different from those of the normal ^6S ground state of manganese. Instead of the zero field splitting between these different spin states being considered as a small perturbation on the main interaction produced by the external magnetic field, the situation is completely reversed, and, as a result, the axis of the internal molecular field is effectively the axis of quantisation, instead of that of the applied magnetic field. When the magnetic field is applied along the axis of the internal field, there is no competition between these, and the observed g-value will then correspond to that of the free spin, as expected for this singlet state in which there is no orbital contribution. On the other hand, there is significant admixture from higher spin levels into the ground state, and the result of the admixture of these higher spin levels is to change the effective g-value observed in directions away from the parallel axis. In the case of axial symmetry, an effective $g_\perp$ of 6.0 is obtained, and the reason for this high effective $g_\perp$ can probably be appreciated qualitatively by a reference to Fig. 3(a). It is seen there that the particular orientation of the total vector of $S = \frac{5}{2}$, which corresponds to the ground state $M_S = \pm\frac{1}{2}$, is making a very wide angle with the molecular axis as it precesses. Thus there will be a large component of magnetization in the directions corresponding to the haem plane (i.e. perpendicular to the axis), and hence a large g-value in these directions is to be expected. The more accurate derivation of this g-value follows from quantum mechanical treatment of the energy levels concerned, and, in particular, from the secular determinants for the cases of the magnetic field parallel and perpendicular to the molecular axis.

It will be seen, in later section, that a second order calculation for the perpendicular g-value, shows that this will diverge from the value of 6.0, when

measurements are made at higher magnetic field strengths. In the earlier
work, however, measurements were made at microwave frequencies in either
the X or Q band region, and in both of these cases, the apparent $g_\perp$, or $g_\perp^{eff}$,
is 6.0. The way in which this g-value for the ground doublet varies with angle
is, therefore, as represented in Fig. 4. It can be seen from this figure that the

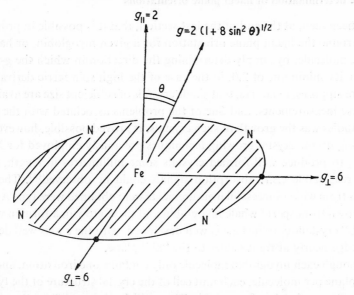

Figure 4 Variation of g value in high spin ferric derivatives. The constant $g_\perp$ equal to 6.0
in the haem plane is a first order approximation valid only at lower microwave frequencies

g-value anisotropy does, in fact, give a highly sensitive method of determining the orientation of the haem, or porphyrin, plane itself. In principle all that
is required in order to determine the orientation of the haem plane, is to
mount a crystal in the cavity resonator and then rotate the crystal in all possible
directions until the g-value equal to 2.0 is obtained. It will then be known
that the magnetic field is being applied along the direction normal to the
haem plane, and hence the orientation of the plane itself is determined. In
practice, however, it is very difficult to move a crystal in three directions at
once, inside a cavity resonator, and the crystals were therefore mounted in
different crystallographic planes in turn, and the g-value variation for these
different planes was then plotted. When these initial electron resonance investigations were made, the X-ray studies of Kendrew[4] and Perutz[5] had
not been completed, and hence no information on the orientation of these
planes was available. The orientation of these planes was, in fact, first deter-

mined by these electron resonance measurements, and this information was
then made available to the X-ray crystallographers and used by them to assist
in the complete analysis of the rest of the molecule.

4   The determination of haem plane orientations

It has been seen, at the end of the last section, that it is possible in principle
to determine the haem plane orientation for a given myoglobin, or haemoglobin, molecule, by simply determining the direction in which the g-value
falls to its minimum of 2.0, in the case of the high spin ferric derivatives.
This presupposes, of course, that single crystals of sufficient size are available
for these measurements, and one of the problems associated with the early
e.s.r. studies was the growth of such crystals. It proved possible, however, by
following up the crystal growing techniques previously employed for X-ray
studies, to produce well formed crystals some millimetres in length, from
which reasonable e.s.r. absorption signals could be observed[6]. The first
crystals from which successful measurements were obtained were type A crystals, grown from sperm whale myoglobin and these grow in a form in which
the "ab" crystallographic face is well developed, with the c axis well defined
as an edge nearly at right angles to the "ab" plane.

Although each myoglobin molecule only contains one iron atom, and one
haem plane per molecule, each unit cell of the crystal structure of the type A
contains two molecules of myoglobin. There will thus be two differently orientated haem planes in such a crystal, and hence two sets of g-value variation are to
be expected in the electron resonance results. In order to obtain a complete
three domensional map of these g-value variations, the type A crystal is first
mounted in a H_{111} cylindrical cavity with the "ab" crystallographic plane
horizontal on the bottom surface. The external magnetic field can then be
rotated conveniently in all directions around the "ab" plane, and thus the
g-value variation in the plane can be measured directly. The results[7] obtained for the "ab" plane of a type A sperm whale myoglobin crystal is shown
in Fig. 5(a). The two g-value variations, corresponding to the two molecules
per unit cell, can be clearly seen, and it is also evident that both of these reach
a maximum value of $g = 6$, since any crystallographic plane must cut a
haem plane in one direction at least, however it is orientated. The minimum
g-value observed does not fall to 2.0 however, since neither of the normals to
the haem plane, corresponding to the $g_{||}$ direction, actually lie in the "ab"
crystallographic plane itself. In fact the minimum g-value observed of 2.64,
which occurs at an angle of 23° to the "a" axis, can be used to determine the

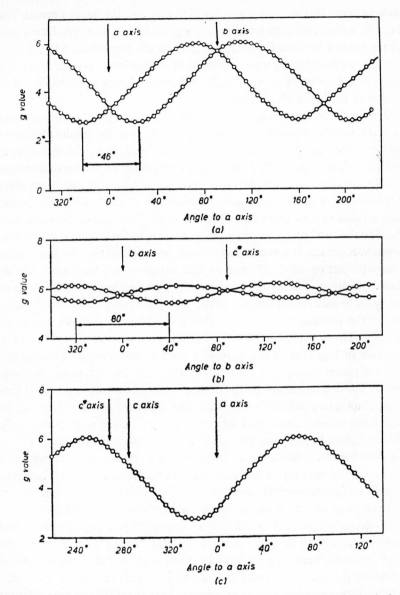

Figure 5 Type *A* myoglobin—*g* value variation for acid met derivative. (a) In ab plane. (b) In bc plane. (c) In ac plane.

angle between this minimum g-value direction and the haem normal. Thus it can be substituted into the expression given in Fig. 4, to produce a value for the angle θ between the haem normal and this direction, of 18°. It therefore follows that this angle measurement of the minimum g-value in the "ab" plane, and its orientation, is actually sufficient to locate the direction of the normal to the haem plane itself.

Measurements are always carried out in other crystallographic planes, however, in order to cross-check the calculation, and the results obtained in the "bc" and "ac" planes are also shown in Fig. 5. It can be seen that the g-value variation in the "bc" plane stays very close to the maximum value of 6.0 in all directions, and this is to be expected since the two haem planes themselves have orientations quite close to that of the "bc" plane. Quite a wide g-value variation is again obtained in the "ac" plane, however, since this, like the "ab" plane, approaches quite close to the direction of the haem normals. Since this is a plane of symmetry, on to which the two haem planes always project equally, only one g-value variation will be obtained in this plane. Quantitative analyses of the g-values observed in both of these planes, can again be used, as for the case of the "ab" plane, to determine the orientation of the two haem planes, and thus a complete cross-check can be obtained[7]. The orientation of the two haem planes, as determined in this way, is shown in Fig. 7(a), in the form of a three dimensional perspective plot, with the planes themselves indicated by the shaded squares. As mentioned earlier, this determination by electron resonance of the orientation of these haem planes, was obtained before the detailed X-ray measurements on the myoglobin were available, and they were, in fact, used as an important piece of information in this final analysis of the myoglobin molecule.

Following this initial work on the myoglobin crystals, electron resonance measurements were then made on crystals of haemoglobin itself[8]. As mentioned earlier, haemoglobin contains four haem planes and iron atoms per molecule, and in the crystals of horse haemoglobin employed there is only one molecule per unit cell. It had been assumed, before the electron resonance measurements were carried out, that the four haem planes in the haemoglobin molecule were probably parallel to one another, but the initial electron resonance measurements indicated immediately that this was not the case. A typical crystal of horse haemoglobin will have the (110) and the (001) faces well developed and readily identifiable, and the g-value variations observed from the acid met derivative in these two planes, are shown in Fig. 6. It is quite clear that there are in fact four separate g-value variations in both of these planes, and these correspond to four differently orientated haem planes

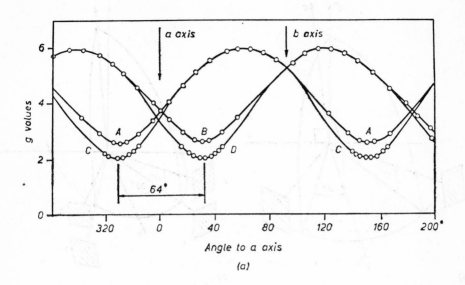

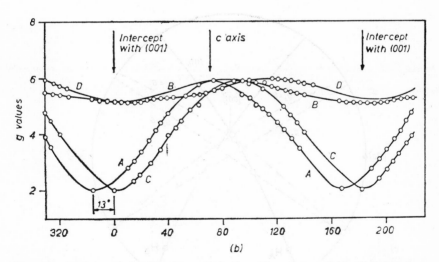

Figure 6 *g*-value variation for acid met derivative of horse haemoglobin. (a) In 001 plane. (b) In 110 plane.

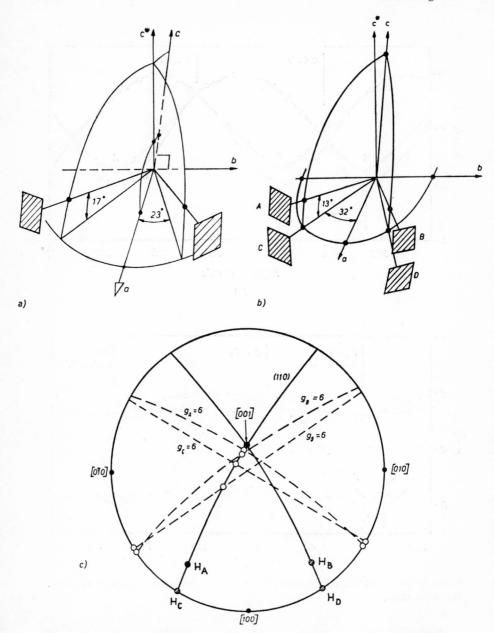

Figure 7 Orientation of haem planes as determined from E.S.R. measurements. (a) For type *A* myoglobin crystals. (b) For horse haemoglobin crystals, as a three dimensional perspective plot. (c) For horse haemoglobin, with complete *g*-value variation on a stereogram

per molecule. In the (110) plane, two of these g-value variations reach their absolute minimum of 2.0 and hence immediately define the actual directions of the haem normals of the two planes with which they are associated. The orientation of the other two planes can be found from the kind of quantitative analysis previously outlined for the myoglobin crystals, and the results can then be summarised as a three-dimensional perspective drawing, or as a stereogram as shown in Fig. 7(b) and (c).

The use of stereograms to represent the directions of different g-values has proved very useful in these single crystal studies, and does enable the electron resonance results to be rapidly correlated with X-ray investigations and other studies. Quite a large number of different types of myoglobin and haemoglobin crystal were in fact analysed in detail by e.s.r. techniques[9] following the initial work on the type A amd haemoglobin molecules, and a typical set of results for the type. F crystals are shown in Fig. 8. It is clear from this, that the orientation of the four haem planes associated with the four different molecules in a unit cell of this particular type of crystal, had been unequivocally identified, and the general g-value variation can be represented in detail by the accompanying stereogram.

5   Structural information from low spin derivatives

It has already been pointed out that the myoglobin and haemoglobin ferric derivatives, which are strongly bound to the central iron atom, would only possess one unpaired spin in their $3d$ orbitals, and hence have resultant $S = \frac{1}{2}$. There will, therefore, be just one electron resonance absorption with g-values spread across the free spin value. Typical variations of such g-values are to be shown in Fig. 9(a) for two different hydroxide derivatives, and can be compared in the same diagram with the g-value for the acid met derivative, since in this particular crystal only the outer half had been converted to the OH derivative, and the angle of variation of all three derivatives could thus be followed at the same time. A summary of the different g-values and their magnitudes along the three principal axes for the azide derivative, is given in Fig. 9(b), and it is seen that there is still a principal value very close to the haem normal at 2.80, but there is now no longer axial symmetry[10]. Thus the g-values around the haem plane itself vary significantly from a minimum of 1.72 to a maximum of 2.22. Three directions of principal g-values, g_x, g_y, g_z, can thus be defined, and it must follow that the variation in g-value in the haem plane reflects some structural features above or below the plane, which are producing the anisotropic effect in the plane itself. The

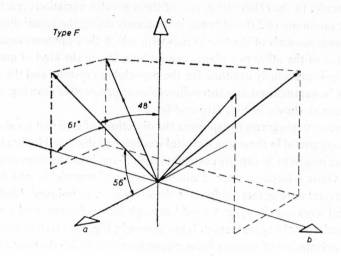

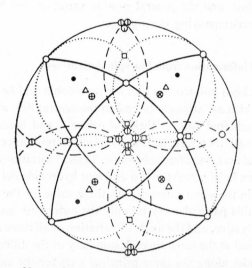

Figure 8   Orientation of haem plane normals in type *F* crystal. The directions of the four normals of the four molecules per unit cell are shown as a three dimensional perspective plot at the top of the figure. The complete *g*-value variation is shown at the bottom on a stereogram, where the circles corresponding to values of 6 ($\leqq$), 5 (---) and 4 (.....) are plotted for all four haem groups

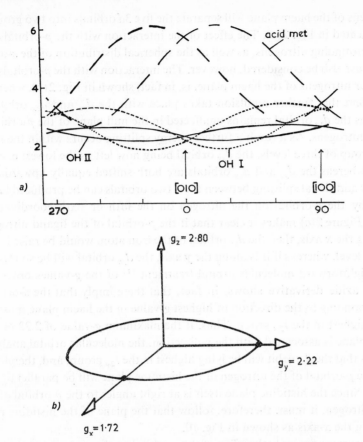

Figure 9 *g*-value variation in low spin derivatives. (a) For the two OH derivatives in the ab plane of type *A* crystal. (b) Summary of principal *g*-values for azide derivative

immediate structural features which might affect the orbitals of the iron atom are, on the one hand, the nitrogen atom on the fifth coordination point belonging to the histidine plane which links the haem plane to the polypeptide chains of the protein. On the other hand, there is a nitrogen atom on the sixth coordination point which is one of the three nitrogen atoms of the N_3 azide group. It would appear, therefore, that some feature of either the nitrogen of the histidine plane below, or of the azide group above, is interacting with the orbitals of the iron atom to produce the anisotropy in the *g*-value around the haem plane itself.

The nature of this interaction is explained diagrammatically in Fig. 2. It has already been seen in an earlier section, that the presence of the four nitrogens of the haem plane will separate the five $3d$ orbitals into two groups, as indicated in Fig. 2(a). The effect of the interaction with the p-orbitals on the surrounding nitrogens, as well as the spherical distribution of the s-orbitals, must also be considered, however. The interaction with the p-orbitals on the four nitrogens of the haem plane. is, in fact, shown in Fig. 2(b), where it is evident that further repulsion takes place with the d_{yz} and d_{zx} orbitals, whereas the d_{xy} orbital remains unaffected in the null plane of the p-orbitals of the nitrogens. As a result a further energy splitting occurs within the bottom group of three levels, the d_{xy} orbital being now left as the lowest energy state, whereas the d_{yz} and d_{zx} orbitals are both shifted equally upwards. A further additional splitting between these two orbitals can be produced however, by the p-orbital on the nitrogen on the fifth or sixth coordination point. Figure 2(c) makes it clear that if the p-orbital of the ligand nitrogen is along the x axis, then the d_{zx} orbital of the iron atom would be raised to a higher level, whereas if it is along the y axis, the d_{yz} orbital will be so raised. A straightforward molecular orbital treatment[11] of the g-values observed in the azide derivative shows, in fact, that these imply that the d-orbital corresponding to the direction of highest g-value in the haem plane must be lying highest in the t_{2g} group. Thus, if the maximum g-value of 2.22 in the haem plane is associated with the y-direction, the molecular orbital analysis implies that the d_{yz} orbit will be lying highest in the t_{2g} group, and, therefore, that the p-orbital of the nitrogen of the histidine plane will be parallel to the y-axis. Since the histidine plane itself is at right angles to the p-orbital of its own nitrogen, it must, therefore, follow that the plane of the histidine ring is along the x-axis as shown in Fig. 10.

However, so far, only the effect of the p-orbital on the nitrogen on the fifth coordination point has been considered, whereas there may well also be an interaction with the p-orbital of the nitrogen on the sixth coordination point, belonging to the azide group. It was initially assumed that the three nitrogens of this azide group were positioned along a projection of the haem normal and this would not cause any asymmetry in the g-value around the plane. Recent X-ray work[12] has shown however that they are, in fact, orientated along a line which makes an angle of 111° with the haem normal. In this case, a significant electron density might well be expected in the p lobes of the nitrogen of the sixth coordination point, at right angles to the line of the azide group, and hence its orientation would also produce an asymmetry in the g-values in the haem plane, in the same way as discussed for the histidine plane above. It is found experimentally[13], that the direction of the minimum g-value in the haem plane lies approximately between the projection

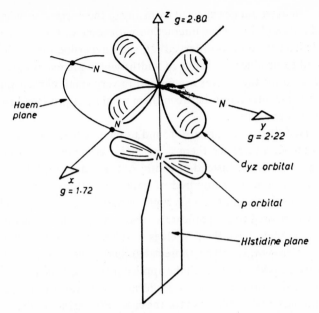

Figure 10 Orientation of histidine plane as related to *g*-value variation. The minimum *g*-value along the *x* axis implies that the d_{yz} orbitals is lying highest and therefore that the *p*-orbital on the attached nitrogen is along the *y* axis, and hence the projection of the histidine plane itself along the *x* axis

of the histidine plane and the azide group, suggesting that both of these structural features are affecting the observed *g*-value variation. Measurements are in hand on the detailed *g*-value variation in cyanide and other covalently bound derivatives of myoglobin and haemoglobin, so that the contributions from the histidine plane and other groups on the sixth co-ordination point can be clearly differentiated.

It is evident from these present results, however, that a determination of the three-dimensional *g*-value variation in these covalently bound derivatives with low spin, together with a small amount of molecular orbital theory, does allow significantly more information to be obtained on both the structure of the haemoglobin molecule, and the binding of the central iron atom.

6   Linewidths and hyperfine interaction

During the last few years, considerable attention has been paid to the widths of the electron resonance absorption line, and the way in which this width varies with angle in the different crystals studied[13]. These widths are in fact much larger than would be expected from any of the normal broadening processes, and although hyperfine interaction with some of the ligands, such as the fluoride iron have been observed[14], such hyperfine interaction is also far too small to produce the range of widths which are encountered. Moreover the iron atoms themselves are present in very small concentration, and hence the dipolar interaction between them would not produce broadenings of more than a few gauss.

The observed linewidths are also found to be independent of temperature from 4° to 90°K, and hence there is no major contribution from the spin lattice relaxation effects. Instead of widths of the order of a few gauss, however, linewidths of several hundred of gauss are observed and these vary quite rapidly with orientation. This rapid variation with angle, and the fact that the line narrowed to a significantly smaller width along the direction of the extreme g-values, gave the first clues to the broadening mechanism that was probably present. It will be appreciated that the resonance field position is very sensitive to the angle which the applied magnetic field makes with the haem normal, since there is such a rapid and large variation of g-value in these compounds, and partially so for the case of the acid-met derivatives. If, therefore, for some reason or other, there was a slight variation in the axial direction of haem normals from molecule to molecule, a major spread of the field positions required for resonance would be produced, and would, moreover, have just the kind of angular variation that was observed for the linewidths of the acid met derivatives. Detailed calculations were therefore undertaken to see if this randomization of the haem plane normals within the crystal might be able to account, both qualitatively and quantitatively for the observed linewidths and their angular variation. In the case of the acid-met derivative only one variable need be considered in such a theory, and this is the mean deviation in angle, associated with such a statistical variation of the orientation of the haem normals. This spread in orientation will produce a corresponding spread in g-values from molecule to molecule, and hence smear out the overall linewidth. The actual expression for the linewidth produced by such an effect can be written down quite explicitly[13], as in the equation below

$$\Delta H = \frac{h\nu}{\beta} \frac{(g_\perp^2 - g_{||}^2)}{2g_\theta^3} \sin 2\theta \sin 2\Delta\theta + \text{const.}$$

where θ is the angle between the applied magnetic field and the haem normal, and $\Delta\theta$ is the standard deviation in the random orientation of the haem group directions. The first term in the above equation thus represents the contribution to the linewidth from the spread in g-values associated with this random orientation, while the second term is a constant which includes the other residual broadening mechanism. The fact that this expression does account for the variation in the linewidths to a very good approximation is shown in Fig. 11, where a comparison is made between the theoretical

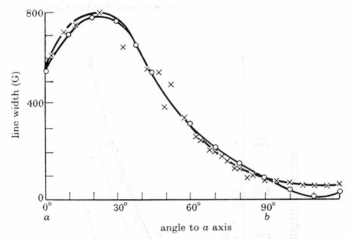

Figure 11   Linewidth variation in acid met myoglobin. The open circles correspond to the theoretical prediction of equation 1, while the experimental results are plotted as crosses

variation in linewidth predicted by equation 1, and the experimentally measured values. The points represented by the open circles correspond to the theoretical variation given by equation 1, assuming a value for $2\Delta\theta$ equal to 0.055 radians, or 3.3°. It would appear, therefore, that a random orientation of only 1.6° in the directions of the haem normals is quite sufficient to account for the line broadening observed in the acid-met myoglobin.

Although the main features of the linewidth variation of these haemoglobin and myoglobin absorptions can be explained in this way, more recent measurements have made it clear that there must also be some other mechanisms present which are not yet fully understood. Thus a systematic study of the linewidth variation with angle at 4 mm wavelengths shows that, as well as the variation which is produced by the g-value spread explained above, there is also an additional component which varies directly with the frequency of the microwaves used in the measurement.

Typical results at 4 mm wavelengths[15], are summarised in Fig. 12, where it is seen that the constant width underneath the angular variation has now risen to a value of about 150 gauss, instead of the 50 gauss observed at the longer wavelengths shown in Fig. 11. Since all the normal mechanisms for line broadening are frequently independent, these results do suggest that there must be another type of broadening mechanism present in these haemo-globin derivatives which has not been observed before.

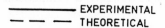

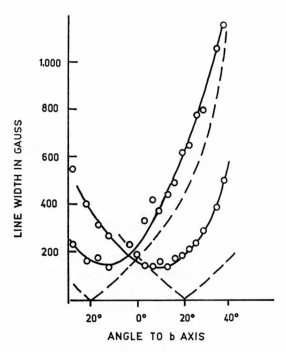

Figure 12   Linewidth variation at 4 mm wavelengths

One possible explanation is that there is not only a random variation in the haem plane orientation, but also a possible variation in the value of the zero field splitting parameter, D, brought about by variation in the distance of the water molecule from the iron atom in the acid met derivative. Such a variation in this zero field splitting parameter, D, would in its turn, produce a randomization in the effective g-value, as given by equation 2 in the next section. Moreover, it is clear from equation 2 that the effective g-value depends on the actual ratio of the splitting parameter, D, to the frequency of the applied microwaves and hence the value of the magnetic field strength required for resonance. This type of variation would, therefore, be one of the few mechanisms that would produce an additional broadening effect which was linearly dependent on the microwave frequency used for measurement. Much more work will have to be carried out, however, before all the problems associated with the linewidths of these compounds have been elucidated. Further details will not be considered here, but the results already quoted should be sufficient to show that systematic studies of e.s.r. line widths in such derivaitives can often give additional information on the basis interactions present in such molecules.

7 Determination of the zero field splitting parameter

It has already been seen that one of the characteristics of the high spin haemoglobin derivatives is the very large splitting which is produced between the energy levels of the spin components by the internal molecular field, and the actual magnitude of this splitting, and the mechanism which produces it, are obviously closely related to the bonding of the iron atom itself. Several attempts have been, and are being made, therefore, to determine this splitting as accurately as possible, in as many different derivatives as are available, so that a consistent picture of the energy level structure of this central part of the molecule can be derived.

If a more accurate second-order perturbation treatment of the energy level system is undertaken, it is found that the levels of the three doublets diverge in high magnetic field strengths, as shown in Fig. 13, instead of the linear divergence indicated in Fig. 3(b). Two effects then become apparent if measurements are undertaken at high microwave frequencies, and high field strengths. On the one hand, it may be possible to determine the value of "D" directly, if it is possible to induce a transition between one of the two levels in the ground state, and one of the levels in the upper $\pm\frac{3}{2}$ component, as indicated by the vertical arrows of Fig. 13. As already explained, the magnitude of the microwave quantum at normal electron resonance wavelengths of 3 cm, and even 8 mm, is not sufficient to induce such a transition, and this is why the shorter wavelength 4 mm and 2 mm resonance spectrometers are being developed. The availability of such wavelengths, together with the high magnetic fields produced by superconducting magnets, should make it possible to observe direct transitions between the $\pm\frac{1}{2}$ and $\pm\frac{3}{2}$ spin levels, and hence direct determination of the value of the splitting parameter "D" will then be available.

The other way in which the value of "D" may be estimated, is from the

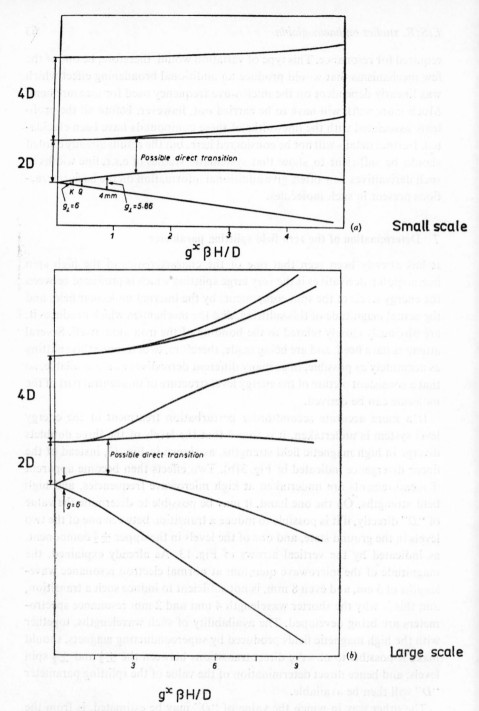

Figure 13 Divergence of energy levels in high magnetic field strengths. The levels are drawn for the magnetic field perpendicular to the crystal field axis and it is seen that $g_\perp$ of 6.0 at low fields will become smaller as the field strength increases

divergence of $g_\perp^{\text{eff}}$ away from 6.0, as measurements are taken at shorter wavelengths, and higher field strengths. When "*D*" is large compared with the microwave quantum, or $g\beta H$, the $g_\perp^{\text{eff}}$ is essentially constant at 6.0, but as H increases, the energy levels cease to diverge linearly, as can be seen from Fig. 13, and a more accurate expression for $g_\perp^{\text{eff}}$ can be derived.

$$\text{i.e.} \quad g_\perp^{\text{eff}} = 3g_\perp^* \left[1 - 2 \left(\frac{g_\perp^* \beta H}{2D} \right)^2 \right] \tag{2}$$

where $g_\perp^*$ is the true $g_\perp$ that would be obtained in infinite magnetic field strengths, and is approximately 2.0. Substitution of typical figures into this expression will show that at microwave frequencies over 72,000 Mc/s (i.e. wavelengths of about 4 mm) there will be significant shifts of $g_\perp^{\text{eff}}$ from 6.0 and hence these will enable the value of "*D*" to be estimated fairly accurately.

This analysis is somewhat more complex than might appear at first sight from equation 2, since the *g*-value variation in the haem plane is not quite isotropic. There is a rhombic component which can be represented by a splitting parameter "*E*", as well as the axial splitting component "*D*". The effect of this rhombic component is to produce anisotropy in $g_\perp^{\text{eff}}$, and hence g_x and g_y now have slightly different values for the acid met, as well as the azide derivatives. Various sets of measurements have been made on the different ionically bound myoglobin and haemoglobin derivatives to obtain values of the parameters "*D*" and "*E*" by this method.

Eisenberger and Persham[16] were the first to deduce values of "*D*" from the divergence of $g_\perp^{\text{eff}}$ from 6.0. They undertook careful measurements at both 13,000 and 35,000 Mc/s, but did not make any allowance for the rhombic term "*E*" in their analysis. Their measurements of $g_\perp^{\text{eff}}$ equal to 5.950 and 5.930 at 13,000 and 35,000 Mc/s respectively then led to an estimate for "2*D*" of 8.76 cm^{-1}. This work was then followed by the measurements of Kotani and Morimoto[14], who measured the anisotropy of $g_\perp^{\text{eff}}$ in the haem plane, as well as its divergence from 6.0. They obtained estimates for "*D*" of 10.0 cm^{-1} for the acid met myoglobin and 6.5 cm^{-1} for the myoglobin fluoride, and the results suggested that "*E*" was about 4% of "*D*". These measurements, however, were made at the relatively low microwave frequencies of 9,000 Mc/s and hence the error in the values of "*D*" are likely to be rather high. Measurements by Ingram and Slade[17] have been made at wavelengths of 4 mm, however, where significantly greater shifts of the $g_\perp^{\text{eff}}$ to 5.915 are obtained, and a detailed analysis of these results gives values of "2*D*" = 8.8 ± 0.8 cm^{-1} for the acid-met myoglobin and a value of "*E*" equal to 0.02 cm^{-1} for the same derivative. Moreover the directions of the g_x and g_y principal axes within the haem plane for the acid-met derivatives are not quite the same as those for the azide derivatives, which suggests that some more general feature of the molecular symmetry, such as the haem plane puckering, is being reflected in these measurements. Further measurements are now being made on other derivatives, and over different temperature ranges, so that a detailed picture of the effect of the internal electrostatic field can be deduced.

It will be clear from all these measurements that considerable work still needs to be carried out in the electron resonance investigations of the different derivatives of myoglobin and haemoglobin before a complete understanding of the energy level system of the iron atom and its surrounding ligands is available. The measurements that have been undertaken so far, however, do indicate the power of electron resonance in probing both the molecular structure and the detailed energy level scheme of such an important biochemical molecule as haemoglobin, and also indicate the extra information that does become available if studies on single crystals of the derivatives can be made.

References

1 Kramers, H.A. *Proc. Acad. Sci. Amst.*, 1930, **33**, 959.
2 Bleaney, B. and Ingram, D.J.E. *Proc. Roy. Soc. A* 1951, **205**, 336.
3 Bleaney, B. and Trenam, R.S. *Proc. Roy. Soc. A* 1954, **223**, 1.
4 Kendrew, J.C., Parrish, R.G., Marrack, J.R., and Orleans, E.S. *Nature*, 1954, **174**, 946.
5 Perutz, M.F. *Proc. Roy. Soc. A* 1954, **225**, 264.
6 Bennett, J.E., Gibson, J.F., Ingram, D.J.E., Haughton, T.M., Kerkut, G.A., and Munday, K.A. *Physics in Medicine and Biology*, April 1957, p. 4.
7 Bennett, J.E., Gibson, J.F., and Ingram, D.J.E. *Proc. Roy. Soc. A*, **240**, 1957, 67.
8 Ingram, D.J.E., Gibson, J.F., and Perutz, M.F., *Nature*, 1956, **178**, 906.
9 Bennett, J.E., Gibson, J.F., Ingram, D.J.E., Haughton, T.M., Kerkut, G.A., and Munday, K.A., *Proc. Roy. Soc. A*, 1961, **262**, 395.
10 Gibson, J.F. and Ingram, D.J.E., *Nature*, 1957, **180**, 29.
11 Griffith, J.S., *Nature*, 1957, **180**, 30.
12 Stryer, L., Kendrew, J.C., and Watson, H.C., *J. Mol. Biol.*, 1964, **8**, 96.
13 Helcke, G.A., Ingram, D.J.E., and Slade, E.F., *Proc. Roy. Soc. B*, 1968, **169**, 275
14 Kotani, M. and Morimoto, H., *Magnetic Resonance in Biological Systems* (Pergamon Press, Oxford 1967), p. 135.
15 Slade, E.F. and Ingram, D.J.E., *Nature*, 1968, **220**, 785.
16 Eisenberger, P. and Pershan, P.S., *J. Chem. Phys.*, 1967, **47**, 3327.
17 Slade, E.F. and Ingram, D.J.E., *Proc. Roy. Soc. A*, 1969, **312**, 85.

A crystal field analysis
of low spin compounds of hemoglobin A

W. E. BLUMBERG and J. PEISACH*

Bell Telephone Laboratories, Incorporated
Murray Hill, New Jersey 07974

FERRIC LOW SPIN FORMS of hemoglobin A are compounds in which the heme iron has six covalently bonded ligand atoms, four from the porphyrin (as in all heme proteins), and two, designated as Z ligands, which are either supplied by amino acids of the protein, or by exogenous ligating groups. EPR spectroscopy conveys information concerning the geometry and strength of the chemical bonds of the non-porphyrin Z ligands.

The forte of the EPR methods in the studies of low spin ferric heme compounds extends only to the immediate ligand environment of the iron, and the technique conveys no information concerning subtle changes of the protein not immediately affecting the ligand field of the iron. Substitutions of one Z ligand for another, accompanying *gross* changes in protein configuration, are readily sensed by EPR. Furthermore, since the number of possible ligand atoms which are endogenous to the protein and which can bind iron is limited, EPR can be used to identify all of them as each has a particular spectroscopic signature. For both endogenous as well as endogenous ligands, EPR is useful in ascribing structures to low spin ferric heme compounds.

All ferric low spin compounds have spin $= \frac{1}{2}$, and the interaction of the spin with an external magnetic field can be completely described by three g values. In the case of low spin compounds of ferric hemoglobin, these three g values are all different, indicating that the heme of these compounds

* Present address: Departments of Pharmacology and Molecular Biology, Albert Einstein College of Medicine, Yeshiva University, Bronx, New York 10461.

has less than axial symmetry. For a frozen sample of hemoglobin in solution, where the molecules are randomly oriented, the three *g* values can be directly determined from three features of the EPR absorption derivative spectrum, an end maximum, an end minimum, and a midpoint crossing (Fig. 1) (1). Since the EPR spectrum is a broadening convolution over a

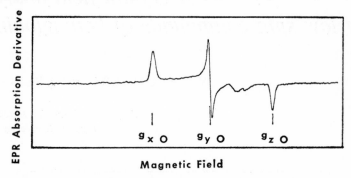

Figure 1   EPR spectrum of isolated ferric alpha chains of hemoglobin, pH 8.7. The three *g* values are 2.56, 2.18, and 1.88. After Ref. (6). The vertical lines show where the three *g* values are measured on the low spin EPR spectrum

second rank tensor interaction, a knowledge of the three *g* values and the broadening function of any one of the features is sufficient to define the whole spectrum, and thus the total number of low spin heme iron atoms in the sample. One can then easily quantitate the concentration of low spin compounds in a sample even if the sample is heterogeneous and contains a mixture of compounds, each with its own set of three *g* values.

For a pure ferric low spin compound, the wave function describing the spin is constituted solely of a combination of the three εd orbitals (2). In such a case the three *g* values are interrelated such that any two determines the third. This can be used to check on the assignment of EPR spectrum or spin state to a specific compound. For these compounds the *g* values are determined completely by a formalism involving two crystal field components. These can be specified arbitrarily as an element of tetragonal symmetry and one of orthorhombic symmetry (3). The *g* values can be used to determine the ratios of the coefficients of these elements (Δ and V, respectively) to the spin orbit coupling energy λ. There is a unique correspondence between sets of three *g* values and sets of two dimensionless ratios Δ/λ and V/λ. Depending on the choice of an axis system there may be several numerically different but geonetrically equivalent pairs of these ratios. We have adopted one self-consistent set of axes for all the compounds studied. The ratio Δ/λ, called the tetragonality (Fig. 2), depends primarily on the charge of the iron atom. This in turn is determined by the sum of the electronegativities of the two Z ligands, the electron donation by the porphyrin ligands remaining constant. The greater the electron donation to the iron atom, the smaller the tetragonality, Δ/λ. The ratio V/Δ, called the rhombicity, is purely a function of the geometrical arrangement of the ligand atoms and the pi bonds between them and the iron atom. Thus, the rhombicity is an indication of whether two different low spin compounds have similar geometry or not. A plot of V/Δ vs. Δ/λ will encompass all possible cases of low spin compounds made

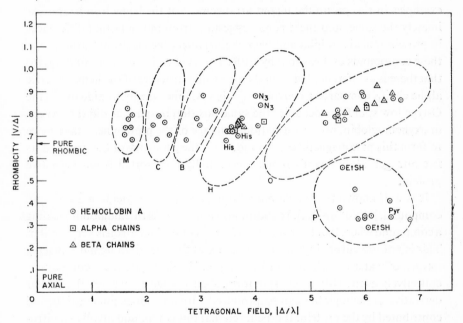

Figure 2 Crystal field parameters for ferric low spin compounds of hemoglobin A and its isolated constituent chains. Unless indicated otherwise, all points in this figure are for compounds formed from ligands endogenous to the hemoglobin molecule. Low spin compounds can be formed using the following reagents referred to in the figure: EtSH, mercaptoethanol; Pyr, pyridine; His, histidine; N_3, azide. The five areas enclosed by the dashed lines define the regions where parameters for the five different compounds may be expected to lie

up of three ε orbitals when the abscissa extends from 0 to infinity and the ordinate from 0 to 2. Thus the location of a point on such a plot specifies uniquely all the determinable crystal field parameters of any low spin compound.

In Fig. 2 we present the results of the analysis of EPR data for a series of low spin derivatives or hemichromes of hemoglobin (4). It can be seen that the low spin forms can be classified by the types P, O, H, B and C, and each type is set off with its own characteristic range of crystal field parameters, all derived from sets of three g values measured by EPR. We have not included any analyses based on only two g values, as these are somewhat less accurate.

The points for the hydroxide low spin forms, designated as O type, $[Hb(d_{5/2}^5)-^\theta OH]$, (5) lie in a well separated group on this kind of plot. In all examples in this group the ligand arrangement of the heme iron is approximately the same, and there is no suggestion from our data that differences in protein moieties (alpha and beta chains) affect the electronic structure of these compounds as measured by EPR. This is in contrast to our findings (6) that there is a difference between the electronic structure of the heme of ferric alpha chains compared to ferrihemoglobin A. Since the hydroxide ion in the O-type low spin form occupies the same position in space as did the oxygen in oxyhemoglobin, no conformational change of the protein need take place to form this paramagnetic compound. Thus the geometry of the iron and its five nitrogen atom ligands remains essentially undistorted from the native protein.

It is well known that azide binds to ferrihemoglobin yielding a low spin compound, $[Hb(d_{1/2}^5)-^\theta N_3]$. Histidine also binds the ferrihemoglobin yielding a compound where the electronic environment of the heme iron is very similar. This is not too surprising, since in both cases the new sixth ligand replacing oxygen of water is an imino nitrogen atom. We designate these compounds collectively as the H-type low spin form. As discussed elsewhere (4) a similar derivative of hemoglobin can be produced with an endogenous sixth ligand contributed by the protein. It is believed that this compound involves a nitrogen atom of the distal histidine (E 7) of hemoglobin, which, in the native protein, is at a distance too far removed to bind to the iron (7). Through a process of hydrogen bond breaking, the tertiary structure of the protein is sufficiently altered so that now this distal histidine can bind to the iron to form a dihistidine low spin form. As the compound formed from ferrihemoglobin H is no different from that formed from isolated ferric alpha chains or from ferrihemoglobin A, here again, EPR is insensitive to differences in protein structure. As the members of the class of H-type low spin compounds nave the same rhombicity as the O-type, it suggests that the five original hitrogen ligands of the heme remain the same and determine the symmetry of the iron ligand field the same way.

Yet another low spin form, designated as C-type, can also be observed in low concentrations from ferrihemoglobin A under appropriate conditions. The EPR parameters for the C-type low spin form are the same as for cytochrome *c* (8), where the fifth and sixth ligands of iron are a nitrogen atom of histidine and sulfur atom of methionine, respectively (9). It is therefore assumed that the ligand environment of the iron in the C-type low spin form of hemoglobin is the same as in cytochrome *c*. In this case also, the five nitrogenous ligands are the same as in the H and O-type low spin forms.

Another low spin form, B-type, which can be prepared as a majority species from hemoglobin incubated with salicylate (4), has the same EPR parameters reported for cytochrome b_2 (10) and b_5 (11). In the case of the B-type, the sixth ligand remains unidentified at present, but from the analysis of EPR data, two facts can be deduced:

1 The five original nitrogen ligand atoms are the same as in the O-, H-, and C-types, and

2 The sixth ligand has an electronegativity lying between that of a nitrogen atom of histidine and the sulfur atom of methionine.

The last group of low spin forms that can be prepared from endogenous ligands of the protein is the P-type. We believe that a necessary Z ligand for the P-type low spin form is a mercaptide sulfur atom. In the case of the compound prepared with urea (4) or at high pH, it is believed that gross denaturation of the tertiary structure of the protein must take place so that a cysteine, on the proximal side of the heme, can now be brought close enough for bonding to iron. Evidence that a sulfur atom is a ligand in the P-type low spin form of hemoglobin is based upon the following:

1 The same compound can be prepared by the addition of a thiol.

2 The P-type low spin form cannot be formed with sperm whale myoglobin, a protein that does not contain cysteine.

3 Mercurial-treated hemoglobin forms the P-type low spin form at a much slower rate.

As can be seen from Fig. 2, the rhombicity of the P-type form is different from all the others. This is not surprising in view of the fact that a sulfur atom has replaced a nitrogen atom on the proximal side of the heme. As sulfur can make δ bonds with iron, this imposes a different geometry on the iron ligand field.

The sum of the electronegativities of the two Z ligands in the P-type low spin compound is approximately the same as that of hydroxide oxygen plus imidazole nitrogen, as in the O-type low spin compounds, giving a tetragonal field of 6λ. As one would expect the electronegativity of a mercaptide group to be greater than that of either hydroxide oxygen or imino nitrogen, it is concluded that the P-type low spin compound does not simultaneously involve both a mercaptide sulfur atom and either of these two other ligand atoms. It is likely that a water molecule or an amine nitrogen atom completes the coordination sphere of the iron in the P-type form.

Although each of these five low spin forms of hemoglobin occupies a small spread in the two crystal field parameters, one can take an average value of each of them over the distribution characteristically typical of the group. The g values of the EPR spectra corresponding to these characteristic crystal fields are then a diagnostic signature for the specific structure involved.

An analysis of the EPR spectra of hemoglobin cyanide (Hb–CN) and cytochrome c cyanide (C–CN) shows that the crystal fields are almost identical, indicating that the same ligand environment of the heme exists in both compounds (11). These points corresponding to these fields are far outside any of the dashed regions of Fig. 2. This fact is not surprising since cyanide is so different from any of the endogenous ligands.

Both hemoglobin $M_{Hyde\ Park}$ (HbM$_{HP}$) and hemoglobin M_{Boston}(HbM$_B$) differ from hemoglobin A in that there is a substitution of tyrosine for histidine in the abnormal beta chain, proximal to the heme in the former and distal to the heme in the latter. Obviously neither protein can produce a dihistidine (H-type) low spin form with endogenous ligands. HbM$_B$, however, can form an H-type low spin form when an exogenous imine nitrogen ligand, such as from azide, is added (12). The azide compound of hemoglobin M_B falls in the same region as hemoglobin azide (Fig. 2). This indicates that the substitution of a tyrosine residue for the distal histidine in this hemoglobin variant does not influence the structure of the low spin azide compound. For HbM$_{HP}$ to form a low spin compound, however, the tyrosine proximal to the heme must be replaced with a more electronegative ligand. By far the most likely replacement is the sulfur of cysteine (β 93), which is rather close to the heme (6), and indeed the only low spin compound reported for hemoglobin M is the P-type or mercaptide form (12).

Thus, we see that EPR is a useful tool in describing the nature of the immediate ligand field environment of heme iron in low spin compounds of hemoglobin, but is insensitive to subtle changes of the configuration of the protein moiety or even substitutions of one protein moiety for another. The number of low spin compounds that can be prepared from a heme protein with endogenous ligands is a small number, as the number of suitable endogenous heme ligands is also a small number. A study of the interconversion of these low spin compounds provides a method for examining gross conformational changes accompanying heme protein reactions. An extension of this analysis to other heme proteins appears elsewhere (13).

The portion of this investigation carried out at the Albert Einstein College of Medicine was supported in part by a Public Health Service research grant to J. Peisach (GM-10959) from the Division of General Medical Sciences. This is Communication No. 192 from the Joan and Lester Avnet Institute of Molecular Biology. The portion of this investigation carried out by J. Peisach was supported in part by a Public Health Service Research Career Development Award (1-K3-GM-31,156) from the National Institute of General Medical Sciences.

The authors would like to thank Dr. Eliezer Rachmilewitz for stimulating discussions, Dr. Helen Ranney for generous samples of hemoglobin H and hemoglobin$_{Riverdale}$, and Miss Rhoda Oltzik for generous samples of hemoglobin A and for technical assistance.

References

1 Kneubühl, F. K., *J. Chem. Phys.* **33**, 1074 (1960).
2 Griffith, J. S., *Nature* **180**, 30 (1957).
3 Griffith, J. S., *The Theory of Transition Metal Ions*, Cambridge University Press, 1961, p. 363.
4 Peisach, J., Blumberg, W. E., Wittenberg, B. A. and Wittenberg, J. B., *J. Biol. Chem.* **243**, 1871 (1968).
5 Peisach, J., Blumberg, W. E., Wittenberg, B. A., Wittenberg, J. B. and Kampa, L., *Proc. Nat. Acad. Sci. U.S.* **63**, 934 (1969).
6 Perutz, M. F., Muirhead, H., Cox, J. M. and Goamann, C. G., *Nature* **219**, 139 (1968).
7 Salmeen, I. and Palmer, G., *J. Chem. Phys.* **48**, 2049 (1968).
8 Dickerson, R. E., Kopka, M. L., Weinzierl, J. E., Vernum, J. C., Eisenberg, D. and Margoliash, E. in *Structure and Function of Cytochromes*, ed. by K. Okunuki, M. D. Kamen and I. Sekuzu, University Park Press, Baltimore, 1968, p. 225.
9 Watari, H., Groudinsky, O., and Labeyrie, F., *Biochem. Biophys. Acta.* **131**, 592 (1967).
10 Bois-Poltoratsky, R. and Ehrenberg, A., *European J. Biochem.* **2**, 361 (1967).
11 Peisach, J. and Blumberg, W. E., unpublished observations.
12 Watari, H., Hayashi, A., Morimoto, H., and Kotani, M., in *Recent Developments of Magnetic Resonance in Biological System.* S. Fujiwara and L. H. Piette, eds. Hirokawa Publishing Co., Tokyo, 1968, p. 128.
13 Blumberg, W. E. and Peisach, J., *Proc. 4th Johnson Foundation Symposium* (in press).

A theoretical survey of EPR spectra due to high spin ferric ions in proteins

MASAO KOTANI*

Science Council of Japan, Tokyo

HIROSHI WATARI**

Faculty of Medicine, Osaka University

1 Introduction

AS IS WELL-KNOWN, there are many kinds of protein molecules which contain iron atoms as constituents. Haem proteins, such as haemoglobin, myoglobin and cytochromes, constitute an important family of iron-proteins, and their EPR spectra have been extensively studied by many scientists. Besides these we have so-called non-haem iron proteins, such as pyrocatechase (catechol 1,2-oxygenase), rubredoxin and ferredoxin. Most of these non-haem iron proteins give EPR signals in the vicinity of $g = 4.28$, as long as the Fe ion is in ferric high spin state.

The arrangement of atoms in the neighbourhood of the Fe ion in the molecule is fairly well known in the case of haemproteins. In this case the Fe atom is situated near the centre of the porphyrin ring, and four N atoms are coordinated to the Fe. Furthermore Fe is coordinated by two more N atoms, or electronegative atoms such as O, or negative ions such as F^-, CN^-, N_3^- etc., in such a way that the Fe atom is coordinated approximately octahedrally by six electronegative ligands. In the case of non-haem iron proteins the surrounding of the Fe atom is not well known, but Fe atoms are considered to be coordinated tetrahedrally by four electronegative ligands,

---

* Present address: Science University of Tokyo, Shinjuiku-ku, Tokyo, Japan.
** Present address: Kyoto Prefectural University of Medicine, Kyoto, Japan.

possibly by negative ions of sulphur. An X-ray analysis of rubredoxin recently carried out by J. R. Heriott *et al.* confirms this structure, although the tetrahedron of sulphur atoms is a little distorted from a regular tetrahedron[1].

In the present paper we discuss the theoretical interpretation of EPR spectra of these iron proteins, covering both haem-proteins and non-haem case, based on the electronic structure of the ferric ion (Fe^{+++}) in ligand fields.

The ferric ion in the high spin state has five electrons outside the A-like closed shell, and in the high spin state spins these electrons are coupled parallel so as to give the resultant spin $S = \frac{5}{2}$. This state is sextet, but under the influence of surrounding atoms the sextet is generally split into three doublets—so-called Kramers' doublets. Splittings of these doublets are measured by EPR spectroscopy.

The orbitals in which five unpaired electrons are accomodated are essentially $3d$ orbitals of the iron atom. It is true that there is evidence, both theoretical and experimental, for hybridization of these orbitals with orbitals of porphyrin ring in the case of haems, and this molecular orbital formation is of vital importance for some purposes, e.g. for interpreting NMR spectra due to protons on the periphery of the porphyrin ring, or even for calculating theoretical energies of these orbitals. But, what is needed in the present paper is the symmetry properties of these orbitals in the group-theoretical sense, i.e., t_{2g} and e_g characters, so that we need not be concerned about the molecular-orbital formation. Hence we use d orbitals for the sake of simplicity, well recognizing that they may not be pure atomic orbitals.

2   Spin Hamiltonian

There is a good theoretical reason to assume that three Kramers' doublets arising from the spin sextet state can be represented by introducing a spin Hamiltonian which is quadratic with respect to S_x, S_y, S_z, components of the resultant spin. Then, by choosing x, y, z axes as the principal axes of the quadratic form, we may write

$$\mathscr{H} = aS_x^2 + bS_y^2 + cS_{z_j}^2. \tag{1}$$

Since $S_x^2 + S_y^2 + S_z^2 = \boldsymbol{S}^2$ is a constant $(= 35/4)$, we may impose the condition $a + b + c = 0$. Then our spin Hamiltonian has only two parameters. The usual choice of these parameters is D, E defined by the following expression:

$$\mathscr{H} = D\left(S_z^2 - \tfrac{1}{3}\boldsymbol{S}^2\right) + E\left(S_x^2 - S_y^2\right) \tag{2}$$

In order to discuss cases corresponding to widely different values of E/D, it is convenient to introduce two new parameter F and α defined by

$$D = F \cos \alpha, \quad E = \frac{1}{\sqrt{3}} F \sin \alpha, \quad (F > 0) \tag{3}$$

Obviously α is a measure of relative magnitudes of E and D:

$$\tan \alpha = \sqrt{3} \frac{E}{D} \tag{4}$$

With the use of these new parameters, our spin Hamiltonian looks like this:

$$\mathcal{H} = F \left\{ \left(S_z^2 - \frac{1}{3} S^2 \right) \cos \alpha + \frac{1}{\sqrt{3}} (S_x^2 - S_y^2) \sin \alpha \right\} \tag{5}$$

Now, we can transform quite elementarily this expression into the following alternate ones:

$$\mathcal{H} = F \left\{ \left(S_x^2 - \frac{1}{3} S^2 \right) \cos \left(\alpha - \frac{2\pi}{3} \right) + \frac{1}{\sqrt{3}} (S_y^2 - S_z^2) \sin \left(\alpha - \frac{2\pi}{3} \right) \right\}$$

$$= F \left\{ \left(S_y^2 - \frac{1}{3} S^2 \right) \cos \left(\alpha - \frac{4\pi}{3} \right) + \frac{1}{\sqrt{3}} (S_z^2 - S_x^2) \sin \left(\alpha - \frac{4\pi}{3} \right) \right\}, \tag{6}$$

or, in terms of notations, whose meanings are quite obvious.

$$\mathcal{H}_z(\alpha) \equiv \mathcal{H}_x \left(\alpha - \frac{2\pi}{3} \right) \equiv \mathcal{H}_y \left(\alpha - \frac{4\pi}{3} \right) \equiv \mathcal{H}_z (\alpha - 2\pi) \tag{7}$$

This means that with the increase of α by $2\pi/3$ we get a cyclic permutation of $x \to y \to z \to x$.

The energies of three Kramers' doublets are obtained by putting matrices of S_x, S_y, S_z for $S = \frac{5}{2}$ into (5), and finding eigenvalues of this 6-dimensional matrix. If we adopt the scheme of representation in which S_z is diagonal, S_x^2 and S_y^2 have non-vanishing elements only for $\Delta M_s = 0$ and ± 2, so that the 6-dimensional secular equation split automatically into two identical 3-dimensional equations. Thiw secular equation is simplified into the following forms:

$$\lambda^3 - 21G^2\lambda - 20G^3 \cos 3\alpha = 0 \tag{8}$$

where $G = \frac{2}{3}F$. The absence of a term proportional to λ^2 corresponds to trace $\mathcal{H} = 0$ in our choice of spin Hamiltonian.

A remarkable feature of equation (8) is found in the fact that α is contained only in the form $\cos 3\alpha$. Accordingly, roots of this equation are periodic functions of α with period $2\pi/3$, and further they are even functions of α. Moreover, if α is increased by $\pi/3$, three roots are obtained simply by changing signs. Therefore, if the roots are calculated as functions of α over the range of $\pi/6$ it is quite simple to find their values for any value of α.

The period of $2\pi/3$ can be forseen from the cyclic relation (7), since eigenvalues should not depend on the naming of three rectangular coordinates in terms of x, y and z. Increase of α by $2\pi/3$ brings the eigenvalues to the same values, but changes eigenfunctions according to cyclic permutation of x, y, z.

Functional dependence of eigenvalues on α, calculated from (8), is illustrated in Fig. 1. For $\alpha = 0$, i.e., $D > 0$, $E = 0$, the roots are $-4G$, $-G$ and $5G$, or $-\frac{8}{3}D$, $-\frac{2}{3}D$ and $\frac{10}{3}D$ in terms of D. The corresponding intervals are $2D$ and $4D$. This is the familiar case described by DS_z^2. For $\alpha = \pi/2$, i.e., $D = 0$, $E > 0$, the roots are $-21G$, 0, $+21G$, or $-2\sqrt{7}E$, 0, $+2\sqrt{7}E$ in terms of E. In this case three Kramers' doublets are equidistantly separated. The same spectrum is obtained for those values of α which are odd multiples of $\pi/6$.

So far we have been concerned with eigenvalues. We can proceed further to find eigenfunctions and to calculate the splitting of each Kramers' doublet caused by the applied magnetic field $|H|$. This gives g values for each doublet. Since g has the tensorial character, it suffices to calculate principal g values

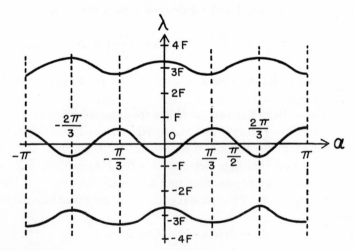

Figure 1 Functional dependence of eigenvalues of Kramers' doublet on α
($\alpha = \tan^{-1}\sqrt{3}\,E/D$)

corresponding to the application of the magnetic field parallel to x, y and z directions. Thus we obtain g_x, g_y and g_z for each of the Kramers' doublets as functions of α. The result is illustrated in Fig. 2.

The region of small α is realized in most haemproteins. They give $g_z = 2$ and g_x, $g_y = 6$. Ferric myoglobin, ferric haemoglobin and their derivatives are examples of this case. In most of these molecules, however, careful experiments show that g_x and g_y are not exactly $= 6$. H. Morimoto succeeded in finding a sinusoidal variation of g depending on the azimuth in the xy-plane, i.e., in the haem plane, on single crystals of ferric myoglobin and its fluoride[2]. Theoretically

$$g_x, g_y = 6 \pm 24\,\frac{E}{D} \qquad (9)$$

for small E/D, and we can estimate E/D. The value of this ratio is 0.0025 and 0.0038 for myoglobin and myoglobin fluoride respectively. Non-vanishing of E indicates that in these molecules symmetry of the ligand field effective on the Fe ion is not strictly tetragonal, but has a small rhombicity. This rhombicity is larger in some of the abnormal haemoglobins. Figure 3 shows EPR spectra measured on unoriented samples of normal haemoglobin (Hb A) and three abnormal haemoglobins Hb M_{Boston}, Hb $M_{Saskatoon}$ and Hb $M_{Hyde Park}$. Spectra of these abnormal haemoglobins clearly show splitting of $g \approx 6$ signals, corresponding to larger values of E/D. The following values have been estimated[3]:

	Hb M_{Boston}	Hb $M_{Saskatoon}$	Hb $M_{Hyde Park}$	
E/D	0.012	0.027	0.013	0.042
			(oxygenated)	(deoxygenated)

Another interesting region is the neighbourhood of $\alpha = \pi/2$. At $\alpha = \pi/2$, i.e., $D = 0$, $E \neq 0$, the intermediate doublet gives

$$g_x = g_y = g_z = \tfrac{30}{7}\,(\doteqdot 4.28) \qquad (10)$$

Since three principal values are equal, g is invariant against variations of direction of applied magnetic field, and a sharp strong signal at $g = 4.28$ is expected. If α is not exactly equal to $\pi/2$, g splits into three different values around $g = 4.28$, such as

$$g_z = 4.28 - \frac{2880}{2401}\left(\frac{\beta H}{E}\right)^2$$

$$g_{x,y} = 4.28 \pm \frac{120}{49}\frac{D}{E} - \frac{1620}{2401}\left(\frac{\beta H}{E}\right)^2, \qquad (11)$$

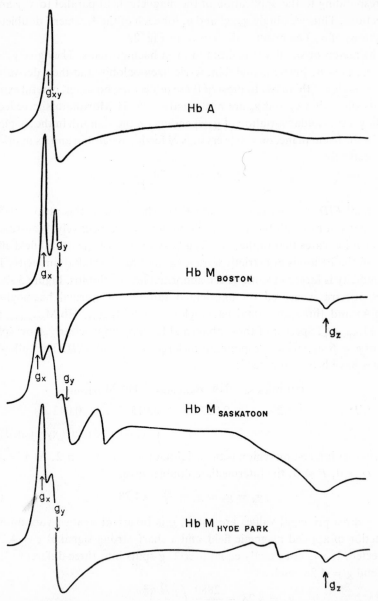

Figure 2   EPR spectra of Hb A, M_{Boston}, Hb $M_{Saskatoon}$ and Hb $M_{Hyde\ Park}$. EPR signal of the last three are only due to the haem iron, which was contained in abnormal subunit

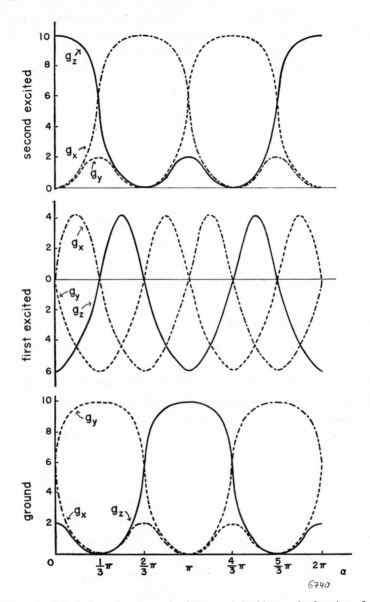

Figure 3 Principal *g*-values for each of Kramers' doublets as the function of α

calculated by Aasa *et al.*[4]. This is actually observed in some cases of non-haem iron proteins, such as inactive pyrocatechase and rubredoxin. Figure 4 gives spectrum of pyrocatechase. In the case of pyrocatechase three g values are 4.41, 4.07 and 3.87, and estimated value of D/E is about $\frac{1}{10}$[5].

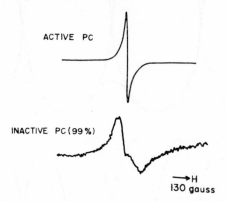

ACTIVE PC

INACTIVE PC (99%)

$\longrightarrow$ H
130 gauss

Figure 4 EPR spectra of active and inactive pyrocatechase (catechol 1,2-oxygenase)

3 Determination of values of D and E

EPR measurement is very effective in determining the relative values of E and D, but their absolute magnitudes are not so easily determined by EPR technique. A possible method is to observe the magnetic field dependence of g; this dependence could be observed at high magnetic field intensity where Zeeman energy $g\beta H$ is not too small compared with distances between Kramers' doublets expressible in terms of D and E. This method has successfully been applied to haemoglobin by Prof. D. J. E. Ingram[6].

Another effective method is measurment of temperature dependence of static magnetic susceptibility. At temperatures where the corresponding thermal energy KT is much higher than the separation between Kramers' doublets, these separations can be disregarded, and the effective Bohr magneton number N_{eff} takes the familiar spin-only value $\sqrt{35} = 5.92$, and this value is independent of the magnetic field $|H|$ relative to x, y, z axes of the spin Hamiltonian. At low temperatures, however, N_{eff} depends on D/kT and E/kT, and moreover, N_{eff} depends also on the direction of $|H|$. Let us consider, as the simplest example, the case $D > 0$, $E = 0$. At extremely low temperature $2D \gg kT$, only the lowest doublet is populated. When $|H|$ is parallel to z, g is $= 2$, and N_{eff} is $\sqrt{3}$, and when $|H|$ is parallel to x, y plane, g and N_{eff} are three times as large: $N_{eff} = 3\sqrt{3}$. Thus there is a large anisotropy in N_{eff}. The average N_{eff}, which is observed by susceptibility measurements on unoriented samples, is given by

$$N_{\text{eff}} = \sqrt{\frac{3 + 27 + 27}{3}} = \sqrt{19}. \tag{12}$$

This is definitely smaller than the corresponding high temperature value $\sqrt{35}$. The decrease of N_{eff} from the high temperature value to the low temperature value is theoretically represented by a simple function of D/kT, so that by fitting observed values by theoretical curve we can determine the value of D. Figure 5 shows results of early measurement on myoglobin done by A. Tasaki *et al.*[7] More recently, effect of saturation, i.e., the non-linear dependence of magnetization on H at very low temperatures has been taken into account and a more satisfactory agreement is obtained.

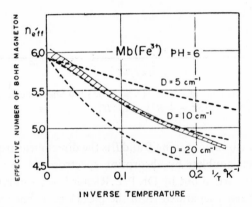

Figure 5 N_{eff} dependency of ferrimyoglobin to the inverse temperature. The shaded area shows the value of N_{eff} and its error limits, calculated from the magnetic susceptibility of ferrimyoglobin. Dotted lines show the theoretical ones of N_{eff} for $D = 5, 10$ and $20\,\text{cm}^{-1}$

An even more sensitive method is the measurement of paramagnetic anisotropy of single crystals. In the uniaxial case $D > 0$, $E = 0$, this measures

$$\Delta(N_{\text{eff}}^2) = N_{\text{eff}}^{\perp 2} - N_{\text{eff}}^{\|2} \tag{13}$$

where suffices $\perp$ and $\|$ indicate that the direction of magnetic field is perpendicular to, and parallel to, the z axis, respectively. $\Delta(N_{\text{eff}}^2)$ is zero at high temperature limit, and approaches $27 - 3 = 24$ at absolute zero. Simple theoretical calculation shows that the curve $\Delta(N_{\text{eff}}^2)$ vs. D/kT has a broad maximum as shown in Fig. 6. H. Uenoyama could fit measured experimental points on myoglobin and its fluoride on this theoretical curve by suitable choices of D [8]. Values of D thus obtained were 10.5 cm^{-1} for myoglobin and 6.5 cm^{-1} for myoglobin fluoride, in good agreements with the corresponding values determined by the preceding method.

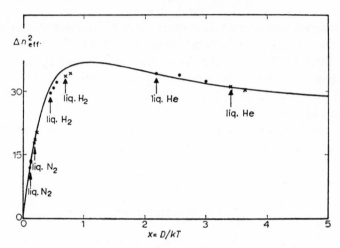

Figure 6   Temperature dependence of $\Delta(N^2_{\text{eff}})$. The solid line was calculated by using the spin Hamiltonian $\mathscr{H} = D\left\{S_z^2 - \tfrac{1}{3}S^2\right\} + 2\beta|H|$

A third method of determining D and E is the direct measurement of energy intervals between doublets by absorption spectroscopy in the far infrared region. This was carried out by Dr. P. L. Richards on a number of different derivations of haems separated from the globin part, but measurement on intact protein containing haems not yet succeeded.

All these methods give consistent results in the case of haemproteins, and D is of the order of 10 cm^{-1} although values are different for different derivatives. This value of ~ 10 cm^{-1} is surprisingly large, since in most inorganic ferric compounds of approximately cubic symmetry D is smaller by one or two orders of magnitude.

4   Interpretation of D and E in terms of orbital energies of d electrons

Since Griffith's early work it is well known that in the low spin case ($s = \tfrac{1}{2}$) g_x, g_y, g_z values are related to orbital energies of d electrons, and from observed g values we can derive information concerning relative orbital energies of three low-lying d orbitals d_{yz}, d_{zx} and d_{xy}. Knowledge about these orbital energies is useful in constructing plausible models of the atomic arrangement of haem and its surroundings.

In the high spin case, with which we are concerned in this paper, results of EPR measurements are described in terms of two parameters, D and E. Our next problem is, can we derive information on orbital energies from the knowledge of D and E? The answer is in principle yes, but in practice it is very different.

First of all, we have to understand the reason why the sextet state of ferric ion is split into three doublets in proteins. Generally speaking, spins can recognize the ligand field only through spin-orbit interaction. Now, in the case of d^5 $S = \frac{5}{2}$, there is no orbital degeneracy, i.e., there is a single wave function Ψ of electronic coordinate associated with this sextet. Therefore, Ψ has no flexibility to instruct the spin system to take different energies according to different orientations of the resultant spin. In technical terms, all matrix elements of spin-orbit interaction vanish among six substates of $^6A_1$. However, the ferric ion has excited quartet states and spin-orbit interaction has non-vanishing matrix elements between the ground sextet $^6A_1$ and excited quartet states of symmetry $^4T_1$. As a consequence of this, the ground sextet gets a small mixing, or "contamination", of quartet states. Now, $^4T_1$ state has three independent wave functions of electronic coordinate ψ_x, ψ_y, ψ_z, which have transformation properties against rotation in space similar to those of coordinates x, y, z themselves (cf. Section 6). Therefore, $^4T_1$ is a superposition of the quartets $^4T_{1x}$, $^4T_{1y}$ and $^4T_{1z}$, corresponding to spatial wave functions ψ_x, ψ_y and ψ_z respectively. As long as the symmetry of the ligand field is cubic, these three quartets have a common energy, but if the symmetry is tetragonal around the z axis T_{1z} will have a different energy and if the symmetry is rhombic, three quartets T_{1x}, T_{1y} and T_{1z} will be energetically separated. Thus $^4T_1$ wave functions have high sensitivity to the ligand field, and through mixing of $^4T_1$ the ground sextet state is endowed with a weak sensitivity to the ligand field. Mixing of $^4T_{1x}, ^4T_{1y}, ^4T_{1z}$ into the ground sextet is not the same for three Kramers' doublets, so that these doublets will have slightly different energies, as described by the spin Hamiltonian.

Now, it is important to note that in the case of ferric ions in proteins, particularly in haemproteins, one quartet excited state of symmetry $^4T_1$ lies unusually low. This gives unusually large mixing of quartet to sextet, and leads to unusually large values of parameters D and E. Theoretical explanation of the unusually low position of $^4T_1$ will be given in section 6, and we shall proceed for the moment assuming this.

5   Derivation of formulae for D and E

What we have just described in the last section can be technically formulated using the second order perturbation theory in quantum mechanics, adopting spin-orbit interaction as perturbing term. Since the spin-orbit interaction is linear in spin components, it can be shown that the resulting spin Hamiltonian is quadratic in spin components.

The explicit form of spin-orbit interaction is

$$V_{so} = \zeta \sum_i l_i s_i \tag{14}$$

where l_i and s_i are orbital and spin angular momenta of the ith d electron, measured in unit of $h/2\pi$, and ζ is an energy called the spin-orbit interaction constant. Since the expression of the theory change due to the second order perturbation consists of terms whose numerators are squares of matrix elements of V_{so} and whose denominators are excitation energies, expressions for D and E must be of the form

$$\zeta^2 \left(\frac{c_x}{E_x} + \frac{c_y}{E_y} + \frac{c_z}{E_z} \right), \tag{15}$$

where E_x, E_y, E_z denote excitation energies to $^4T_{1x}$, $^4T_{1y}$, $^4T_{1z}$ from $^6A_1$ respectively (see Fig. 7), and c_x, c_y, c_z are pure numbers. These numbers are obtainable from perturbation calculation, but the following symmetry arguments give their relative values in an elementary way.

It is evident that

(i)   when $E_x = E_y = E_z$, then $D = E = 0$, and

(ii) if x and y are interchanged, D should remain invariant and E should change its sign.

Condition (i) gives

$$c_x + c_y + c_z = 0 \tag{16}$$

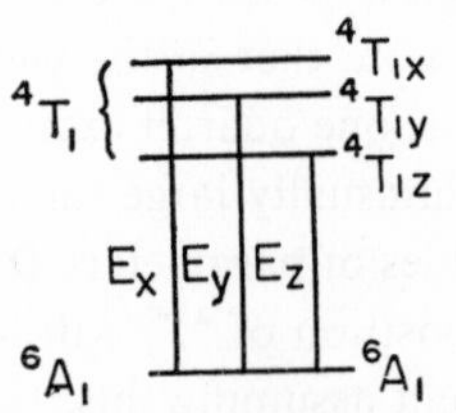

Figure 7   Energy splitting of $^4T_1$ by lower symmetry

for both D and E, while condition (ii) gives

$$c_x = c_y \quad \text{for } D \quad \text{and} \quad c_x = -c_y \quad \text{for } E \tag{17}$$

From (16) and (17), we have

$$D = c\zeta^2 \left(\frac{2}{E_z} - \frac{1}{E_x} - \frac{1}{E_y} \right)$$

$$E = c'\zeta^2 \left(\frac{1}{E_x} - \frac{1}{E_y} \right) \tag{18}$$

Now, we can prove $c = c'$ as follows. Our spin Hamiltonian $D\,(S_z^2 - \frac{1}{3}S^2)$ $+ E\,(S_x^2 - S_y^2)$ can be written as

$$\bar{D}\,(S_x^2 - \tfrac{1}{3}S^2) + \bar{E}\,(S_y^2 - S_x^2) \tag{19}$$

with the use of

$$\bar{D} = \tfrac{2}{3}E - \tfrac{1}{2}D, \quad \bar{E} = -\tfrac{1}{2}E - \tfrac{1}{2}D \tag{20}$$

On the other hand expressions for $\bar{D}$ and $\bar{E}$ must be obtained from (17) by cyclic permutation of x, y, z. For instance

$$\bar{E} = c'\zeta^2 \left(\frac{1}{E_y} - \frac{1}{E_z} \right). \tag{21}$$

Compatibility between (18), (20) and (21) is obtained only when $c = c'$. Thus finally we have

$$D = c\zeta^2 \left(\frac{2}{E_z} - \frac{1}{E_x} - \frac{1}{E_y} \right)$$

$$E = c\zeta^2 \left(\frac{1}{E_x} - \frac{1}{E_y} \right) \tag{22}$$

c is $\frac{1}{10}$ if configuration interaction can be neglected.

6 Some characteristic features of electronic structure of ferric ions in proteins

In section 4 we mentioned that the excitation energy to $^4T_1$ is unusually low in the case of the ferric ion in haemproteins. In order to explain this and other features of electronic structures of iron in proteins, we start our discussion assuming the symmetry of the ligand field as cubic in the first approximation.

We take x, y, z axes in the directions of the cubic axes, and denote the five independent pure d orbitals as follows

$$\left.\begin{aligned} d_{yz} &= 2yzR\,(r)\\ d_{zx} &= 2zxR\,(r)\\ d_{xy} &= 2xyR\,(r) \end{aligned}\right\} \tag{23}$$

$$\left.\begin{aligned} d_{x^2-y^2} &= (x^2 - y^2)\,R(r)\\ d_{z}^{2} &= \frac{1}{\sqrt{3}}\,(2z^2 - x^2 \cdot y^2)\,R(r) \end{aligned}\right\} \tag{24}$$

The first three orbitals (23) are called $d\varepsilon$ orbitals and the last two orbitals (24) $d\gamma$ orbitals. From the equivalence of x, y and z in cubic symmetry, it is obvious that the three $d\varepsilon$ orbitals have the same energy. Thus $d\varepsilon$ is triply degenerate. By writing

$$d_{z}^{2} = \frac{1}{\sqrt{3}}\,(d_{z^2-x^2} - d_{y^2-z^2})$$

it is easy to see that the orbital energies of d_{z^2} and $d_{x^2-y^2}$ are the same. Hence $d\gamma$ is doubly degenerate. These degeneracies of $d\varepsilon$ and $d\gamma$ are not lifted as long as the ligand field retains the cubic symmetry.

As mentioned earlier, d orbitals are more or less delocalized in iron complexes. Actual orbitals are molecular orbitals, but transformation properties of these molecular orbitals against cubic symmetry operations (rotations) remain the same as pure d orbitals, as proved rigorously on group-theoretical considerations. In particular, the degeneracies of $d\varepsilon$ and of $d\gamma$ remain unaffected.

The effect of a cubic ligand field on these orbitals can be characterized by a single parameter Δ, where Δ is the difference of orbital energies of $d\gamma$ and $d\varepsilon$:

$$\Delta = (\text{orbital energy of } d\gamma) - (\text{o.e. of } d\varepsilon) \tag{25}$$

The charge distributions of $d\gamma$ orbitals are more or less concentrated along coordinate axes, but those of $d\varepsilon$ orbitals have vanishing density there. Electronegative ligands are situated on the coordinate axes in the case of octahedral coordination, while they are located on space diagonals in the case of tetrahedral coordination. From these considerations it is reasonably understood that in octahedrally coordinated complexes Δ is positive and in tetrahedrally coordinated complexes Δ is negative.

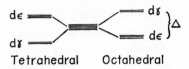

Tetrahedral   Octahedral

Figure 8   Electronic levels of d^5 (cubic approximation)

This splitting of d level into two levels $d\varepsilon$ and $d\gamma$ gives as a consequence splitting of d^5 configuration into five sub-configurations $(d\varepsilon)^{5-n}(d\gamma)^n$, $n = 0, 1, 2, 3, 4$. If we disregard interelectronic Coulomb interaction for the moment, energies of these sub-configurations are simple multiples of Δ. Taking the sub-configuration $(d\varepsilon)^3(d\gamma)^2$ as standard, energies of $(d\varepsilon)^5$, $(d\varepsilon)^4(d\gamma)$ and $(d\varepsilon)^2(d\gamma)^3$, $(d\varepsilon)(d\gamma)^4$ are given by -2Δ, $-\Delta$ and Δ, 2Δ respectively. This is illustrated in Fig. 9. This diagram gives a common single energy value for $\Delta = 0$, which corresponds to the free ion. It is well known, however, that states of different spin multiplicities arising from d^5 configuration are widely separated in energy. Quartet states ^4G, ^4P, ^4D, ^4F are above the ground state ^6S by several electron volts, and the doublet states are located still higher. This separation comes from interelectronic Coulomb interaction. This Coulomb interaction can be taken into account semi-qualitatively in the following way. Generally speaking, effective contribution of Coulomb interaction is larger as the resultant spin is smaller, and the states with smaller spin values are pushed up relatively to those with larger spin values (Hund's rule). Now, the states of highest spin values arising from five sub-configurations are

$(d\varepsilon)^5$ and $(d\varepsilon)(d\gamma)^4$: $^2T_2$

$(d\varepsilon)^4(d\gamma)$ and $(d\varepsilon)^2(d\gamma)^3$: $^4T_1$

       ($^4T_2$, which arises from these sub-configurations is omitted

       since this state is not important for our subsequent discussions)

$(d\varepsilon)^3(d\gamma)^2$: $^6A_1$

Assuming that contributions from Coulomb interaction are independent of Δ, we may include its effect in the diagram of Fig. 10 simply shifting straight lines in the vertical direction by different amounts for different spin values. Since we are interested only in relative heights of energy levels, we may keep the horizontal line for $(d\varepsilon)^2(d\gamma)^2$ $^6A_1$ fixed, and shift upwards two straight lines for $^4T_1$ by a certain amount, and finally shift upwards two straight lines for $^2T_2$ by a larger amount. The result is shown in Fig. 10.

Now, an interesting feature of this diagram is the crossing of lines for different spin values. In particular, doublet lines cross sextet line at Δ_c nad Δ_c'

($\fallingdotseq -\Delta_c$). Therefore the sextet state is the ground state only in a limited range of weak ligand field $\Delta'_c < \Delta < \Delta_c$, and doublet states become the ground state outside this range. Quartet states never become lowest provided that the vertical shift of doublets is less than twice that of quartets. This is verified by more detailed calculation.

d^5-system

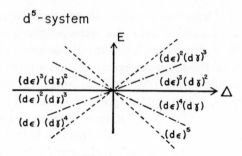

Figure 9   Energy scheme of subconfiguration of d^5 system

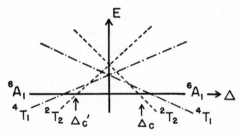

Figure 10   Relative heights of energy levels accuming the contributions of Columb interaction

From the thermal coexistence of high spin (sextet) and low spin (doublet) states observed on many derivatives of haemproteins and other evidences, actual effective Δ for the ferric ion in haemproteins seems to be very close to Δ_c. Δ for ferric ions coordinated tetrahedrally in proteins may not be very far from Δ'_c although data for this case are very limited. Then we may reasonably expect from the diagram that the excitation energy to $^4T_1$ is fairly small. The presence of quartet $^4T_1$ state with unusually low excitation energy has been explained.

In real molecules the symmetry of the ligand field is lower than cubic, and the energy diagram such as in Fig. 10 must be constructed in multidimensional space. This remark is important particularly with regard to $^2T_2$, as the stable state of $^2T_2$ seems to have much larger rhombicity than the stable state of $^6A_1$. If atomic arrangements are exactly the same for $^6A_1$ and $^2T_2$, there must be quantum-mechanical mixing of these states in the close neighbourhood of the crossing point, but this has never been observed.

7 Case of tetrahedral coordination

As mentioned earlier, EPR signals in the neighbourhood of $g = 4.28$ are often observed in non-haem iron proteins in which ferric ions are believed to be tetrahedrally coordinated. In this section we discuss how tetrahedral coordination is favourable for giving signals around $g = 4.28$.

As discussed in section 2, $g = \frac{30}{7} = 4.28$ is obtained when the spin Hamiltonian can be written as

$$E(S_i^2 - S_j^2), \tag{26}$$

where i, j are any two of x, y, z. If (i, j, k) is a permutation of (x, y, z), this condition requires

$$D' = c\zeta^2 \left(\frac{2}{E_k} - \frac{1}{E_i} - \frac{1}{E_j} \right) = 0 \tag{27}$$

or in other words, $1/E_k$ should be the arithmetical mean of $1/E_i$ and $1/E_j$. If deviation from the cubic symmetry is not too large, T_{1x}, T_{1y}, T_{1z} are not very widely separated, and E_i, E_j, E_k are nearly the same. Then the above condition may be replaced by the following one:

"Three sublevels of $^4T_1$ are spaced equidistantly, i.e.

$$E_i - E_k = E_k - E_j" \tag{28}$$

Then our problem is: why does tetrahedral coordination favour this relation (28)?

In order to discuss this problem, we assume that the ligand field V to which d electrons are subjected can be expressed as a sum of contributions due to four ligands A, B, C, D:

$$V = V_a + V_b + V_c + V_d \tag{29}$$

V_a denotes contribution due to ligand A, and so on. We assume that V_a is axially symmetric around the line connecting iron and atom A, and similarly for V_b, V_c, V_d. These assumptions may appear too restrictive, but essential features of the tetrahedral ligand field may be reasonably represented by this model.

First we examine orbital energies of a d-electron under a component field V_a. Taking a rectangular coordinate system $x_a y_a z_a$ with z_a coinciding with the direction Fe$_{\cdot}$—A, five independent d orbitals may be written as $d_{y_a z_a}, d_{z_a x_a}, d_{x_a^2 - y_a^2}$ and $d_{z_a^2}$. Since V_a is axially symmetric around z_a axis, these five functions are eigenfunctions of V_a, and $(d_{y_a z_a}, d_{z_a x_a})$ and $(d_{x_a y_a}, d_{x_a^2 - y_a^2})$ are two degenerate pairs. Thus V_a is represented by a diagonal matrix

$$\bar{V}_a = \begin{pmatrix} V_2 & & & & \\ & V_2 & & 0 & \\ & & V_1 & & \\ & 0 & & V_1 & \\ & & & & V_3 \end{pmatrix} \tag{30}$$

in the 5-dimensional space spanned by $d_{y_a z_a}, d_{z_a x_a}, d_{x_a y_a}, d_{x_a^2 - y_a^2}, d_{z_a^2}$. It is reasonable to assume $V_1 < V_2 < V_3$.

We note here that eigenfunctions of V_a may not be pure d atomic orbitals, but may be linear combinations with atomic orbitals of A. (30) is obviously valid in such a case too.

In order to construct the matrix of total V, it is necessary to transform $\bar{V}_a$, $\bar{V}_b$, $\bar{V}_c$ and $\bar{V}_d$ to a common scheme. For this purpose we take the common rectangular axes x, y, z in coincidence with the cubic axes. Ne set up the transformation matrix T_a between $d_{y_a z_a}, d_{z_a x_a}, d_{x_a y_a}, d_{x_a^2 - y_a^2}, d_{z_a^2}$ and $d_{yz}, d_{xz}, d_{x^2 - y^2}, d_{z_a^2}$:

$$T_a = \begin{array}{c} \\ d_{y_a z_a} \\ \\ d_{z_a x_a} \\ \\ d_{x_a y_a} \\ \\ d_{x_a^2 - y_a^2} \\ \\ d_{z_a^2} \end{array} \begin{Vmatrix} \overset{d_{yz}\quad d_{zx}\quad d_{xy}\quad d_{x^2-y^2}\quad d_z^2}{} \\ -\dfrac{1}{\sqrt{6}} & \dfrac{1}{3\sqrt{2}} & -\dfrac{1}{\sqrt{3}} & -\dfrac{1}{3} & \dfrac{1}{\sqrt{3}} \\ \dfrac{1}{\sqrt{6}} & \dfrac{1}{3\sqrt{2}} & \dfrac{1}{\sqrt{3}} & -\dfrac{1}{3} & \dfrac{1}{\sqrt{3}} \\ 0 & -\dfrac{\sqrt{2}}{3} & 0 & \dfrac{2}{3} & \dfrac{1}{\sqrt{3}} \\ \sqrt{\dfrac{2}{3}} & 0 & -\dfrac{1}{\sqrt{3}} & 0 & 0 \\ 0 & \sqrt{\dfrac{2}{3}} & 0 & \dfrac{1}{\sqrt{3}} & 0 \end{Vmatrix} \tag{31}$$

and transform $\bar{V}_a$ into the common scheme by

$$V_a = T^{-1}\bar{V}_a T \tag{32}$$

The result of this transformation is as follows

$$
V_a = \left\| \begin{array}{ccccc}
p & q & q & -s & \dfrac{1}{\sqrt{3}}s \\[2ex]
q & p & q & s & \dfrac{1}{\sqrt{3}}s \\[2ex]
q & q & p & 0 & -\dfrac{2}{\sqrt{3}}s \\[2ex]
-s & s & 0 & r & 0 \\[2ex]
\dfrac{1}{\sqrt{3}}s & \dfrac{1}{\sqrt{3}}s & -\dfrac{2}{\sqrt{3}}s & 0 & r
\end{array} \right\|
\tag{33}
$$

where $p = \frac{4}{9}V_1 + \frac{2}{9}V_2 + \frac{1}{3}V_3$, $q = -\frac{2}{9}V_1 - \frac{1}{9}V_2 + \frac{1}{3}V_3$, $r = \frac{1}{3}V_1 + \frac{2}{3}V_2$ and $s = \frac{1}{3}(V_2 - V_1)$. In similar ways, we obtain

$$
V_b = \left\| \begin{array}{ccccc}
p & q & -q & s & -\dfrac{1}{\sqrt{3}}s \\[2ex]
q & p & -q & -s & -\dfrac{1}{\sqrt{3}}s \\[2ex]
-q & -q & p & 0 & -\dfrac{2}{\sqrt{3}}s \\[2ex]
s & -s & 0 & r & 0 \\[2ex]
-\dfrac{1}{\sqrt{3}}s & -\dfrac{1}{\sqrt{3}}s & -\dfrac{2}{\sqrt{3}}s & 0 & r
\end{array} \right\|
\tag{34}
$$

$$
V_c = \left\| \begin{array}{ccccc}
p & -q & -q & -s & \dfrac{1}{\sqrt{3}}s \\[2ex]
-q & p & q & -s & -\dfrac{1}{\sqrt{3}}s \\[2ex]
-q & q & p & 0 & \dfrac{2}{\sqrt{3}}s \\[2ex]
-s & -s & 0 & r & 0 \\[2ex]
\dfrac{1}{\sqrt{3}}s & -\dfrac{1}{\sqrt{3}}s & \dfrac{2}{\sqrt{3}}s & 0 & r
\end{array} \right\|
\tag{35}
$$

$$V_d = \begin{Vmatrix} p & -q & q & s & -\dfrac{1}{\sqrt{3}}s \\[2mm] -q & p & -q & s & \dfrac{1}{\sqrt{3}}s \\[2mm] q & -q & p & 0 & \dfrac{2}{\sqrt{3}}s \\[2mm] s & s & 0 & r & 0 \\[2mm] -\dfrac{1}{\sqrt{3}}s & \dfrac{1}{\sqrt{3}}s & \dfrac{2}{\sqrt{3}}s & 0 & r \end{Vmatrix} \tag{36}$$

Since these matrices are represented in a common scheme, they can be added simply element by element. The result matrix is diagonal:

$$V = \begin{pmatrix} 4p & & & & \\ & 4p & & 0 & \\ & & 4p & & \\ & 0 & & 4r & \\ & & & & 4r \end{pmatrix} \tag{37}$$

Thus $4p$ is the orbital energy of $d\varepsilon$ and $4r$ is that of $d\gamma$.

In this calculation we have assumed that the four ligands are completely equivalent, so that the resulting field V is strictly cubic, and both parameters D and E of the spin Hamiltonian vanish. To have a non-vanishing spin Hamiltonian we have to introduce some deviation from the cubic symmetry.

There are many ways of introducing deviation from the cubic symmetry. Here we discuss the case on which V_a, V_b are equivalent as are V_c and V_d, but these two sets of ligands are slightly inequivalent. To describe this situation, we add primes to p, q, r, s appearing in V_c and V_d. Then the resulting matrix of V' has small non-diagonal elements:

$$V' = \begin{Vmatrix} 2(p+p') & 2(q-q') & 0 & 0 & 0 \\[2mm] 2(q-q') & 2(p+p') & 0 & 0 & 0 \\[2mm] 0 & 0 & 2(p+p') & 0 & \dfrac{2}{\sqrt{3}}(s-s') \\[2mm] 0 & 0 & 0 & 2(r+r') & 0 \\[2mm] 0 & 0 & \dfrac{2}{\sqrt{3}}(s-s') & & 2(r+r') \end{Vmatrix} \tag{38}$$

On the assumption that $|s - s'|$ is much smaller than the energy difference between $d\gamma$ and $d\varepsilon$, e.q. $2(p + p') - 2(r + r')$, this matrix V' shows that $d\varepsilon$ is split into the three levels as follows:

$$2(p + p') + 2(q - q'), \quad 2(p + p'), \quad 2(p + p') - 2(q - q'),$$
$$2(r + r'), \quad 2(r + r').$$

Therefore, energies of three orbitals of $d\varepsilon$ are equidistantly spaced provided that the second order terms of off-diagonal elements can be neglected.

The $^4T_1$ state has three orbital configurations as discussed in section 6. If d_{yz}, d_{zx}, d_{xy} are equidistantly spaced, three substates $^4T_{1x}, ^4T_{1y}, ^4T_{1z}$ are also equidistantly separated. This corresponds to the situation $D \fallingdotseq 0$, $E \neq 0$, and $g = 4.28$ signal in EPR spectra. Thus, we have shown that a small deviation from tetrahedral symmetry may give $D \neq 0$, $E = 0$.

It may be interesting to study various modes of deviations from the cubic, and to examine conditions under which $D \fallingdotseq 0$, $E \neq 0$ can be expected.

8 Conclusion and summary

EPR signals due to high spin ferric ions in proteins can normally be described in terms of two parameters D and E, by using a quadratic spin Hamiltonian. For this purpose it is convenient and instructive to plot energies and g-values of three Kramers' doublets as functions of angle α, which indicates relative magnitudes of D and E. In the case of haemproteins D is much larger than E, but the reverse situations $E \gg D$ usually applies in the case of non-haem iron proteins. Parameters D and E are related to orbital energies of d electrons subject to ligand fields. A convenient semi-qualitative method is given for estimating relative values of these orbital energies. This method is applied to the case of tetrahedral coordination, and it is shown that the relation $E \gg D$ can be reasonably explained.

Discussion

D.J.E. INGRAM Does the expression for the D splitting depend on any other factors than the spin-orbit coupling parameter of the free ion, and the E_x, E_y and E_z of the free energy levels, in your theory?

M. KOTANI In my theory D depends on E_x, E_y and E_z and the spin-orbit coupling parameter, but the effective value of the spin-orbit coupling parameter of Fe^{+++} in the haem will be somewhat smaller than its value for the

free ion. Furthermore, more detailed analysis using the higher order perturbation theory we get the following expression for the spin Hamiltonian:

$$DS_z^2 + FS_z^4,$$

and D will depend also on excitation energies to higher energy levels than
those expressed by E_x, E_y and E_z.

References

(1) J.R.Herriott, L.C.Sieker, L.H.Jensen and W.Lovenberg, to be submitted.
(2) H. Morimoto and M. Kotani, *Magnetic Resonance in Biological Systems* (ed. by A. Ehrenberg *et al.*, Pergamon Press, London 1967) p. 135.
(3) H.Watari, A.Hayashi, H.Morimoto and M.Kotani, *Recent Developments of Magnetic Resonance in Biological System* (ed. by Shizuo Fujiwara *et al.*, Hirokawa Publishing Comp., Tokyo 1968) p. 128.
(4) R.Aasa, B.G.Malmström, P.Saltman and T.Vänngard, *Biochim. Biophys. Acta*, **88** (1964) 430.
(5) H.Watari, T.Nakazawa and T. Yamano, *Biochim. Biophys. Acta*, **146** (1967) 409.
(6) D.J.E.Ingram, Lecture given at this conference.
(7) A.Tasaki, J.Otsuka and M.Kotani, *Biochim. Biophys. Acta*, **140** (1967) 284.
(8) H.Uenoyama, T.Iizuka, H.Morimoto and M.Kotani, *Biochim. Biophys. Acta*, **160** (1968) 159.

NMR studies of the role of the heme group during the cooperative oxygenation of hemoglobin

S. OGAWA, R. G. SHULMAN, K. WÜTHRICH*

and T. YAMANE

Bell Telephone Laboratories, Incorporated Murray Hill, New Jersey 07974

Abstract

High resolution NMR studies of hemoglobin have been made at 220 MHz. Two conclusions have been reached. First there are changes in the protein structure upon oxygenation. Second there are no changes at one heme group when its neighbors are alternately oxygenated and deoxygenated. From these two results we conclude that the free energy of interaction responsible for the well-known cooperative oxygenation of hemoglobin is created by the dependence of the intersubunit binding energy upon the degree of ligation.

HEMOGLOBIN CONSISTS of two pairs of two different subunits, α and β chains. Each chain contains one heme group which can reversibly bind one oxygen molecule. Hemoglobin increases its affinity towards oxygen during the course of oxygenation. This cooperative oxygen binding, sometimes called "heme-heme" interaction, which shows up as the well-known sigmoidal binding curve, has been studied in many laboratories[1], since hemoglobin, aside from its intrinsic interest, can be a model system for regulatory functions of various enzymes.

---

* Present address: Laboratorium für Molekularbiologie, E.T.H. Zurich, Switzerland.

In spite of the detailed knowledge accumulated for years about hemoglobin the understanding of the molecular basis for the cooperative oxygen binding has not yet been established. The increase of oxygen affinity in hemoglobin during the sequential oxygenation of four heme groups is believed to be associated with the conformational changes in hemoglobin, in particular the quaternary structural differences between deoxy and oxyhemoglobin which have been determined by the X-ray studies of Perutz[2] and his colleagues. However, it has not yet been shown how the information of oxygenation of one heme group is transmitted through the molecule in order to increase the strength of the subsequent oxygen binding.

High resolution proton NMR can be very useful in delineating these changes[3]. The magnetic hyperfine interaction between the protons of the heme group and the unpaired electrons of the heme iron when it is paramagnetic, such as in deoxyhemoglobin and in cyanoferrihemoglobin, shifts the resonance lines of those protons to higher and lower fields. Hence they are well separated from the resonances arising from the amino acid residues. Therefore, resonance lines in two different forms of hemoglobin, which have low and high oxygen affinity respectively, could provide information on the extent of heme participation in the affinity changes. Deoxyhemoglobin is the low affinity form which we have studied. The high affinity form, which we have prepared is a mixed-state hemoglobin, in which the heme irons of the chains of one kind, for example the α-chains, are in the ferric state and those of the other kind, in this case the β-chains, are in the ferrous state. This mixed-state hemoglobin has been shown[5] to have high oxygen affinity with Hill constant $n \approx 1$.

High resolution proton NMR can also show conformational changes in proteins, especially when ring current shifts[6] are observable. These shifts are very sensitive to the relative positions of the proton to be observed and the aromatic ring which is responsible to the shift. The ring current shifted lines are temperature independent, in contrast to the lines shifted by magnetic interactions, which are strongly temperature dependent. In this way those two kinds of shifted lines are distinguishable.

Experimental methods and materials

The NMR spectra were measured at 220 MHz with a Varian HR-220 spectrometer. The signal-to-noise ratios were improved by averaging with a Fabritek 1062 computer. The volumes of the samples studied by NMR were 0.3 to 0.5 millimeter, and the heme concentrations were approximately

6×10^{-3} M. The resonance positions are relative to the internal standard, DSS (2,2-dimethyl-2-silapentane-5-sulfonate). Human oxyhemoglobin A was prepared from freshly drawn blood by the method described by Bunn and Jandl[7]. Oxyhemoglobin was separated into individual chains with para-chloromercuribenzoate (PMB) by the method of Bucci and Fronticelli[8]. The PMB was removed from the oxy-α-chains by absorption on a carboxy-methylcellulose (CMC) and by washing with mercaptoethanol. The oxy-α-chains free from PMB were then eluted from the column. Hemoglobin samples were deoxygenated by alternating purified nitrogen gas with a vacuum. The deoxy chain solutions were prepared by addition of dithionite, followed by anaerobic gel filtration to eliminate the reaction products. The state of oxygenation was mentioned by means of optical absorption spectra. After deoxygenation, the NMR samples were stored at 4°C until used.

Mixed-state hemoglobin was prepared from aged oxyhemoglobin with Huishman's procedure[9], using bio-rex 70 ion exchange column. After deoxygenation and several hours at room temperature, the sample was oxygenated and rechromatographed. The chromatographic elution pattern was the same as before, indicating that negligible amounts of oxyhemoglobin A (HbO_2) or ferric hemoglobin A [$Hb(H_2O)$] tetramers were generated through heme, subunit, or electron exchange.

Experimental results

Figure 1 shows the low-field region of the NMR spectra of deoxygenated α-chains and β-chains and compares them with the spectrum of deoxyhemoglobin A (Hb). The spectra of the α-chains and β-chains differ to some extent from each other, and both differ greatly from the spectrum of the tetrameric form. The positions of these resonances are all temperature-dependent and thus evidently arise from hyperfine interactions. In the cyanoferric state the differences between the isolated chains and the tetrameric form are smaller[10] than in the deoxygenated state. It has been reported elsewhere[11] that the hyperfine shifted resonances from the ferric heme cyanide in a mixed-state hemoglobin composed of oxy- and ferri-cyanide chains were the same as those in ferric HbCN. This indicates that the distribution of unpaired electrons in the ligated heme of one chain was unchanged when the ligands of the other chains were changed from cyanide to oxygen. This is expected from the well-known similarity in protein structure between oxy- and cyano-derivatives of hemoglobin. Figure 2 shows that the shifted resonances in mixed-state hemoglobins composed of deoxy- and ferric chains, are a superposition of the resonances of Hb and Hb(H_2O), while other experiments have shown that those composed of deoxy- and ferric cyanide-chains are a similar superposition of Hb and HbCN. Similar results are obtained at 18 °C, so evidently the temperature dependence of the resonances is the same as that of the homogeneous tetramers.

A comparison of the central regions of the spectra of Hb and HbO_2 is presented in Fig. 3. In the spectrum of Hb, a well-resolved line with intensity corresponding to approximately six protons per subunit is observed close to DSS. The position of this line was independent of temperature between

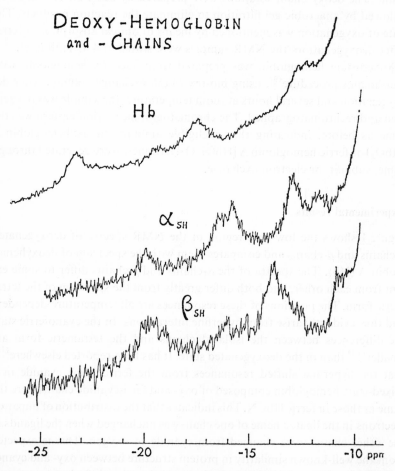

Figure 1 A comparison of the down field shifted NMR lines at pD 7.0 of (top spectrum) deoxy hemoglobin and (middle and bottom spectra) deoxy α_{SH} and deoxy β_{SH} chains.

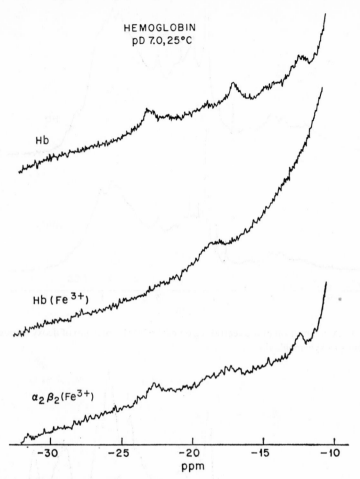

HEMOGLOBIN
pD 7.0, 25°C

Hb

Hb (Fe^{3+})

$\alpha_2 \beta_2$(Fe^{3+})

-30   -25   -20   -15   -10
ppm

Figure 2 A comparison of the NMR spectra of (top spectrum) deoxy hemoglobin (middle spectrum) ferrihemoglobin and (bottom spectrum) deoxy mixed-state hemoglobin. Temperature, 25°C; pD, 7.0.

10° and 35°C. Hence this line is evidently shifted by ring currents and its position is sensitive to conformational changes in the molecule. The resonance observed at this position in Hb is not observed in HbO$_2$. The most likely explanation of its disappearance is that the subunits in HbO$_2$ and Hb have different tertiary structures.

Because NMR spectra of smaller proteins are better resolved, it is easier to obtain details of tertiary structural changes upon oxygenation from myoglobin than from tetrameric hemoglobin. Evidence for conformational

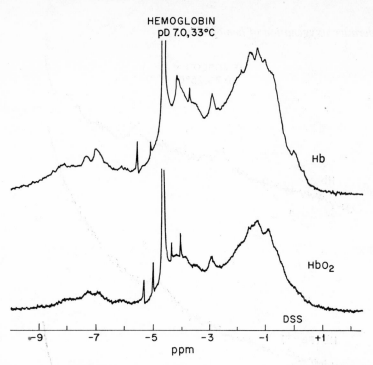

Figure 3 A comparison of the central region of the NMR spectrum of deoxy hemoglobin with that of oxyhemoglobin

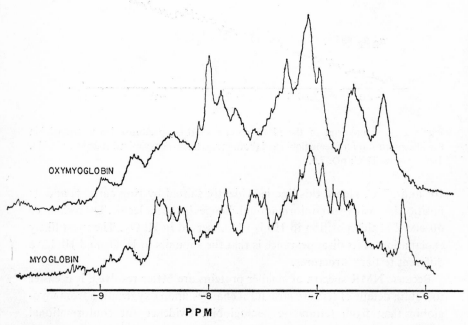

Figure 4 A comparison of the aromatic region of the NMR spectrum of deoxy myoglobin with that of oxy myoglobin

changes upon oxygen binding within the single chain of myoglobin has been obtained by a comparison of the ring-current-shifted resonances in Mb and MbO_2. For example, a well-resolved temperature-independent resonance with the intensity of two protons is observed at -6.13 parts per million in Mb but not in MbO_2, as shown in Fig. 4. This resonance line was assigned[3] to the two protons in the meta positions of phenylalanine CD1. The movement of the position of this phenylalanine was estimated to be at least 0.2 Å away from the iron when deoxymyoglobin was converted to the oxy form. In addition to this particular resonance, there are many differences between the NMR spectrum of Mb and that of MbO_2 in the aromatic region shown in Fig. 4, as well as other spectral regions—differences which appear to indicate extensive structural changes between the two states.

Discussion

The NMR experiments on mixed-state hemoglobin show that the heme groups and their immediate surroundings in the α-chains are not affected by the state of ligation of the neighbouring β-chains and vice versa. The peaks shifted by the paramagnetic iron have positions quite similar to those of corresponding homogeneous deoxy or ferric cyanide tetramers. In the high and low affinity forms of hemoglobin studies, there were no differences in heme environments, in both the ligated and unligated states. Therefore, it can be concluded that there are no energy changes at the heme groups which can account for the free energy of cooperative interaction. According to Wyman's[12] analysis, this interaction energy ΔF_I for oxygenation of mammalian hemoglobin is 3 kcal/mole of oxygen out of total of ~ 25 kcal/mole,[13] which is the free energy of binding four oxygen molecules. Hence, if any appreciable fraction of ΔF_I had shown up at the other heme group as an enthalpy change, this should have been detected in the NMR spectra. Furthermore, if the interaction energy had shown up in the immediate heme surroundings as conformational changes, it should also have been detected, because of the high sensitivity of such NMR measurements to structural changes near the heme[14, 15]. Hence we conclude that heme groups do not make an appreciable contribution to ΔF_I and therefore "heme–heme" interaction does not occur at the heme site. This conclusion is consistent with the well-known observation that ΔF_I is almost the same for many different ligands which have a wide range of heme–ligand binding energies[16, 17]. In the absence of heme–heme interaction at heme sites, ΔF_I must be explained by changes in the protein part of the molecule.

The present NMR study indicates that there are ligand-induced conformational changes within a subunit of hemoglobin and that the tertiary structure changes at the heme are sequential, since they depend only upon the state of ligation of that subunit. NMR shows conformational differences between the oxy and deoxy form of myoglobin. There are also many reported[18-20] observations of ligand-linked conformational changes in hemoglobin. The present mixed state NMR experiments show that the heme groups, and their immediate environments, are not affected by ligation of the neighbouring subunits.

The sequential model proposed and applied to hemoglobin by Koshland *et al.*[21] is consistent with the above-mentioned observations that the tertiary structural change at the hemes is independent of its neighbors and hence sequential. In this model, cooperativity comes from differences in the inter-subunit interactions at various stages of ligation. There must be ligand-induced conformational changes within a subunit which modify the interface interaction between subunits. The difference between the free energies of intersubunit interaction of unligated and partially ligated forms of hemoglobin is not the same as that between partially and totally ligated hemoglobin, and therefore the affinity towards ligands changes during the course of ligation. It should be pointed out that for cooperativity to occur in this model the tertiary structural change within a subunit has to propagate to the interface and also to interact or to overlap with the change in structure induced by the ligation at neighbouring subunits. If the overlap of the conformational changes at the interfaces is absent, there is no cooperativity, even though there can be a heterogeneity among the subunits in their affinity towards ligands. Evidence for this overlapping of the conformational changes is not as strong as for the presence of ligand-induced conformational changes within a subunit, although there is some indication of this overlapping from the work[18] on spin-labelled hemoglobin.

The allosteric model for hemoglobin proposed by Monod, Wyman and Changeux[22] finds no support in our mixed state NMR experiment. In this model, hemoglobin, conserving its molecular symmetry, can exist in two states of distinctly different conformation. One state has lower affinity towards oxygen than the other, the difference of affinity supposedly being due to differences in the heme surrounding. Our NMR results showed that there was no difference between the heme surrounding in the low and high affinity forms of hemoglobin which presumably correspond to the two states mentioned above, and that the tertiary structure changes sequentially, breaking the molecular symmetry.

Although the sequential model is favored from our NMR experiments and other reported observations in hemoglobin, an alternative description, in which the tertiary structure changes sequentially but the quarternary structure does not, cannot be ruled out. More detailed discussion of this alternate model will be given elsewhere[23]. In order to distinguish among these models, it is necessary to know how the quaternary structural change proceeds and to look for some conformational changes which are affected by ligation of more than one subunit. It will be interesting to compare the tertiary structural difference which X-ray study should show between oxy and deoxy hemoglobin, when the structure of deoxy hemoglobin, as that of oxy hemoglobin, is obtained at the resolution of 2.8 Å. Then it will be possible to see which ligand-induced conformational changes are interacting with each other and to see how important those interactions are.

Discussion

M. BRUNORI I would like to point out a problem which is of relevance to the experiments presented by Ogawa *et al.* and which has received our attention since we made the first experiments with artificial intermediates (Antonini, Brunori, Noble and Wyman, 1966, *J. Biol. Chem.*, *241*, 3236).

The understanding of structure–function relationships in hemoglobin would advance greatly if it was possible to study the functional and conformational properties of the intermediates occurring in the reaction with ligands. However, naturally occurring intermediates are short-lived and therefore their characterization is not easily achieved, if possible at all. Therefore artificial intermediates, in which one type of chain is "frozen" in the ligand bound form, have been prepared. These have the form: $(\alpha X \beta)_2$ or $(\alpha \beta X)_2$, where X may be a difficultly dissociable ligand of the ferrous (NO) or the ferric (CN^-) form, and only the partner (ferrous) chain undergoes reversible reactions, binding with O_2 or other ligands. A primary problem in this connection is to know how far the properties of these intermediates depend on the nature of the fixed ligand, X; therefore we have compared the functional and conformational properties of the CN-met and the NO intermediates. We have studied the ligand binding equilibrium, the kinetics of combination with CO and, for the CN-met intermediates, the changes in reactivity of the β_{93} SH groups toward PMB and the changes in optical activity in the far ultraviolet. Comparison of the functional properties of the two types of intermediates shows that, although certainly different in some details, they are very similar in their overall properties. Thus both types of intermediates show a large reduction in the heme–heme interactions ($n = 1$–1.2), the presence of a large alkaline Bohr effect and a low rate of combination with CO. Therefore it seems possible to conclude that to a first approximation the functional properties of the artificial intermediates are relatively independent of the nature of the fixed ligand (X), and they are probably good models for the transient intermediate occurring in the reaction of hemoglobin with a single type of ligand.

References

1   A. Rossi-Fanelli, E. Antonini and A. Caputo, *Adv. Protein Chemistry* **19**, 73 (1964).
2   W. Bolton, J. M. Cox and M. F. Perutz, *J. Mol. Biol.* **33**, 283 (1968).
3   R. G. Shulman, S. Ogawa, K. Wüthrich, T. Yamane, J. Peisach and W. E. Blumberg, *Science* **165**, 251 (1969).
4   A. Kowalsky, *Biochemistry* **4**, 2382 (1965): K. Wüthrich, R. G. Shulman and J. Peisach, *Proc. Natl. Acad. Sci. U.S.* **60**, 373 (1968).
5   Y. Enoki and S. Tomita, *J. Mol. Biol.* **32**, 121 (1968): R. Banerjee and R. Cassoly, *J. Mol. Biol.* **42**, 351 (1969); M. Brunori, E. Antonini, G. Amiconi and K. H. Winterhalter (to be published).
6   C. C. McDonald and W. D. Phillips, *J. Am. Chem. Soc.* **88**, 6332 (1967).
7   H. F. Bunn and J. H. Jandl, *J. Biol. Chem.* **243**, 465 (1968).
8   E. Bucci and C. Fronticelli, *J. Biol. Chem.* **240**, P.C. 551 (1965).
9   T. H. Huishman, A. M. Dozy, B. F. Horton and C. M. Nechtman, *J. Lab. Clin. Med.* **67**, 355 (1966). G. Guidotti, personal communication.
10 S. Ogawa, R. C. Shulman, T. Yamane, in preparation.
11 K. Wüthrich, R. G. Shulman, T. Yamane, *Proc. Nat. Acad. Sci. U.S.* **61**, 1199 (1968).
12 J. Wyman, *Adv. Protein Chem.* **19**, 323 (1964).
13 R. Noble, *J. Mol. Biol.* **39**, 479 (1969).
14 D. G. Davis, N. L. Mock, V. R. Laman and C. Ho, *J. Mol. Biol.* **40**, 311 (1969).
15 K. Wüthrich, R. G. Shulman, T. Yamane and S. Ogawa, to be published.
16 Q. H. Gibson, *Biochem. J.* **77**, 519 (1960).
17 R. C. C. St. George and L. Pauling, *Science* **114**, 629 (1951).
18 S. Ogawa, H. M. McConnell and A. Horwitz, *Proc. Natl. Acad. Sci. U.S.* **61**, 401 (1968); S. Ogawa and H. M. McConnell, *ibid.* **58**, 19 (1967).
19 E. Antonini and M. Brunori, *J. Biol. Chem.* **244**, 3909 (1969).
20 S. R. Simon and C. R. Cantor, *Proc. Natl. Acad. Sci. U.S*, **63**, 205 (1969).
21 D. E. Koshland, G. Némethy and D. Filmer, *Biochemistry* **5**, 365 (1966).
22 J. Monod, J. Wyman and J. P. Changeux, *J. Mol. Biol.* **12**, 88 (1965).
23 S. Ogawa, R. G. Shulman and T. Yamane (to be published).

NMR studies of hemoglobins

CHIEN HO, DONALD G. DAVIS, NANCY H. MOCK
and SAMUEL CHARACHE*

*Department of Biophysics and Microbiology, University of Pittsburgh
Pittsburgh, Pennsylvania*

Introduction

THE CHIEF PHYSIOLOGIC FUNCTION of hemoglobin is to transport molecular oxygen from lung to tissues by virtue of its ability to combine reversibly with oxygen. Hemoglobin is a protein molecule consisting of four subunits (normally two α chains and two β chains). Each of the four subunits has a heme group and these heme groups are the "active centers" of hemoglobin. Many of the functional properties of hemoglobin are highly sensitive to the quarternary structure of the molecule[1,2]. Perutz and Lehmann[3] have recently pointed out that although these properties are, in general, insensitive to replacements of most amino-acid residues on its surface, they may be profoundly altered by relatively small changes in internal non-polar contacts, such as those near the heme and those contacts between the subunits. Hence, a knowledge of the reactivity as well as the equivalence or non-equivalence of the heme groups in the individual chains of hemoglobins toward ligands and the detailed environment of these groups will be of value in our understanding of structural-functional relationship in hemoglobins.

Because the nuclear magnetic resonances of the protons in paramagnetic hemin derivatives are shifted by electron hyperfine interactions to regions of the spectrum outside the range for diamagnetic compounds[4,5], it is possible

* Address: Special Hematology Laboratory, Johns Hopkins Hospital, Baltimore, Maryland.

to probe the heme environment of hemoproteins by NMR spectroscopy[6-9]. Proton NMR data reported by us[6,7,10] and others[8,9,11] on various high and low spin forms of metmyoglobin and methemoglobins reveal that the heme proton resonance shifts depend on the heme environment in these proteins.

Results and Discussion

The NMR spectra were measured by means of a Varian HA-100 spectrometer operated at a frequency of 100 MHz and at a temperature of 27°–31 °C. The hemoglobin samples were prepared in 0.1 M deuterated phosphate buffer at pD 7. Details of the preparation of the samples are given elsewhere[7,12]

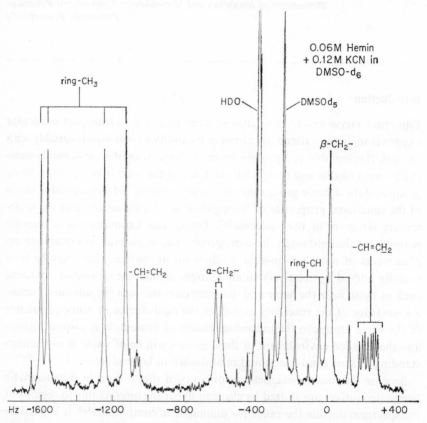

Figure 1 100 MHz proton NMR spectrum of 0.06 M hemin plus 0.12 M KCN in DMSO-d_6 at 27°C. The assignments for the heme protons are given in the figure. The frequency scale is measured with respect to tetramethylsilane (TMS)

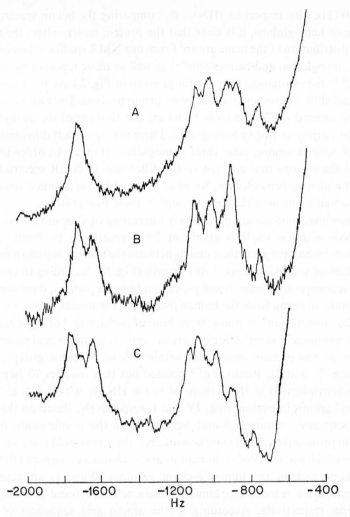

Figure 2 100 MHz proton NMR spectra of cyanomethemoglobins in 0.1 M deuterated phosphate and in 0.05 M KCN at pD 7 and at 27 °C. The frequency scale is measured with respect to HDO. A, human adult hemoglobin; B, human fetal hemoglobin; and C, horse hemoglobin

Figure 1 gives a typical 100 MHz proton NMR spectrum of the low spin cyanide complex of the hemin in DMSO-d_6 at 27 °C. The assignments for the heme protons are also given in Fig. 1. Figure 2 shows the 100 MHz proton NMR spectra of the low spin state of human Hb A ($\alpha_2\beta_2$), Hb F ($\alpha_2\gamma_2$) and horse Hb ($\alpha_2'\beta_2'$) in the cyanomet form at pD 7 over the range from -700 to

−1800 Hz with respect to HDO[7]. By comparing the hemin spectrum with those of hemoglobins, it is clear that the protein moiety alters the electron spin distribution of the heme group. From our NMR studies of hemin derivatives, myoglobin, and hemoglobin[6,7] as well as those reported by Wüthrich et al.[8,9], the resonances over the range given in Fig. 2 have been assigned to contact shifted resonances of the heme group protons. The resonances in the region around −1700 Hz from HDO are due to some of the methyl groups on the porphyrin ring in hemoglobin. There are significant differences in the NMR spectra among these three hemoglobins (Fig. 2). In order to understand the reasons that give rise to the differences in NMR spectra between adult and fetal hemoglobins, we need to consider the detailed structure or conformation around the heme groups in these two proteins.

The above findings are particularly interesting in view of the recent x-ray analysis of horse oxyhemoglobin at 2.8 Å resolution by Perutz and coworkers[2] who have given the contacts between the heme group and amino-acid residues of α- and β-chains in this protein (Fig. 3). According to the amino-acid sequences of human β- and γ-chains and horse β-chain, there are 39 substitutions in going from the human β-chain to the human γ-chain and there are 26 substitutions in going from human β-chain to horse β-chain[13]. Of these amino-acid substitutions, there are only two amino-acid replacements in the β- and γ-chains which are within 4 Å to the heme group[2], namely residues 70 and 71. Perutz et al.[2] pointed out that residues 70 (serine) and 71 (phenylalanine) in the β-chain of horse Hb are within 4 Å of the two methyl groups in pyrrole rings IV and I, respectively. Based on the amino-acid sequences of human β- and γ-chains[13] and the atomic model of hemoglobin proposed by Perutz and coworkers[2], the amino-acid residues near the heme which are replaced in human β- and γ-chains as compared to those of the horse β-chain are: human β-chain, residues 70 and 71 are alanine and phenylalanine, respectively; human γ-chain, residues 70 and 71 are serine and leucine, respectively. According to the amino acid sequences of human α-chain and horse α-chain[13] and the atomic model of hemoglobin[2], there is no amino-acid replacement which is within 4 Å to the heme group. Furthermore, according to Perutz[14], the serine oxygen in the residue 70 can make a contact with the methyl group of the pyrrole ring IV only if the serine oxygen is cis to the amide hydrogen (NH) in the polypeptide chain and would not make a contact with the methyl group if the serine oxygen is cis to the carbonyl oxygen (CO) in the polypeptide chain. However, at the present resolution of the x-ray analysis of hemoglobin, the position of the serine oxygen is indeterminate[14].

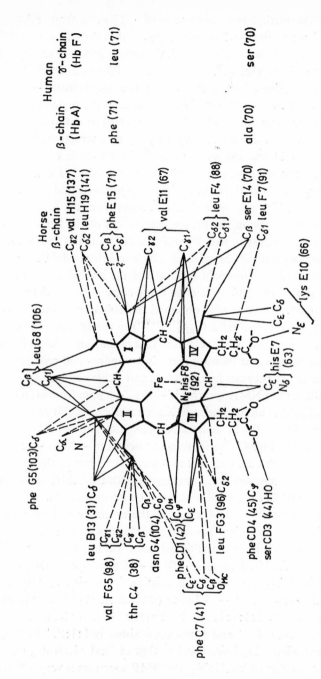

Figure 3 Contacts between the heme group and residues of the globin in horse and human β-chain and human γ-chain. All contacts of approximately 4 Å or less are listed. Plain lines indicate contacts on the side of distal histidine; broken lines, contacts on the side of proximal histidine. This figure is taken from the article by Perutz (Proc. Roy. Soc. (London), B **173**, 113 (1969)).

The fact that the methyl resonance around -1700 Hz from HDO in human Hb F and horse Hb (Fig. 2) is split suggests that the serine oxygen in residue 70 does, in fact, make a contact with the methyl group of the pyrrole ring IV in these two hemoglobins[7]. There are also differences in the NMR spectra among these three hemoglobins over the range from -800 to -1200 Hz from HDO. In particular, the resonances in the region of about -900 Hz from HDO in human Hb F are quite different from those in horse and human Hb A. Our NMR results suggest that the interactions between the leucine in residue 71 and the methyl group of the pyrrole ring I in the γ-chain are different from those of phenylalanine in residue 71 with corresponding methyl group in both horse and human β-chains and that those amino-acid residues which are more than 4 Å away from the heme group may also have some effects on the proton resonances of the heme group. Our prediction is also consistent with Perutz's atomic model of hemoglobin, namely that the replacement of phenylalanine by leucine in residue 71 reduces the distance between the amino-acid residue and the methyl group of the pyrrole ring I so that the $CH_3(\delta)$ of leucine could come closer to the methyl group of the heme[14]. We have also obtained the 100 MHz proton NMR spectrum of human sickle-cell Hb S ($\alpha_2\beta_2^{6\,Glu \rightarrow Val}$) in the cyanomet form at pD 7. This spectrum is indistinguishable from that of the human adult hemoglobin. This is to be expected because there is no amino-acid replacement around the heme group between human adult and sickle-cell hemoglobins.

A portion of the proton NMR spectra[12] (-30 to -8 ppm from HDO) for the azide forms of human adult, fetal, and Zürich methemoglobins as well as that of metmyoglobins in 0.1 M deuterated phosphate and in 0.05 M KN_3 at pD 6.9 are shown in Fig. 4. The prominent lines from -23 to -15 ppm from HDO in these spectra are assigned to some of the heme methyls on the basis of relative line intensities and comparison with the NMR spectra of $MetMbN_3$ and MetHbCN recorded under similar experimental conditions. Above -12 ppm intense, low field wings of the aromatic residue proton resonances of the protein moiety prevent accurate assessment of the line intensities and possibly obscure additional heme proton resonances. No resonances were observed in the region -50 to -30 ppm from HDO. In the human adult $MetHbN_3$ spectrum, the relative intensities of the lines at -22.4, -21.3, -16.5, and -15.7 ppm (labeled A, B, C, and D respectively) are approximately $1:1:1:1$. For human fetal $MetHbN_3$, the relative intensities of the lines at A, B, and C are approximately $1:1:2$. The NMR spectrum of horse $MetHbN_3$ is identical to that of fetal $MetHbN_3$ in this region. In the case of Zürich $MetHbN_3$, the NMR spectrum is very different

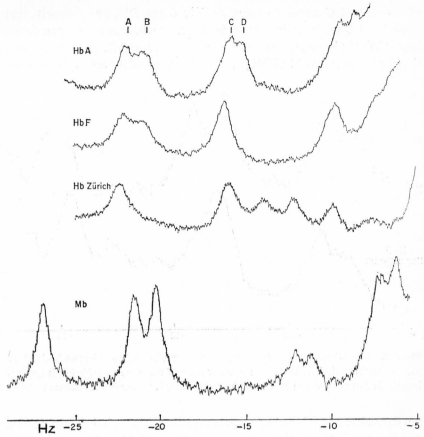

Figure 4 100 MHz proton NMR spectra of human adult (Hb A), human fetal (Hb F), and Zürich (Hb Zürich) methemoglobin azide derivatives and metmyoglobin azide (Mb) in the range −28 to 5 ppm from HDO at pH 7 and at 31 °C

from that of the normal adult MetHbN$_3$; the relative intensities of the lines at A and C are about 1:1. While it is essentially impossible to measure the absolute intensities of these lines, it was found that in protein solutions with equal concentration of heme, the intensities (i.e. total area under the resonance line) of the heme methyl resonances in MetMbN$_3$ spectrum were approximately twice the intensities of lines A, B, C, and D in the human adult MetHbN$_3$ spectrum. This implies that any one of these lines belongs to half the total number of hemes in the hemoglobin molecule.

The azide ion titration experiment (Fig. 5) which ahows that in the spectra of solutions containing less than saturating amounts of N$_3^-$, the intensities of lines A and C decrease relative to lines B and D^{12}. This suggests that lines A and C can be assigned to a heme group whose affinity for azide ion is less than the heme group associated with lines B and D. The conclusion is that for human adult $MetHbN_3$, the heme groups of the α and β chains are

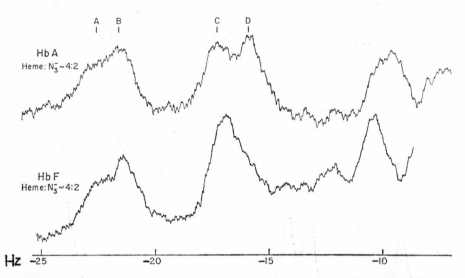

Figure 5   100 MHz proton NMR spectra of human adult (Hb A) and human fetal (Hb F) methemoglobin solutions in which the molar ratio of the heme to azide ion is approximately 4 : 2 at pD 7 and at 31 °C. The frequency scale is measured with respect to HDO

not equivalent and this non-equivalence is reflected in both the NMR spectra and the reactivity of azide ion toward methemoglobins. Because the titration experiments show that the lines at A and C belong to the same heme group and the same decrease in relative intensity is found between the pair of lines at -22.4 and -21.3 ppm in fetal methemoglobin titrations with azide ion, the lines at -22.4 and -16.5 ppm belong to the α-chain heme and this heme group binds azide ion less strongly than the β- or γ-heme. It is also inferred that the lines B and D belong to the β-heme.

These conclusions are given additional weight by the result of the proton NMR spectrum of the azide form of MetHb Zürich (Fig. 4). Qualitatively, the NMR spectrum of this hemoglobin is quite different from that of Hb A. This is to be expected because the abnormality in Hb Zürich is due to the substitution of the distal histidine residue at the β-63 position by a bulky arginine residue[15]. This group can alter the nature of the heme-azide interaction and therefore the NMR spectrum of the β chain heme protons. Our present NMR results indicate that the heme environment in the abnormal β chain is different from that of the normal one. Quantitatively, the lines at -22.4 and -16.5 ppm from HDO are essentially the same as those for adult and fetal MetHbN$_3$. This is consistent with the assignment of these lines to α-heme protons. The absence of the lines at -21.3 and -15.7 ppm is equally consistent with the assignments for the adult MetHbN$_3$ spectrum and reflects the altered environment of the β-heme azide complex in Hb Zürich. Another implication of these results is that the abnormal β-chain does not appear to affect the heme proton resonances of the α-chain (i.e. at least the methyl resonances at -22.4 and -16.5 ppm). The present NMR results are in agreement with the recent kinetic studies by DiPasquale and Ho[16] as well as by Gibson *et al.*[17] that α- and β-chains in methemoglobin have different reactivities toward N$_3^-$.

We have obtained the 100 MHz NMR spectrum of oxyhemoglobin Chesapeake ($\alpha_2^{92\,\text{Arg}\to\text{Leu}}\beta_2$) and have found that the aromatic proton resonances of HbO$_2$A and HbO$_2$ Chesapeake are not identical. These differences are particularly interesting in view of the fact that there is only one amino acid substitution between these two hemoglobins and that Hb Chesapeake is characterized by a high oxygen affinity, decreased or absent "heme–heme interaction" and a normal Bohr effect[18]. Perutz *et al.*[2] pointed out that residue 92 in the α-chain is located in α_1–β_2 contact region. Our NMR results suggest that there is a difference in the conformation between HbO$_2$A and HbO$_2$ Chesapeake and that this difference in the structure between these two hemoglobins may be related to their difference in the functional properties.

By combining the information obtained from our NMR results and that from other physico-chemical techniques, we hope to have a better understanding of the structural-functional relationship in human hemoglobins.

Acknowledgement

This work is supported by grants HE-02799, HE-10383 and FR-00292 from the National Instititutes of Health and by grant GB-8596 from the National Science Foundation and by a Grant-in-Aid from the American Heart Association. We wish to thank Mrs. Verna R. Laman and Mrs. Esther E. Gayle for their excellent technical assistance.

References

1   Antonini, E., *Science* **158**, 1417 (1967).
2   Perutz, M.F., Muirhead, H., Cox, J.M., and Boaman, L.C.G., *Nature* **219**, 131 (1968); Perutz, M.F., *Proc. Roy. Soc. (London)* **B173**, 113 (1969).
3   Perutz, M.F., and Lehmann, H., *Nature* **219**, 902 (1968).
4   Bloembergen, N., *J. Chem. Phys.* **27**, 595 (1957).
5   Eaton, D.R., and Phillips, W.D., *Advan. Magnet. Resonance* **1**, 119 (1965).
6   Kurland, R.J., Davis, D.G., and Ho, C., *J. Am. Chem. Soc.* **90**, 2700 (1968).
7   Davis, D.G., Mock, N.L., Laman, V.R., and Ho, C., *J. Mol. Biol.* **40**, 311 (1969).
8   Wüthrich, K., Shulman, R.G., and Peisach, J., *Proc. Natl. Acad. Sci. U.S.* **60**, 373 (1968).
9   Wüthrich, K., Shulman, R.G., and Yamane, T., *Proc. Natl. Acad. Sci. U.S.* **61**, 1199 (1968).
10 Kurland, R.J., Little, R.G., Davis, D.G., and Ho, C., Results on NMR studies of low spin hemin derivatives to be published.
11 Kowalsky, A., *Biochemistry* **4**, 2382 (1965).
12 Davis, D.G., Charache, S., and Ho, C., *Proc. Natl. Acad. Sci. U.S.*, in press.
13 Eck, R.V., and Dayhoff, M.O., *Atlas of Protein Sequence and Structure*, 1967.
14 Personal communication with Dr. M.F.Perutz.
15 Hitzig, W.H., Frick, P.G., Betke, K., and Huisman, T.H., *Helv. Paediat. Acta* **15**, 499 (1960); Muller, C.J., and Kingma, S., *Biochem. Biophys. Acta* **50**, 595 (1961).
16 DiPasquale, V.J., and Ho, C., manuscript submitted to *Biochemical and Biophysical Research Cummunications*.
17 Gibson, Q.H., Parkhurst, L.J., and Geraci, G., *J. Biol. Chem.* **244**, 4668 (1969).
18 Nagel, R.L., Gibson, Q.H., and Charache, S., *Biochemistry* **6**, 2395 (1967).

E.P.R. and N.M.R. spectra
of some haemin derivatives

H. A. O. HILL, K. G. MORALLEE and A. RÖDER

Inorganic Chemistry Laboratory, South Parks Road, Oxford

WE HAVE INVESTIGATED (1–3) the possibility that the oxidised form of the microsomal oxidase, cytochrome P-450, contains an iron(III)-haeme mercaptide group. It has long been known (4) that it is possible to form complexes of metmyoglobin and met-haemoglobin with hydrogen sulphide. We have examined the optical and e.p.r. spectra of these complexes and those with other thiols and have shown (1–3) that they resemble those of P-450 in a number of respects. In addition, it has been possible to form mercaptide complexes of iron(III) protoporphyrin IX which are stable in solution below $-60\,^{\circ}$C. All the complexes exhibit (3) e.p.r. spectra very like that of P-450 with g_1 and g_3 dependent on the mercaptide, g_2 remaining almost constant. There (3) is a linear correlation between g_1, g_3 and the logarithm of the acid dissociation constant of the coordinated mercaptide. The addition of a nitrogeneous base gives, we assume, the mixed ligand complex. With a constant mercaptide ligand, g_1 and g_3 again depend (Fig. 1) on the basicity of the nitrogenous base though the dependence is less marked and, for heterocyclic ligands, in the reverse order from that of the mercaptide. As Dr. Blumberg mentioned (5), the mercaptide ligand dominates the e.p.r. spectrum in the mixed ligand complex.

These results indicate the importance of the axial ligands in determining the properties of the complex as a whole. This is more apparent in another series of complexes (6), the *bis*-pyridine iron(III) protoporphyrin IX complex cations (Fig. 2). The ^{1}H n.m.r. spectrum of the *bis-d_5*-pyridine complex in

117

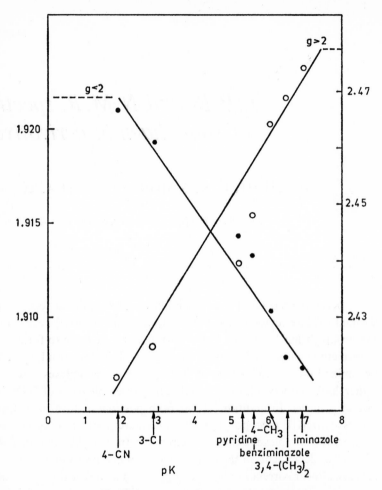

Figure 1 The g-values derived from the e.p.r. spectra of solutions of iron (III) proto-porphyrin IX $^-SCH_2COOCH_3$ in $CHCl_3$ at 100°K in the presence of excess substituted pyridines, benziminazole or iminazole (3)

the presence of a large excess of pyridine is shown in Fig. 3; the assignment is as indicated in Fig. 2. The shift to low-field for the peripheral methyl groups (a_3) and to high-field for the meso-hydrogens (e_1 and e_3) suggests that the origin of the shifts lies in the contact term with positive spin density present on the relevant adjacent carbons. As expected, therefore, the shifts are temperature dependent but *not* linearly with T^{-1}, as shown for the low-field resonances in Fig. 4. We interpret the observed temperatures dependence as

Fe protoporphyrin IX py$_2^+$

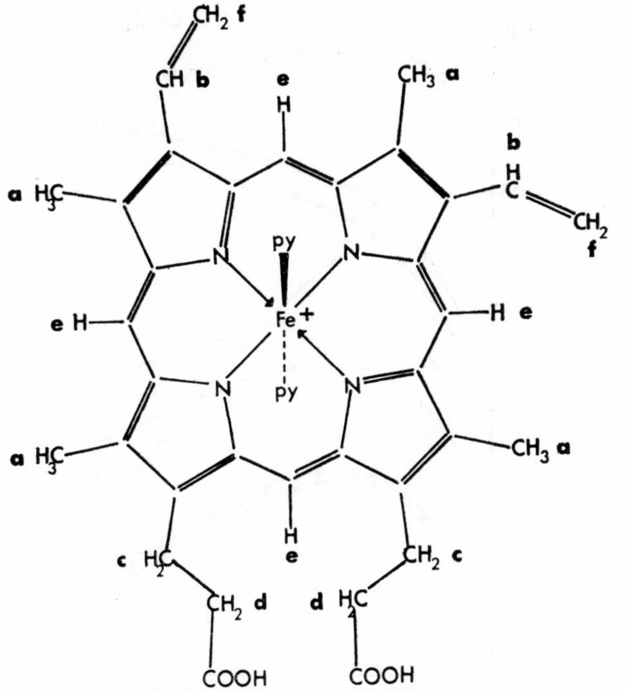

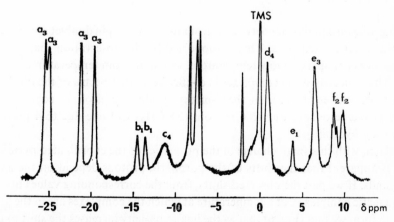

Figure 2 Structural formula of *bis*-pyridine iron(III) protoporphyrin IX complex cation

60 MHz P.M.R. spectrum of haemin + D$_5$ pyridine in CDCl$_3$ at 209°K

Figure 3 ^{1}H n.m.r. spectrum at 60 MHz of *bis*-d$_5$-pyridine iron(III) protoporphyrin IX chloride in CDCl$_3$ in the presence of excess d$_5$-pyridine

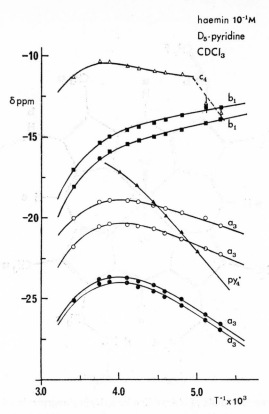

Figure 4   Temperature dependence of chemical shifts of low-field resonances in spectrum at 60 MHz of *bis*-d_5-pyridine iron(III) protoporphyrin IX chloride in presence of excess d_5-pyridine in $CDCl_3$

reflecting an equilibrium between a ground-state low-spin doublet and a high-spin sextet excited-state. This is supported by the observation that the resonances are sharp at low temperature, broad at room temperature.

The structure of the complexes is confirmed in the spectra of the *bis*-picoline derivatives (Fig. 5). In the 4-methylpyridine (γ-picoline) complex we observe a resonance due to 6 protons ($2 \times CH_3$) and one due to 4 protons (2×2 β-hydrogens).

Even with this small change in the properties of the coordinated pyridine, it is apparent that the shifts of the porphyrin protons depend on the axial ligands. If we plot the observed shifts, from the corresponding values in diamagnetic complexes, at a temperature at which the complexes are in the low-spin form we find (Fig. 6) that as the ligand basicity increases the shift of the peripheral methyl hydrogens (a_3) decreases. The effect on the *meso*-hydrogens is much smaller and, in the opposite direction.

To ensure that the major contribution to the shifts resulted from the contact term, an estimate was made of the pseudo-contact contribution taking $g_{11} - g_1 = 0.38$ with $T_{1e} \ll \tau/c$, the following pseudo-contact shifts are predicted: methyl protons 1.4 ppm, meso protons 4.0 ppm, which compare very well with these deduced for the cyanohaemes by Wüthrich *et al.* (7). This suggests that pseudo-contact interactions are of minor importance compared with the contact interactions.

The increasing shift with decreasing electron density on the iron suggests that the main mechanism for spin delocalization can be described in terms

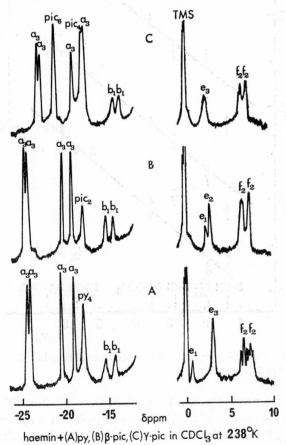

haemin+(A)py,(B)β-pic,(C)γ-pic in CDCl₃ at 238°K

Figure 5 ^{1}H n.m.r. spectra at 60 MHz of haemin in the presence of excess (A) pyridine, (B) β-picoline and (C) γ-picoline in CDCl₃

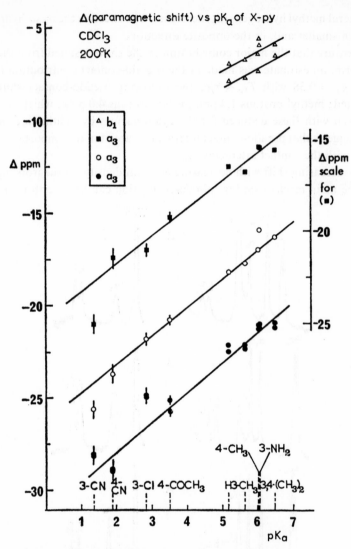

Figure 6 Δ (paramagnetic shift) of low-field resonances of haemin in the presence of excess substituted pyridines at 200°K

of charge transfer from ligand π to metal. M.O. calculations (8) have shown that for the low-spin cyano iron(III) haeme complexes the e_g^* orbital, though mainly metal, has considerable ligand character, consistent with the above explanation. This would account for the dependence of the shifts of the methyl hydrogens. Why are the shifts of the *meso*-hydrogens axial-ligand independent? As Eaton has pointed out (9) if spin-density is present in the ligand nitrogens π-orbitals it is impossible to write resonance forms which place spin density on the meso-hydrogens. Using the McConnell formula $^aCH_3 = {^Q}CH_3 \, \varrho_c^{\pi}$ carbons ϱ_c^{π}, doubles from ~1.5; ~1.00 (% of one unpaired electron) for the 4-aminopyridine derivative to ~2.4; ~2.00 for the 4-cyano-derivative for the inequivalent peripheral methyls.

References

1   E. Bayer, H. A. O. Hill, A. Röder and R. J. P. Williams, *Chem. Commun.* (1969), 109.
2   H. A. O. Hill, A. Röder and R. J. P. Williams, *Naturwissenschaften* (1969).
3   E. Bayer and A. Röder, *European J. Biochem.* (1969).
4   D. Keilin, *Proc. Roy. Soc.*, (1933), **113 B**, 393.
5   W. Blumberg and J. Peisach, this volume, p. 67.
6   H. A. O. Hill and K. G. Morallee, to be published.
7   K. Wüthrich, R. G. Shulman, B. J. Wyluda and W. S. Caughey, *Proc. Nat. Acad. Sci. U. S.* (1969), **62**, 636.
8   M. Zerner, M. Gouterman and H. Kobayashi, *Theoret. Chim. Acta* (1966), **6**, 363.
9   D. R. Eaton and E. A. LaLancette, *J. Chem. Phys.* (1964), **41**, 3534.

On the nature of hemoglobin as found in Heinz bodies of red cells from patients with alpha thalassemia

W. E. BLUMBERG, J. PEISACH*

*Bell Telephone Laboratories, Incorporated
Murray Hill, New Jersey 07974*

E. A. RACHMILEWITZ

Department of Hematology, Hadassah Hospital, Jerusalem, Israel

T. B. BRADLY

*Department of Medicine, Albert Einstein College of Medicine
Bronx, New York 10461*

INCLUSIONS of precipitated hemoglobin (Heinz bodies) can be demonstrated in the red cells of patients with alpha thalassemia (1). In thalassemia a defect in the synthesis of specific globin polypeptide chains results in an unbalanced production of hemoglobin subunits. In α thalassemia excess β chains form hemoglobin tetramers (hemoglobin H), which can be demonstrated in hemolysates by electrophoresis (2, 3). The subunits which are produced in relative excess are unstable on isolation and tend to precipitate (4). In hemoglobinH disease, which appears to result from heterozygosity for each of two variants of α thalassemia, hemolysates contain up to 30 per cent hemoglobin H. In this disease prominent inclusions (H-bodies) are seen after exposure of peri-

---

* Present address: Departments of Pharmacology and Molecular Biology. Albert Einstein College of Medicine, Yeshiva University, Bronx, New York 10461.

pheral red cells and bone marrow normoblasts to oxidative dyes (3, 5). Inclusion bodies form *in vivo* in hemoglobin H disease but are largely removed by the spleen (6). After splenectomy inclusion bodies are present in the circulating red cells and can easily be detected (7). We have shown that the red cell inclusion bodies in hemoglobin H disease are heme-containing β chains with the optical and electron paramagnetic resonance (EPR) properties of material commonly called "hemichrome" (8).

Hemichromes are low spin forms of ferric heme proteins in which both the fifth and sixth coordination positions of heme iron are bound to electron donating groups, such as imidazole (9). Hemichromes of hemoglobin A can be made by a wide variety of denaturing agents and have been studied by EPR (10). When they derive two ligand atoms from the protein chain, their formation requires changes in tertiary structure near the heme. In one form of hemichrome (H-type) of hemoglobin the histidine at helical position E7 (11) is thought to be bound to the sixth coordination position of heme iron while the bond between the proximal histidine F8 and the fifth coordination position of iron is maintained as in the native protein (12). The new histidine-iron bond requires some change in the tertiary structure of the hemoglobin, for the nitrogen of the histidine imidazole must move closer to the iron (13). Conversion of hemoglobin to hemichrome takes place after oxidation of the heme iron to its ferric form and is enhanced when the interaction between dissimilar α and β subunits in the hemoglobin A tetramer is absent. The hemichromes have a distinct absorption peak at 530–535 mμ, a shoulder at 565 mμ and characteristic low spin EPR absorptions.

The molecular instability of isolated hemoglobin H and its relation to the formation of inclusion bodies have not been adequately explained. Our own studies have shown that oxidation of hemoglobin H, as well as isolated heme-bearing α subunits of hemoglobin A, leads to the formation of hemichromes after the transient appearance of ferrihemoglobin, as monitored by optical and EPR absorption spectra (12). Hemichrome formation in vitro is followed by progressive precipitation, which increases with increasing temperature.

In order to study these precipitated hemichromes, we performed the following experiments. Fresh red cells were lysed with water and the ghosts examined for inclusion bodies by phase contrast microscopy. In addition, hemoglobin-free ghosts were prepared for visible and EPR spectral analyses in 20 milliosmolar phosphate buffer pH 7.5, by the method of Dodge *et al.* (14). Urea/starch gel electrophoresis was used for identification of the globin chains precipitated within the ghosts (15).

We studied Heinz bodies from two unrelated patients with hemoglobin H disease. The first patient was a 59 year old woman who had undergone splenectomy 25 years ago. The second patient was a 20 year old woman with a 4 cm splenomegaly. Phase contrast microscopy revealed inclusion bodies of various sizes in lysed red cells of the splenectomized patient, but only occasional inclusions in preparations from the non-splenectomized patients. Ghosts of normal red cells and cells of the non-splenectomized patient, prepared at pH 7.5, were colorless and showed only minute traces of oxyhemoglobin in the visible spectrum. A similar preparation of cells from the splenectomized patient remained brownish-red in spite of repeated washes with the hypotonic phosphate buffer. These latter ghosts showed a characteristic hemichrome optical spectrum although the major absorption peak was shifted to higher wavelengths attributable to the presence of a small amount of oxyhemoglobin.

When whole blood of the non-splenectomized patient was incubated at 37 °C for 12 hours before lysis of the red cells, the ghosts prepared at pH 7.5 were brownish-red as in the case of the splenectomized patient and had inclusion bodies and the spectral characteristics of hemichrome. Similar incubation did not alter the properties of the normal red cells. In all the ghost preparations, phase contrast microscopy revealed intact membranes without evidence of stromotolysis.

Definitive evidence for the presence of hemichromes in the inclusion bodies of hemoglobin H disease were obtained by EPR spectroscopy. The iron in both ferrihemoglobin and hemichromes is paramagnetic and each exhibits characteristic EPR absorption spectra. On the other hand, diamagnetic oxyhemoglobin produces no EPR absorption. The presence of oxyhemoglobin does not interfere with the EPR study of ferrihemoglobin or hemichromes when solutions containing both oxygenated and oxidized hemoglobin derivatives are examined by this method. In addition, the high and low EPR spin absorption spectra, which are characteristic of ferrihemoglobin and hemichromes respectively, do not interfere with one another and allow an identification and quantitation of the various paramagnetic compounds present. This is in contrast to optical studies where the optical absorption spectra of all three derivatives of hemoglobin, oxyhemoglobin, ferrihemoglobin and hemichromes overlap. Another advantage of EPR spectroscopy over optical spectroscopy is that it allows measurements in turbid and even solid materials. Thus neither the presence of oxyhemoglobin in the suspensions of ghosts nor their turbidity interferes with accurate detailed study of high and low spin forms of hemoglobin. Only ghosts which contained numerous inclusion bodies exhibited EPR absorptions attributable to heme iron, and its low spin forms were predominant. High spin heme iron, either ferrihemoglobin or free heme, was present in only trace amounts. Low spin EPR spectra were observable 45 minutes after a solution of hemoglobin H was oxidized with ferricyanide at pH 7.0 (12). Oxidation of hemoglobin A in the same conditions chiefly produces a high spin EPR spectrum with the major absorption at $g = 6$ (16).

The nature of the hemoglobin protein comprising the inclusion bodies was determined by urea/starch gel electrophoresis. The ghosts from both patients which contained inclusion bodies had a protein band in the position of normal β globins; α chains were not detected. No protein bands were apparent in hemoglobin-free ghosts, prepared from normal red cells at pH 7.5.

The present findings, indicating that the inclusion bodies in the red cells of patients with hemoglobin H disease are largely present as hemichromes, provide further evidence that the instability of hemoglobin H is directly related to hemichrome formation. The finding that the inclusion bodies in hemoglobin H disease involve primarily the β subunits of hemoglobin confirms the belief that the inclusions bodies result from the precipitation of hemoglobin H.

The EPR and optical spectra of inclusion bodies in hemoglobin H disease indicate that heme remains attached to globin in the precipitated hemichromes. The chemical nature of the hemichromes described in this paper depends on the time of aging of the material. Thus ferrihemoglobin H incubated at room temperature for 45 minutes yields predominantly the H type hemichrome, the hemichrome found in the red cell of the splenectomized patient after 12 hours incubation is predominantly the B type while the hemichromes from the red cells of the nonsplenectomized patient is primarily the P type. The chemical nature of each of these is discussed elsewhere in this symposium (16). Because hemichromes of hemoglobin H in the inclusion bodies are similar by EPR criteria to those that can be formed from hemoglobin A, (9) the structure in the vicinity of the heme is probably the same in both. It is thought then that the predominant cause of hemichrome formation in hemoglobin H is the absence of interchain contacts between dissimilar subunits. Whether hemichromes play a part in the precipitation of other unstable hemoglobins remains to be investigated.

The portion of the investigation carried out at the Albert Einstein College of Medicine was supported in part by a Public Health Service Research Grant to J. Peisach (GM-10959) from the Division of General Medical Sciences, and a grant from the Cooley's Anemia Blood and Research Foundation for Children to E. A. Rachmilewitz and as such is communication No. 191 from the Joan and Lester Avnet Institute of Molecular Biology. The portion of this investigation carried out by J. Peisach was supported by a Public Health Service Career Development Award (1K3-GM-31,156) from the National Institute of General Medical Sciences.

We thank Dr. Helen M. Ranney for her help and encouragement in performing these studies.

References

1 Dacie, J.V., Grimes, A.J., Meisler, A., Steingold, L., Hemsted, E.H., Beaven, G.H., and White, J.C., *Brit. J. Haematol.* **10**, 388 (1964).

2 Jones, R.T., Schroeder, W.A., Balog, J.E. and Vinograd, J.R., *J. Amer. Chem. Soc.* **81**, 3161 (1959).

3 Gabzuda, T.G., *Blood* **27**, 568 (1966).

4 Nathan, D.G. and Gunn, R.B., *Amer. J. Med.* **41**, 815 (1966).

5 Fessas, P. and Yataghanas, X., *Blood* **31**, 323 (1968).

6 Wennberg, E. and Weiss, L., *Blood* **31**, 778 (1968).

7 Rigas, D.A. and Koler, R.D., *Blood* **18**, 1 (1961).

8 Rachmilewitz, E.A., Peisach, J., Bradley, T.B., and Blumberg, W.E., *Nature* **222**, 248 (1969).

9 Blumberg, W.E. and Peisach, J. in *Proceedings of the 4th Johnson Foundation Symposium* (in press).

10 Hollocher, T.C., *J. Biol. Chem.* **241**, 1958 (1966).

11 Dickerson, R.E., in *The Proteins* (second ed.) (edit. by Neurath, H.), **2**, 603 (Academic Press, New York, 1964).

12 Rachmilewitz, E.A., *Ann. N.Y. Acad. Sci.* **165**, 171 (1969).

13 Perutz, M.F., Muirhead, H., Cox, J.M., and Goaman, L.C.G., *Nature* **219**, 131 (1968).

14 Dodge, J.T., Mitchell, C., and Hanahan, D.J., *Arch. Biochem. Biophys.* **100**, 119 (1963).

15 Chernoff, A.I. and Petit, jun., N.M., *Blood* **24**, 750 (1964).

16 Blumberg, W.E. and Peisach, J., this book, p. 67.

A negative outcome of the ^{17}O N.M.R. approach to oxygen bonding by haemoglobin

S. MARIČIČ,

Institute of Biology, University of Zagreb, Yugoslavia

GRETA PIFAT, and M. PETRINOVIČ,

"Rudjer Bošković" Institute, Zagreb, Yugoslavia

V. KRAMER, and J. MARSEL

"Jožef Stefan" Institute, Ljubljana, Yugoslavia

K. BONHARD

*Wissenschaftliche Abteilung, Biotest-Serum-Institut GmbH
Frankfurt, Germany*

IN THE PAST FEW YEARS attempts have been made to determine, by x-ray crystallographic analysis, the exact mode in which oxygen is bound to haemoglobin, but no definite answer has yet emerged*. ^{17}O N.M.R. spectra of oxygen bound reversibly to haemoglobin may contribute towards an understanding of the stereochemistry and the chemical nature of this complex, although, of course, the method does not yield the coordinates of oxygen atoms. However, it could distinguish the symmetrical structure from the asymmetrical one. In the first, sometimes called the Griffith structure, the

---

* The full paper has been published in *Croat. Chem. Acta* **41** (1969/No. 4).

oxygen molecule is aligned parallel to the haem group[2] in such a way that the two oxygen nuclei are magnetically equivalent so that one single [17]O N.M.R. peak is expected. The other case referred to as the Pauling structure shows the two nuclei in magnetically different environments because the molecular axis is slanted at 30° towards the haem plane[3]. Thus, one would expect two [17]O N.M.R. lines chemically shifted in respect to each other.

Three years ago one of us (S.M.) reported briefly[4] the recording of the [17]O N.M.R. spectrum from oxygen bound to haemoglobin and later a paper was published[5]. In a continuation of this work we have now come to the conclusion that the interpretation of the first experiment was wrong and the purpose of this report is to show why it is so.

In order to increase the signal to noise ratio very concentrated haemoglobin solutions ought to be prepared -15 to 20×10^{-3} M (in haem), which is in fact 23 to 30 w-% in protein. We were unable to get a [17]O N.M.R. signal from the ordinary solvent water in such haemoglobin solutions under air. A linear dependence of the [17]O line width on Hb-concentration could be measured up to about 12×10^{-3} M. This broadening is due to an increase of the correlation time of water molecules when attached to the protein surface and to their fast exchange with bulk water[6]. Thus, opposite to our former fears[5], in a search for [17]O N.M.R. signal from oxygen bound to haemoglobin the [17]O line from ordinary water would not show up in such highly concentrated solutions.

However, we prepare our samples by sealing them off the vacuum line under 300 mm Hg of pure oxygen enriched to 62% in [17]O. Initially there is no [17]O signal at all, but a faint one usually appears within a day or two and increases progressively up to three times in a couple of weeks. The observed signal is not shifted from the position in the spectrum where the water [17]O line would be. Spectra taken with such samples extend the straight line dependence of line width on Hb-concentration from some 400 m Gauss (uncorrected for any instrumental broadening) for 23% up to 550 mGauss for a 32% Hb-solution (all measurements at 7.33 MHz).

By the time the signal intensity attains a final value haemoglobin turns from its oxy- into the deoxy-form, but the [17]O N.M.R. signal persists. It does so even if the sample is deliberately deoxygenated by evacuation before itself turns spontaneously into the deoxy-form.

In the spontaneous deoxygenation the pressure above the Hb-solution diminishes continuously until a steady state is reached coinciding with the oxy-deoxy conversion. The mass-spectroscopic analysis showed a two- to threefold enrichment in ^{17}O of the solvent water after this conversion. In the gas-phase above such samples oxygen was replaced almost completely by carbon dioxyde (which, of course, explains the deoxygenation of haemoglobin).

We conclude from all this data that what we observe is indeed the ^{17}O N.M.R. line from solvent water enriched in ^{17}O by a reduction of the gaseous isotope above the solution, into it.

As our first sample[5] was 11 months old at the time of the N.M.R. measurement, and the recorded line agrees in its position and saturation characteristics with the present measurements there is every reason to believe that the same conclusion holds in that case, too.

There is but one discrepancy left. The main argument for observing the O-17 line from oxygen bound to haemoglobin in the first experiment was the gradual disappearance of the signal in contact with air with a half-time of about six hours. This was ascribed to the $^{17}O/^{16}O$ exchange of bound oxygen. We could not repeat this observation now.

In looking for a possible explanation of this discrepancy we see two possibilities:

a)   the decrease in intensity observed formerly was of a trivial cause, i.e. the overall sensitivity of the spectrometer was being slowly degraded.

b)   the only preparative difference between the two sets of experiments is in the presence of $MnCl_2$ added to the sample in the first experiment before it was exposed to air. (It was added with deliberate intention to broaden the supposedly underlying water line, which was not indeed necessary because that line is anyway broadened beyond detectibility by high protein concentration.) One of the referees of the first paper doubted the validity of our argument by supposing that the line was only due to ^{17}O from water but was slowly broadened. Against this is our observation of instantaneous broadening, on adding $MnCl_2$, of the ^{17}O water line in ordinary solution of Hb, though in a less concentrated solution (7 mM). This point has yet to be checked.

At the moment, it is of more concern to establish the origin of the oxy-deoxy conversion and find ways of circumventing it in order to enable further N.M.R. measurements.

The pattern of change of the isotope content in the liquid phase and the replacement of O_2 by CO_2 in the gas phase in this system suggest strongly that the reduction of oxygen is being acomplished via the bacterial metabolism. This is further corroborated by preparing sterile haemoglobin with which no such effects were observed. The haemoglobin solution stayed in its oxy-form for 6 months and in two vials (two months after sealing) the air-composition did not change at all. On balance we believe that bacterial contamination of ordinarily prepared samples does indeed cause the whole trouble. (However, the kinetics of pressure change in the course of approaching the oxy-deoxy conversion; its similarity at room temperature and near zero centigrade, as well as between preparations with vastly different contamination grades, require further consideration.)

We found that addition of sodium azide to the concentrated haemoglobin solution (in amounts greater than equimolar in respect to haem) while indeed not affecting the oxyhaemoglobin[7] does preserve this form for months. For practical reasons of doing N.M.R. measurements this is helpful because it eliminates the need for the rather involved sterile preparation.

As for recording the ^{17}O spectrum of oxygen bound to haemoglobin the main conclusion of the present report is, apart from invalidating the first experiment, that it seems indeed almost hopeless to look for the signal in this classical way. At the moment, having found the way of stabilizing the oxy-haemoglobin samples during long periods of time, the main problem of sensitivity remains. In all probability the N.M.R. line of ^{17}O bound to haemoglobin will be wider than the H_2O ^{17}O line broadened by high protein concentration.

At least there is a guide now in searching for this ^{17}O N.M.R. signal: A. Velenik and R. M. Lynden-Bell[8] calculated the chemical shifts for the two stereochemical structures. In both cases the spectrum is expected to appear approximately 3000 ppm downfield from the H_2O ^{17}O line, with a separation between the two peaks in the Pauling structure of about 1500 ppm.

Even in case of a successful location of oxygen reversibly bound to haemoglobin, by x-ray (or neutron?[9]) diffraction, the ^{17}O N.M.R. spectra, if ever recorded, will prove useful in a comparative investigation of the phylo- an ontogenesis of haemoglobins, as well as in relating other aspects of the physical chemistry of this unique reversible binding of oxygen to the nature of its bond with haemoglobin. Perhaps the time is now to make a proper start (instead of the jump attempted several years ago!) by using suitable model compounds of reversible oxygen carriers where the deleterious broadening of the ^{17}O N.M.R. line should be smaller than in case of haemoglobin.

References

1   H. C. Watson and C. L. Nobbs: "The Structure of Oxygenated and Deoxygenated Myo-globin", in *Biochemie des Sauerstoffs*, ed. by B. Hess and Hj. Staudinger, Springer-Verlag Berlin–Heidelberg–New York, 1968, p. 37.

2   J.S.Griffith, *Proc. Roy. Soc.* **A235** (1956) 23.
3   L.Pauling, *Nature* **203** (1964) 182.
4   See the discussion, pp. 282–284 in *Hemes and Hemoproteins*, ed. by B.Chance, R.Estabrook and T.Yonetani, Academic Press New York–London, 1966.
5   S.Maričić, J.S.Leigh, Jr., and D.E.Sunko, *Nature* **214** (1967) 462.
6   J.A.Glasel, *Nature* **218** (1968) 953.
7   D.Keilin, *Proc. Roy. Soc.* **B121** (1936) 165.
8   A.Velenik and R.M.Lynden Bell, *Croat. Chem. Acta* **41** (1969/No. 4), in press.
9   B.P.Schoenborn, *Nature* **224** (1969) 143.

Iron-nitrosyl paramagnetic complexes with biological and organic ligands: an EPR study[*]

L. BURLAMACCHI, G. MARTINI and E. TIEZZI

Institute of Physical Chemistry, University of Florence, Florence, Italy

Abstract

The EPR spectra of a variety of iron-nitrosyl compounds with several inorganic and organic ligands have been investigated.

Particular interest has been devoted to the behaviour of the iron-nitrosyl system where the $-SH$ group may or may not be present in the ligand. On the basis of the nuclear hyperfine structure of the EPR spectra, it is possible to distinguish two different kinds of complexes with two NO groups bonded to the iron atom. They both have an almost tetrahedral structure.

Two other kinds of iron-nitrosyl complexes have been analyzed in which only one NO group is present. The structure of these paramagnetic systems in one case is pyramidal with the nitrosyl group at the vertex, while in the other case it is octahedral.

Sometimes the same ligand gives rise to two different complexes, thus causing changes in the EPR spectra. These variations can be correlated to the establishment of an equilibrium between the two different types of complexes structures.

All the paramagnetic iron-nitrosyl complexes which are EPR detectable can be related to the following formulations: $Fe(NO)L_2$, $Fe(NO)L_5$, $Fe(NO)_2L$ and $Fe(NO)_2L_2$, where L is a monodentate or bidentate ligand.

[*] This research was supported by the Italian National Research Council (CNR).

137

Introduction

THE STUDY of the complexes of iron and nitric oxide has received increasing attention in the last few years.

The biological implications of this class of paramagnetic compounds have been reported by several authors. Dobry-Dulcaux[1] have shown that Roussin's black salt (Roussinate ion: $Fe_4S_3(NO)_7{}^-$) effectively inhibits the enzyme alcohol dehydrogenase and other enzymes. Paramagnetic complexes are also formed between NO and iron containing protein[2]. Finally, Fe–NO complexes were found to be involved in carcinogenetic processes, as studied by Commoner's group. Recently Kayushin[3] and Woolum *et al.*[4] have given fundamental contributions on this subject.

In this report the electronic and chemical structures of various types of Fe–NO complexes are discussed and correlated to their activity and chemical behaviour. Some new complexes are then described with particular attention given to the interactions between different species and to the equilibria which may be established between complexes of different structure.

Sometimes very complicated patterns were present in EPR spectra. In these cases we used isotopic substitution (mostly ^{15}N and ^{57}Fe) and computer simulated spectra in order to clarify the nuclear hyperfine structure.

Experimental

Apparatus

EPR spectra were recorded by means of Varian E-3 or V 4502 X-band 100 Kc field modulation EPR spectrometers equipped with temperature controller. In some cases a special EPR spectrometer for aqueous solutions was used. It was designed and constructed at the laboratories of the Department of Biology, Washington University, operating with a microwave frequency of 8956 Mc. Klystron frequency was measured by a Hewlett-Packard model X532B frequency meter. A double cavity in the V 4502 spectrometer was used for the measurements of *g*-factors and isotropic hyperfine constants, by comparison with Fremy salt as an external reference standard. In some cases, complicated hyperfine structures were resolved with the aid of computer simulated spectra.

Samples and chemicals

Water and water–ethanol solutions of $FeSO_4$ or $FeCl_2$ $5 \cdot 10^{-4}$ to $5 \cdot 10^{-3}$ M were deoxygenated by stirring under a nitrogen stream. After addition of the proper amount of ligand, the solutions were saturated with gaseous NO. This was obtained from ferrous sulfate, sulfuric acid and sodium nitrite, as described by Blanchard *et al.*[5]. An alternative method of obtaining NO, was provided by the reaction of $NaNO_2$ and L-ascorbic acid dissolved in the solution, as described in ref. 4. The pH of the solutions was adjusted by adding small amounts of NaOH, HCl or $HClO_4$.

All chemicals were reagent grade and used without further purification. $Na^{15}NO_2$ was obtained from Tracerlab, Waltham, Mass. (USA) and $^{57}FeSO_4$ was obtained from New England Nuclear, Boston, Mass. (USA).

Results and discussion

It is possible to distinguish four most important types of Fe–NO paramagnetic complexes. They can be represented by the following general simplified formulas:

$$Fe(NO)L_5 \qquad Fe(NO)L_2 \qquad Fe(NO)_2L_2 \qquad Fe(NO)_2L$$

$$\text{I} \qquad\qquad \text{II} \qquad\qquad \text{III} \qquad\qquad \text{IV}$$

where L is a simple organic or inorganic ligand as well as a biologically interesting ligand. The ligands are monodentate in structure I and III and bidentate in structures II and IV. We will describe these complexes separately.

I Fe(NO)L_5 complexes

They present sixfold octahedral coordination around the central iron atom. Two main classes of $Fe(NO)L_5$ complexes are known:

a)   high-spin d^7 complexes, described as "brown ring" type compounds, such as $[Fe^I(H_2O_5)NO]^{+2}$ and $[Fe^I(NH_3)_5NO]^{+2}$. Infrared spectra and magnetic susceptibility measurements have shown[6] that NO is bonded as a nitrosium NO^+ ion. Iron can be considered formally in the +1 oxidation state; the compound being formulated as an outer-orbital complex (sp^3d^2). In water solution the equilibrium:

$$[Fe(H_2O)_6]^{+2} + NO \rightleftharpoons [Fe(H_2O)_5NO]^{+2} + H_2O$$

is present. The brown colour of these complexes has been attributed to charge transfer.

The high-spin d^7 complexes usually display a large orbital contribution to the g-factor and very short relaxation times. For these reasons these complexes are EPR undetectable.

b) low-spin d^7 complexes. The most celebrated and discussed compound of this class is the blue complex $[Fe(CN)_5NO]^{-3}$. This species presents narrow EPR lines. The spectrum consists of a triplet at a g-value of 2.026, due to the interaction with the ^{14}N nucleus in the NO^+ group ($a_N = 14.8$ gauss). For many years there have been conflicting speculations on the molecular orbital schemes for this complex. The electronic structure proposed by Manoharan and Gray[7] is now the most widely accepted. These authors have also presented a molecular calculation of the red-brown nitroprusside ion $[Fe(CN)_5NO]^{-2}$ which formally contains Fe^{+2} and NO^+, the electronic configuration of iron thus being $3d^6$. A distorted octahedral structure with approximate C_{4v} symmetry, was suggested for the blue complex. The Fe–N–O grouping was found to be bent.

On these theoretical bases Van Voorst and Hemmerich[8] performed EPR experiments using ^{57}Fe enriched $[Fe(CN)_5NO]^{-3}$. They identified two species in the equilibrium:

$$[Fe(CN)_5NO]^{-3} + H^+ \rightleftharpoons [Fe(CN)_5NOH]^{-2}$$

The triplet pattern mentioned above has been attributed to the proton adduct species, whose EPR parameters suggest that the compound has high spin density at the iron atom and a low spin density at the nitrogen atom. In the unprotonated species the unpaired electron is almost completely localized on the nitrosyl ligand ($\cdot NO$).

II Fe(NO)L$_2$ complexes

Five coordinated iron-nitrosyl complexes have been extensively studied by means of infrared spectroscopy, X-ray analysis and EPR spectra. Generally these complexes display pyramidal structure with ligands chelating the central ion through sulfur atoms.

We may distinguish among several types of compounds:

a) 1,2 dithiolene type complexes exist in three different chemical configurations. Following the nomenclature of McCleverty and coworkers[9], they are:

   i) $[Fe(NO) (—S_4)]^{-1}$: diamagnetic; iron in the formal $+2$ oxidation state; NO^+; five-membered chelate ring.

   ii) $[Fe(NO) (—S_4)]^{-2}$: three hyperfine lines; $g = 2.027$; $a_N \simeq 15$ gauss; $a_{57_{Fe}} = 9.4$ gauss; five-membered chelate ring.

   iii) $Fe(NO) (—S_4)^0$: one EPR line; $g = 2.009$; five membered ring.

We can see that species ii) shows EPR parameters almost identical to those of the blue complex (I). Thus we suggest the formal oxidation state $+1$ for the iron atom, considering NO as nitronium NO^+ group, coordinated along the axis perpendicular to the plane of the four sulfur atom. The electronic structure of the ground state can be considered the same as in (I). Referring to C_{2v} symmetry, McCleverty and coworkers[9] proposed a molecular orbital scheme which, in a broad sense, can be applied to all three differently charged types of complexes. The Fe–N–O system is assumed to be linear in i) and iii) and non-linear in the dianionic species. The positive g-shift in ii) is due to spin-orbit coupling which allows admixture of excited configurations.

b)   Dithiocarbamate complexes exhibit a three-line EPR spectrum[10] with $g \simeq 2.04$ and $a_N \simeq 12$–13 gauss. The complexes form a four-membered chelate ring, with iron in the formal $+1$ oxidation state. In both dithiolene and

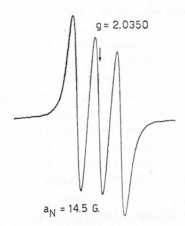

$g = 2.0350$

$a_N = 14.5$ G.

Figure 1   EPR spectrum of Fe–NO-dithioxamide complex

dithiocarbamate complexes, the nitrogen hyperfine splitting is remarkably sensitive to the size of the chelate ring.

The EPR analogies between ii) and dithiocarbamate species are supported by X-ray studies. They attributed a rectangular-based pyramidal structure with the nitric oxide molecule at the apex to dithiocarbamate complexes[11] and a square pyramidal structure to ii)[12]. Finally EPR data suggest that they both are low-spin d^7 complexes with the unpaired electron mainly localized at the iron atom.

The Fe–NO complex with dithioxamide as ligand is of particular interest. The spectrum given by the dark-green complex formed in water-ethanol solution is shown in Fig. 1. By analogy with spectra of ii) we can suggest two structures:

(chemical structures of two Fe–NO dithioxamide complexes)

A four-membered chelate ring involving S and N atoms is less probable due to the high value of the nitrogen coupling constant ($a_N = 14.5$ gauss).

c) Cysteine reacts with the Fe–NO system giving rise to a brown-violet complex in aqueous solution at pH $\simeq 9$. The related EPR spectrum displays a triplet with a nitrogen coupling constant $a_N \simeq 14$ gauss or, in the presence of $Na^{15}NO_2$, a doublet ($a_{15_N} \simeq 19.5$ gauss). After about five minutes the colour of the solution changes from brown-violet to light green and a III type of complex[13] is formed. It is noteworthy that even sulfide ions in the presence of ferrous ion and nitric oxide give rise to a triplet ($g = 2.021$; $a_N \simeq 5$ gauss) as does the diamagnetic Roussin's black salt on raising the pH to 11[13]. Finally cystine gives broad EPR spectrum quite distinct from that of cysteine. We can see that the —SH group, the disulfide group and the sulfide ions, present a different behaviour in relation to the bonding with the Fe–NO group. This could be used as a basis of enzyme studies.

Glutathione displays an unusual behaviour by giving rise to three different kinds of complexes as shown in Fig. 2. At first a dark-purple complex with structure II is formed ($g = 2.0436$; $a_N \simeq 14.6$ gauss; $a_{15_N} \simeq 20$ gauss). Then after a few minutes a different EPR spectrum arises. As we will see in the next section, this is attributable to structure III. Finally, after several hours, a triplet appears at a higher field with the EPR characteristics of the sulfide complexes ($g = 2.022$; $a_N \simeq 5$ gauss). This last complex is light yellow and its formation is favored at higher pH. On the bases of these experimental results, we can suggest a five-membered structure for the II-type cysteine and glutathione complexes:

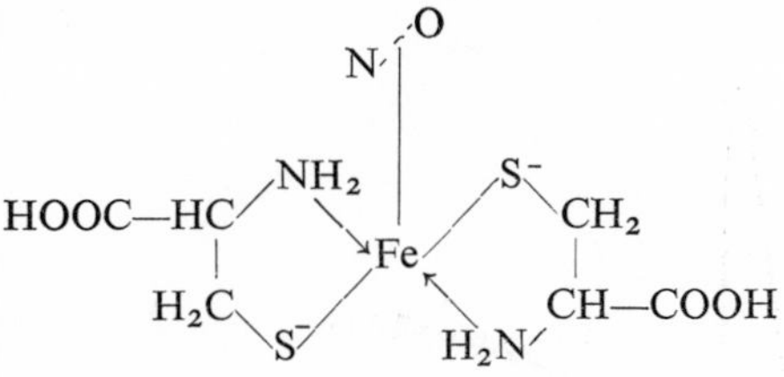

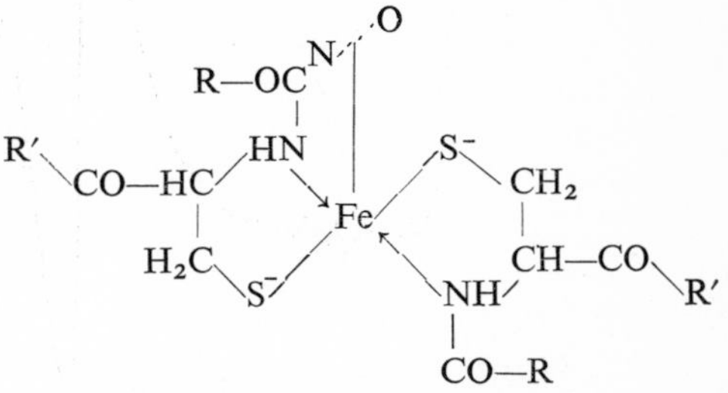

A chelate structure involving carboxyl group can be excluded because: (i) the chelate ring would be six-membered in cysteine, (ii) glutathione has no carboxyl group available in β-position with respect to the sulfur atom, (iii) further experiments showed that thioglycolic acid, $HS—CH_2—COOH$, only gives rise to the III type complex with no triplet present in its EPR spectrum. This supports the hypothesis that in the chelate type complexes, one amino (or amide) group has to be involved together with the —SH group.

d) Diphenylthiocarbazide gives a very stable triplet with g-factor $= 2.0391$ and $a_N = 11.5$ gauss. The spectrum is shown in Fig. 3a. The most probable structure seems to be a five-membered chelate ring complex:

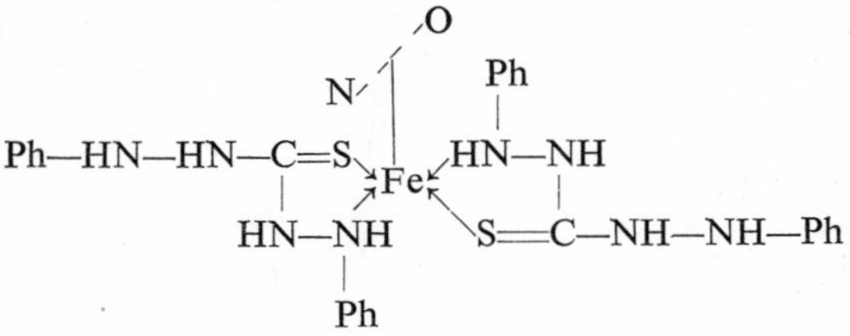

This dark-violet complex is in equilibrium with another one which may be a III type complex, whose spectrum is reported in Fig. 3b. However, because of the poor resolution of the spectrum, the structural attribution is ambiguous. It is noteworthy that diphenylcarbazide and diphenylcarbazone only have EPR spectra which can be interpreted in terms of a chelate IV-type structure.

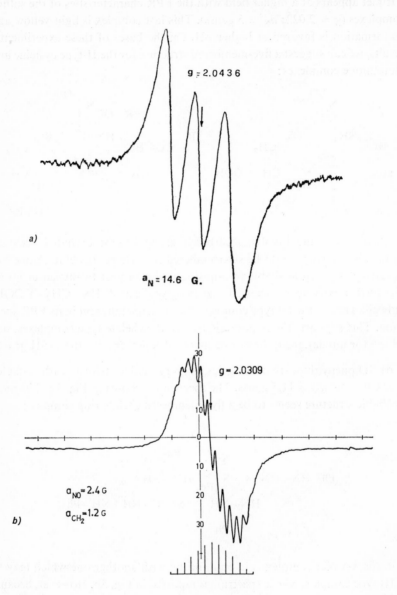

a)

g = 2.0436

$a_N = 14.6$ **G.**

b)

g = 2.0309

$a_{NO} = 2.4$ G

$a_{CH_2} = 1.2$ G

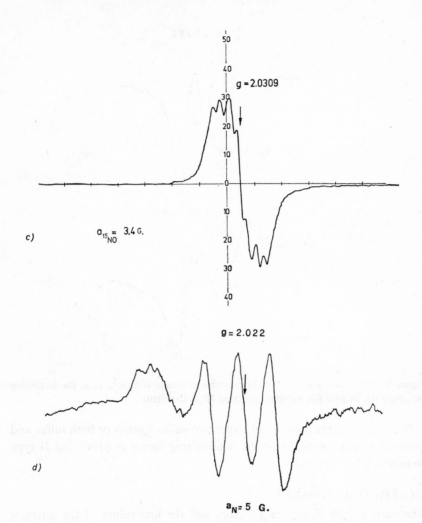

Figure 2 EPR spectra of Fe–NO-glutathione complexes. a) at the beginning (structure II). b) after about five minutes (structure III). c) after about five minutes with $Na^{15}NO_2$ (structure III). d) after 10 hours (sulfide structure)

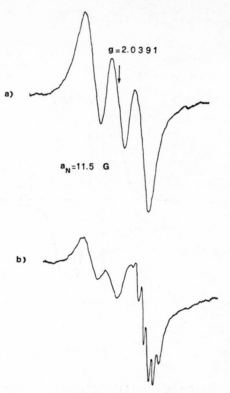

Figure 3 EPR spectra of Fe–NO-diphenylthiocarbazide complex. a) at the beginning (structure II). b) after five minutes (structure III at the right)

In conclusion, the presence of either two sulfur ligands or both sulfur and nitrogen ligands seems to be the determining factor in producing II type iron-nitrosyl compounds.

III   Fe(NO)$_2$L$_2$ complexes

The narrow EPR lines, the g-factors and the low values of the nitrogen coupling constants strongly suggest a low-spin d^7 configuration and a low degree of covalent character for the iron atom.

By analogy with X-ray structure of Roussin's red salt,[14] it is possible to assume a roughly tetrahedral arrangement for the bonds about the iron atom, the degeneracy of the d orbitals being completely destroyed.

The most interesting feature of this class of complexes is that iron participates in the complex formation in the +1 oxidation state. The NO groups are present as ·NO, with their unpaired electrons coupled in a molecular orbital. When formed from ferrous ion, the reduction of the iron atom must occur in the stage preceding the ligand association. Two possible reduction mechanism may be suggested:

a) $$2Fe^{+2} \rightleftharpoons Fe^{+3} + Fe^{+}$$

b) $$Fe^{+2} + 3NO \rightleftharpoons [Fe(NO)_2]^{+} + NO^{+}$$

We have observed that, after a few minutes, the formation of a II-type complex is very often followed by the appearance of a new spectrum corresponding to a III-type compound. Obviously redox processes must be involved in this equilibrium.

As mentioned above, cysteine, glutathione (Fig. 2b,c) and diphenylthiocarbazide (Fig. 3b) display this behaviour. Figure 4a represents both kinds of ^{57}Fe–NO cysteine complexes contemporaneously present in solution. Figure 4b shows the pure III-type complex. Interactions with the $I = \frac{1}{2}$ nuclear spin of ^{57}Fe splits the spectrum unto two groups of lines. Each group shows a nuclear hyperfine structure (relative intensity of the 13 lines: $1:4:8:12:16:20:22:20:16:12:8:4:1$) due to the two nitrogen atoms and the four methylene protons. The high value of the ^{57}Fe coupling constant proves that our assignment of the $+1$ oxidation state (low-spin d^7) to the iron atom is correct.

Thioglycolic acid gives a similar pattern (Fig. 5). As outlined above, it does not give rise to the triplet spectrum. A structure similar to that of cysteine can be attributed to this complex:

$$\begin{array}{c} ON \\ \diagdown \\ ON \diagup \end{array} Fe \begin{array}{c} \diagup S—CH_2—COOH \\ \diagdown S—CH_2—COOH \end{array}$$

Surprisingly, using thiolactic acid $HS—CH(CH_3)—COOH$ as ligand no paramagnetic absorption was observed. This can be explained in terms of steric hindrance, which could be due to the contemporaneous presence of methyl and carboxyl groups in the molecule.

As outlined by McDonald and coworkers[13] some of the $Fe(NO)_2L_2$ complexes can be considered as the monomeric form in equilibrium with the diamagnetic Roussin's red salt:

$$\begin{array}{c} ON \\ \diagdown \\ ON \diagup \end{array} Fe \begin{array}{c} \diagup L \diagdown \\ \diagdown L \diagup \end{array} Fe \begin{array}{c} \diagup N \\ \diagdown N \end{array} + 2L \rightleftharpoons 2 \begin{array}{c} ON \\ \diagdown \\ ON \diagup \end{array} Fe \begin{array}{c} \diagup L \\ \diagdown L \end{array}$$

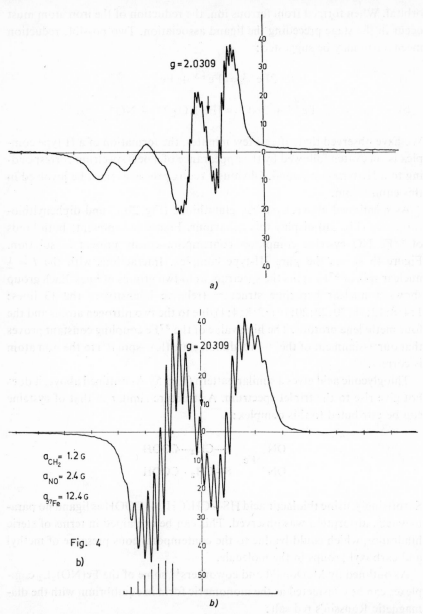

Figure 4   EPR spectra of Fe–NO-cysteine complexes. a) structure II and III; with $^{57}FeSO_4$. b) after about five minutes; with $^{57}FeSO_4$ (structure III)

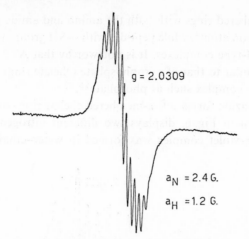

g = 2.0309

a_N = 2.4 G.

a_H = 1.2 G.

Figure 5 EPR spectrum of Fe–NO-thioglycolic acid complex (brown colour; pH $\simeq$ 10)

Also, in the absence of any ligand, a single line was observed in water solution. It was tentatively attributed to the $[Fe(NO)_2(H_2O)_2]^+$ complex. On raising the pH to 11, the spectrum changed into a quintet, attributed to $[Fe(NO)_2(OH)_2]^-$. As an alternative interpretation, the complex present in neutral and acid pH could be considered as the proton adduct $[Fe(NOH)_2 (H_2O)]^{+3}$.

Recently we observed similar effects in a novel class of iron-nitrosyl complexes with halide ligands[15]. They can be classified as III-type compounds. It is noteworthy that fluoride ion forms a "sui generis" complex, with six ligands bonded to iron: $[Fe(NO)_2F_4]^{-3}$.

Imidazole[4], phosphate, and other anionic ligands[13] give rise to nitrosyl complexes which belong to this class.

IV Fe(NO)₂L *complexes*

These compounds do not differ substantially from type III complexes, except for the fact that two monodentate ligands are substituted by a chelate ligand. They are therefore approximately tetrahedral, low-spin d^7 complexes. The same considerations about iron reduction apply here, as in the case of the type III complexes.

Several ligands may be involved, with different chelating groups, namely amino, amide, sulfur, carboxyl, phosphate etc. α-aminoacids have been[4] observed to form five-membered chelate ring complexes, whereas β-aminoacids do not give detectable EPR spectra. This can be explained in terms of unstability of six-membered chelate rings. Peptides without —SH grousp give rise to five-membered rings with both the amino and amide groups bonded to the central iron atom[4], while peptides with —SH groups (as glutathione) form II- and III-type complexes. It is noteworthy that ATP and ADP give EPR spectra similar to that of pyrophosphate (chelate ring) and that AMP forms a III-type complex such as phosphate[13].

Diphenylcarbazide forms a five-membered chelate ring complex. Its EPR spectrum, shown in Fig. 6, displays two different nitrogen splitting constants. The deep-violet complex was formed in water–ethanol solution at

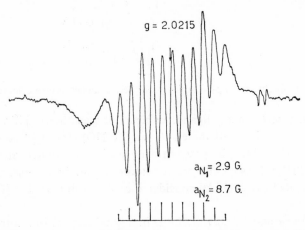

$g = 2.0215$

$a_{N_1} = 2.9$ G.

$a_{N_2} = 8.7$ G.

Figure 6   EPR spectrum of Fe–NO-diphenylcarbazide complex (structure IV)

$pH \simeq 7$–8. The g-factor and the nitrogen hyperfine constants ($a_{NH_2} = 8.7$; $a_{NO} = 2.9$ gauss) are very close to the respective values found for the α-amino-acid complexes[4], thus suggesting the following structure:

$$Ph—NH—NH—C=O$$
$$NH—NH \overset{}{\nearrow} Fe \overset{NO}{\underset{NO}{<}}$$
$$Ph$$

Conclusions

Table I summarizes and compares the EPR parameters of the four classes of the iron–nitrosyl complexes.

It was already noted that proteins give rise to at least two different EPR signals. This depends on whether the protein is rich or poor in —SH groups.

Table I g factors and isotropic hyperfine coupling constants of Fe–NO complexes

Type	Complexes	g-factor	a_{NO} (gauss)	$a_{15_{NO}}$ (gauss)	$a_{57_{Fe}}$ (gauss)	a_N (gauss)	a_H (gauss)	Ref.
I: Fe(NO)L$_5$	[Fe(CN)$_5$NOH]$^{-2}$	2.026	14.8	—	—	—	—	8
	[Fe(CN)$_5$NO]$^{-3}$	2.00	—	—	—	—	—	8
II: Fe(NO)L$_2$	[Fe(NO)(—S$_4$)]$^{-2}$	2.027	15	—	9.4	—	—	9
	Fe(NO)(—S$_4$)	2.009	—	—	—	—	—	9
	Fe(NO)(S$_2$CNR$_2$)$_2$	2.040	12.8	—	—	—	—	9
	Fe(NO)[S$_2$CC(CN)$_2$]$_2$	2.041	12.8	—	8.4	—	—	13
	Fe(NO)(sulfide)	2.021	5	—	—	—	—	13
	Fe(NO)(dithioxamide)$_2$	2.035	14.5	19.5	—	—	—	°
	Fe(NO)(cysteine)$_2$	2.043	14	20	—	—	—	°
	Fe(NO)(glutath.)$_2$	2.0436	14.6	—	—	—	—	°
	Fe(NO)(sulfide type)	2.022	5	—	—	—	—	°
	[Fe(NO)(DPTCZ)$_2$]$^{+2}$	2.0391	11.5	—	—	—	—	°
III: Fe(NO)$_2$L$_2$	[Fe(NO)$_2$(H$_2$O$_2$)]$^+$	2.033	—	—	—	—	—	13
	[Fe(NO)$_2$(OH$_2$)]$^-$	2.027	2.2	—	—	—	—	13
	[Fe(NO)$_2$Cl$_2$]$^-$	2.034	—	—	—	—	—	15
	[Fe(NO)$_2$Br$_2$]$^-$	2.045	—	—	—	—	—	15
	[Fe(NO)$_2$I$_2$]$^-$	2.070	—	—	—	—	—	15
	[Fe(NO)$_2$F$_4$]$^{-3}$	2.033	—	—	—	—	—	15
	[Fe(NO)$_2$(cysteine)$_2$]$^-$	2.0309	2.4	3.4	12.4	—	1.2	°
	[Fe(NO)$_2$(glutath.)$_2$]$^-$	2.0309	2.4	3.4	—	—	1.2	°
	[Fe(NO)$_2$(thiglycolic)$_2$]$^-$	2.0309	2.4	—	—	—	1.2	°
	[Fe(NO)$_2$(imidazole)$_2$]$^+$	2.033	—	—	—	—	—	4
IV: Fe(NO)$_2$L	Fe(NO)$_2$(aminoacid)	2.020	3.5	—	15.6	6.7	—	4
	Fe(NO)$_2$(peptide)	2.016	3	—	—	3.6–(7.3)	—	4
	Fe(NO)$_2$(ATP)	2.038	—	—	—	—	—	13
	Fe(NO)$_2$(ADP)	2.040	—	—	—	—	—	13
	[Fe(NO)$_2$(DPCZ)]$^+$	2.0215	2.9	—	—	8.7	—	°

°: this work; DPTCZ: diphenylthiocarbazide; DPCZ: diphenylcarbazide.

A more complete understanding of protein complexes should include the knowledge of different kinds of amino-acids-iron-nitrosyl complexes. We have seen from Woolum's data[4] and from our own work that the following structures exist:

a) —SH containing aminoacids (cysteine type): structures II and III.
b) —S—S— containing aminoacids (cystine type): unresolved structure II.
c) imidazole containing aminoacids (hystidine type): structure III.
d) α-aminoacids (only —NH$_2$ and —COOH groups): structure IV.
e) β-aminoacids (only —NH$_2$ and —COOH groups): no EPR detectable
                                                              Fe–NO complexes.

Of course other groups can be involved in the metal-protein bond. A deeper analysis with simple organic ligands, at different temperatures and pH, would be useful to check the existence of possible equilibria. We believe that a future program studying the relaxation times of both electron and nuclear T_1 and T_2, can throw light on the metal-protein bond in different complexes.

Acknowledgements

The authors are grateful to Prof. B. Commoner and to Dr. J. Woolum, department of Biology of the Washington University, St. Louis (USA) for basic ideas, suggestions and discussions. One of us (E.T.) thanks Prof. B. Commoner for the opportunity to spend one year at the Wahington University and for the facilities and instruments available, with which part of this work was carried out.

References

1   A. Dobry-Dulcaux, *Biochim. Biophys. Acta*, **39**, 33 (1960) and **39**, 44 (1960).
2   W. Gordy and H. N. Rexroad, in *Free Radicals in Biological Systems*, M. S. Blois *et al.*, Academic Press, New York, 1961, p. 268.
3   Ya. I. Axhipa and L. Kayushin, this book, p. 183.
4   J. C. Woolum, E. Tiezzi, and B. Commoner, *Biochim. Biophys. Acta*, **160**, 311 (1968).
5   A. A. Blanchard, *Inorganic Synthesis*, vol. II, p. 126, W. C. Fernelius edit. (1946).
6   W. P. Griffith, J. Lewis, and G. Wilkinson, *J. Chem. Soc.*, 3993 (1958).
7   P. T. Manoharan and H. B. Gray, *J. Amer. Chem. Soc.*, **87**, 3340 (1965).
8   J. D. N. Van Voorst and P. Hemmerich, *J. Chem. Phys.*, **45**, 3914 (1966).
9   J. A. McCleverty *et al.*, *J. Amer. Chem. Soc.*, **89**, 6082 (1967).

10 H.B.Gray, I.Bernal, and E.Billig, *J. Amer. Chem. Soc.*, **84**, 3404 (1962).

11 G.R.Davies, R.H.B.Mais, and P.G.Owston, *Chem. Commun.*, 81 (1968).

12 A.I.M.Rae, *Chem. Commun.*, 1245 (1967).

13 C.C.McDonald, W.D.Phillips, and H.F.Mower, *J. Amer. Chem. Soc.*, **87**, 3319 (1965).

14 J.T.Thomas, J.H.Robertson, and E.G.Cox, *Acta Cryst.*, **11**, 599 (1958).

15 L.Burlamacchi, G.Martini, and E.Tiezzi, *Inorg. Chem.*, **8**, 2021 (1969).

EPR studies of some aspects of copper-thiols interaction

GIUSEPPE ROTILIO, CARLO DE MARCO and

SILVESTRO DUPRÉ

*Institute of Biological Chemistry, University of Rome
and Molecular Biology Center of C.N.R.*

WE SHALL REFER here to some preliminary results obtained in extending our previous studies on copper-catalyzed thiols oxidation (1–4) and on the structure of the copper-thiols complexes involved in the reaction (5, 6).

It has been shown (5) that the Cu^{++} catalyzed oxydation of cysteine proceeds with the metal firmly complexed by the thiol. Recently (6) data have been obtained which strongly suggest that one copper is bound to two moles of cysteine in a complex recognizable by a characteristic EPR spectrum. Moreover, this complex has been indicated as being responsible for the typical absorption band centered at 330 nm, which appears immediately after mixing cysteine and cupric copper in alkaline solutions. The amount of bis-cysteine-copper complex remains practically unchanged in the presence of oxygen until an excess of cysteine is present, while in anaerobiosis it disappears slowly. It is clearly an intermediate in the oxidation of cysteine to cysteine, and quantitative estimation of paramagnetic copper by EPR indicated that during the thiol oxidation almost all the copper is present as cupric copper.

We have extended the optical and EPR analyses to other thiol compounds, namely cysteamine, cysteine methyl ester and mercaptosuccinic acid.

As in the case of cysteine, immediately after mixing with cupric copper in 0.1 N NaOH, these thiols give rise to the appearance of yellow compounds, whose absorption spectra are reported in Fig. 1. It is evident that the spectra of cysteamine and cysteine methyl ester with copper are similar to that of cysteine with copper; that of the mercaptosuccinate-copper is different, the peak being shifted to 340 nm. Moreover cysteamine differs from the other

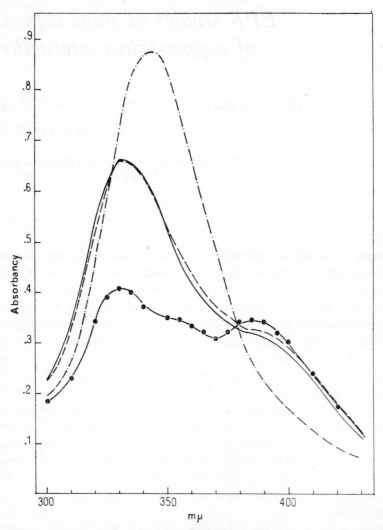

Figure 1 Absorption spectra of some thiols with Cu^{++}. The solutions contained 10^{-3} M thiol and 10^{-4} M $CuCl_2$ in 0.1 N NaOH. Light path, 1 cm. ●—● cysteamine, –– cysteine, —— cysteine methylester, –·– mercaptosuccinic acid. The spectrum of cysteamine was recorded at 5 seconds; the other spectra at 30 seconds after mixing

thiols in that the yellow compound formed after the addition of copper disappears very quickly, in either the presence of oxygen or anaerobiosis. This indicates that either the oxidation of cysteamine to the disulfide by molecular oxygen, or the reduction of copper in anaerobiosis is a very rapid reaction. Therefore the spectrum of cysteamine–copper complex reported in Fig. 1 has been obtained by determining at different wavelengths the O.D. of various solutions 5 sec. after mixing copper and cysteamine; and the absolute absorbancy values, lacking the O.D. at zero time, cannot be directly compared with the other two spectra of Fig. 1.

On the other hand mercaptosuccinate with copper gives rise to a yellow intermediate more stable in either oxygen or anaerobiosis: even on bubbling oxygen in the solution the oxidation of a 10^{-3} M solution of mercaptosuccinate in the presence of 10^{-4} cupric chloride requires about 20 min.

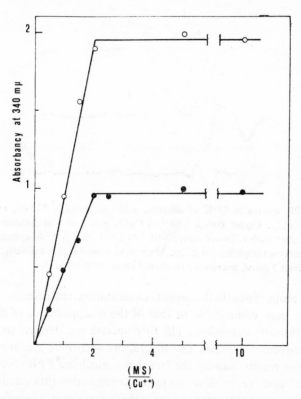

Figure 2 Maximal optical density at 340 mm as a function of the mercaptosuccinate over copper ratio. NaOH final concentration 0.1 N. Full circles: $CuCl_2$ 10^{-4} M, open circles: $CuCl_2$ $2 \cdot 10^{-4}$ M

During this time the absorbancy at 340 nm remains unchanged, until a sudden decrease is observed at the end of oxidation. With mercaptosuccinate it is therefore easy to demonstrate that the absorption band is a function of the copper concentration, and that the maximal O.D. at 340 nm is obtained when thiol and copper are in 2:1 ratio (Fig. 2). This indicates that, as in the case of cysteine[6], the stoichiometry for the formation of the complex is two moles of mercaptosuccinate for one of Cu^{++}.

The EPR analyses of the different thiol-copper solutions have shown that the absorption bands are strictly related to the presence of paramagnetic copper.

The EPR spectra of cysteamine and cysteine-methyl-ester copper solutions frozen immediately after mixing demonstrate that distinct complexes between cupric copper and thiol are present in the solutions (Fig. 3). The signal

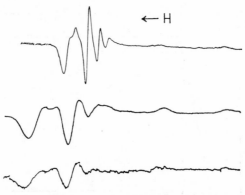

← H

Figure 3 EPR spectra at 77°K of different solutions in 0.1 N NaOH. Upper curve: 2×10^{-4} M $CuCl_2$. Center curve: $2 \cdot 10^{-4}$ M $CuCl_2$ plus $2 \cdot 10^{-3}$ M cysteine methylester, frozen 5 sec. after mixing. Lower curve: $2 \cdot 10^{-4}$ M $CuCl_2$ plus $2 \cdot 10^3$ M cysteamine, frozen 5 sec. after mixing (amplification $\times 2.5$). Microwave power was 6.4 mwatts, modulation amplitude about 5 gauss, microwave frequency about 9170 MHz

intensity is proportional to the copper concentration, and decreases in anaerobiosis at a rate comparable to that of the disappearance of the peak at 330 nm in the same conditions. The EPR spectra are identical to that of the bis-cysteine-copper complex ($g_{||} = 2.13$, $A_{||} = 570$ Mc/sec, for the three thiols). These results, namely the fact that optical and EPR spectra are the same as with cysteine, confirm our previous suggestion (6) that nitrogen and sulfur are the ligand atoms for copper in these complexes. The only difference is the different rate of the disappearance of the intermediate complex formed with cysteamine. This suggests that the rate limiting step for the reduction of copper by sulfhydryl is the redox dissociation of one sulfur atom, yielding sulfur radicals, and that the rate of this dissociation is related to the presence in the molecule of another group beside the ligand ones. More detailed studies are in progress to obtain a more quantitative correlation between kinetics of oxidation and structure of cysteine derivatives.

In the case of mercaptosuccinate-copper solutions the EPR spectrum is somewhat different (Fig. 4). Although the shape of the EPR spectra, and particularly the hyperfine pattern, which is not very well resolved, raises some doubt on the presence of only one species of copper at the steady state in air, it is evident that a complex between copper and the thiol has been formed in this case also.

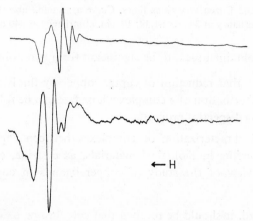

Figure 4 EPR spectra at 77°K of different solutions in 0.1 N NaOH. Upper curve: 2×10^{-4} M CuCl$_2$. Lower curve: $4 \cdot 10^{-4}$ M CuCl$_2$ plus $4 \cdot 10^{-3}$ M mercaptosuccinate, frozen 5 sec. after mixing (amplification $\times 2$)

Additional evidence for that comes from the identity between the EPR spectrum of CuCl$_2$ in 0.1 N NaOH and that of copper in the presence of a tenfold excess of succinate. With mercaptosuccinate however, the presence of SH groups affords the formation of a distinct complex, which is responsible for the optical absorption at 340 nm, as is clearly shown in Fig. 5, where the disappearance of the EPR signal in anaerobiosis, parallels fairly well the decrease of the 340 nm peak. Also in this case the spectrophotometric titration suggests the presence of a 2 : 1 complex which should have sulfur and a carboxyl as ligands for copper. The carboxyl group involved is most probably that near the sulfhydryl, as preliminary EPR experiments on copper chelates in 0.1 NaOH have shown that a tenfold excess of β-alanine does not form a complex with copper at an appreciable extent, whereas glycine does. In the case of the sulphur- and carboxyl-containing ligand, the redox dissociation of sulfur atom is slower as indicated by the much slower rate of oxidation of mercaptosuccinate.

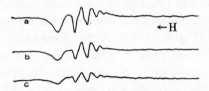

Figure 5 EPR spectra at 77°K of a solution containing $2 \cdot 10^{-4}$ M $CuCl_2$ and $2 \cdot 10^{-3}$ M mercaptosuccinate in 0.1 N NaOH, prepared in anaerobiosis and from which aliquots were taken out and frozen at various times. Curve a): 1 min., absorbancy at 340 nm: 1.76; 5 min., absorbancy at 340 nm: 1.39; 10 min. absorbancy at 340 min.: 0.78.

The results obtained seem to be significant from two points of view:

1) they indicate that reduction of cupric copper by thiols occurs in alkali through the formation of a complex whose fate can be followed owing to its well defined spectral properties.

2) the spectral chracterization of complexes between copper and sulfur ligands occurring in biological materials, as cysteine, can be of some interest in view of the study of copper ligands in copper-containing proteins.

In this regard, it should be recalled that cytochrome oxidase, as already pointed out by Blumberg and Peisach (7), has g values very similar to those of the copper-cysteine complex (i.e. $g_{||} = 2.17$ and $g_\perp = 2.03$). Moreover, some blue copper proteins (fungal laccase, ceruloplasmin, oxyhaemocyanin) show optical absorption bands in the region around 330 nm (8). Malkin *et al.* (9) attribute the 330 nm band of fungal laccase to a cupric-ligand charge transfer band; its extinction coefficient is 3 mM^{-1} cm^{-1}, which is the same order of magnitude of that of cysteine-copper complex (6, Fig. 1). Nakamura and Ogura (10) have emphasized the fact that the copper proteins characterized by an absorption band at 330 nm have a high rate of aerobic oxidation of the copper atom. These authors suggest that the site responsible for this band is essential in bringing about a specific mode of interaction between the copper atom and molecular oxygen.

References

1 D. Povoledo, C. De Marco and D. Cavallini, *G. Biochim.* **7**, 78 (1958).
2 C. De Marco, B. Mondovi and D. Cavallini, *G. Biochim.* **7**, 361 (1958).
3 C. De Marco, *G. Biochim.* **7**, 375 (1958).
4 C. De Marco and B. Mondovi, *Boll. Soc. It. Biol. Sper.* **36**, 260 (1960).
5 D. Cavallini, C. De Marco and S. Dupré, *Archiv. Biochem. Biophys.* **124**, 18 (1968).
6 D. Cavallini, C. De Marco, S. Dupré and G. Rotilio, *Arch. Biochem. Biophys.* **130**, 354 (1969).
7 W. E. Blumberg and J. Peisach, *J. Chem. Phys.* **49**, 1793 (1968).
8 K. E. Van Holde, *Biochemistry* **6**, 93 (1967).
9 R. Malkin, B. G. Malmström and T. Vänngård, *European J. Biochem.* **10**, 324 (1969).
10 T. Nakamura and Y. Ogura, *J. Biochem.* (Tokyo), **64**, 267 (1969).

A comparison of the information derivable from ESR and paramagnetic Mössbauer spectra of biological macromolecules

Nuclear Physics Division
Atomic Energy Research Establishment, Harwell, England

Abstract

Biological molecules containing both high and low spin ferric iron have been the subject of extensive investigations by both electron spin resonance and Mössbauer spectroscopy. It has proved possible to relate these measurements to the extent that the results of one kind of measurement can be used to predict, at least roughly or in part, the results of the other. It is thus important to examine the relation closely in order to determine to what extent the methods are redundant and, of more interest, to determine how they may complement each other and thus provide a more accurate picture of the state of the iron complex and its electronic level scheme. In the low spin ferric case ESR provides the means of determining with some precision the rhombic field splitting of the t_{2g} electronic levels of the iron, subject to some possible ambiguity. Mössbauer paramagnetic spectra provide considerably less accuracy but may often remove the ambiguity. As a simple example, a g tensor with components clustered closely about the free spin value may indicate either a very symmetric or a very asymmetric crystal field. Mössbauer spectra readily distinguish the two cases and determine in the latter the nature (pi or sigma) of the orbital containing the unpaired spin. Mössbauer spectroscopy of polycrystals is sensitive to the relative orientation of rhombic and cubic fields, while the ESR method can produce such information only through measurement of single crystals of known structure. Finally the sensitivity of Mössbauer spectra to the electric field gradient can result in information about paired as well as unpaired electrons. Calculated Mössbauer spectra are shown which demonstrate the relative precision of the methods. These are compared with examples of experimental Mössbauer spectra. In summary it could be stated that of two methods ESR provides the more precise information, while Mössbauer spectroscopy provides a wider range of types of information. The measurements are complementary rather than redundant and the combination may reasonably be expected in the coming few years to provide extremely valuable information on the class of materials to which they are applicable.

I Introduction

PARAMAGNET BIOLOGICAL MOLECULES have for a number of years been the subjects of intensive study by electron spin resonance. More recently some of the iron-containing members of this group have been examined as well by Mössbauer spectroscopy. The separation between magnetic centres afforded by the large size of biological molecules serves to suppress spin-spin interactions. At low temperature spin-lattice interaction is also very small, and the resulting long spin relaxation time makes possible the observation of rather well defined paramagnetic Mössbauer spectra. These spectra are determined largely by the magnetic interactions between unpaired electrons and the iron nuclear moment. ESR spectra on the other hand reflect the interaction of unpaired electrons with external fields. It is not surprising that they sense different features of unpaired spin distribution.

In the absence of hyperfine interaction, the Hamiltonian which determines the ESR spectrum is $\mathscr{H} = \beta H_a \cdot \sum_k (l_k - 2s_k)$. This is sensitive to the spin and orbital angular momentum of the electrons, but not to their spatial distribution. Hyperfine interactions can provide more information in the ESR spectra, but these are usually not obervable in biological molecules.

If there are unpaired electrons in the vicinity of the iron nucleus, and if their spins are not being flipped too rapidly by spin–spin or spin–lattice interactions, the hyperfine Hamiltonian which determines the Mössbauer spectrum may be written as

$$\mathscr{H} = P \sum_k \{l_k \cdot I + 3\,(\hat{r}_k \cdot s_k)\,(\hat{r}_k \cdot I) - (s_k \cdot I) - \mathscr{K}\,(s_k \cdot I)\}$$

$$- g_N \beta_N I \cdot H_a + \mathscr{H}_Q$$

$$P = 2\tfrac{1}{N}g\beta\beta_N \langle r^{-3}\rangle_{\text{effm}} \tag{1}$$

Here I is the nuclear spin operator, β_N is the nuclear magneton, β is the electron magneton, l_k is the orbital angular momentum of the kth electron, s_k is its spin, and the summation is taken over all unpaired electrons. The first term represents the interaction of the nucleus with the electrons regarded as current loops. The next two come from ordinary magnetic dipole–dipole interactions and are sensitive to the spatial distribution of spin. The term involving $\mathscr{K}$ results from the Fermi contact interaction, and takes account of two effects: (1) the core s electrons are polarized through exchange effects with the unpaired 3 d electrons; (2) in suitable symmetries the d electrons have a small amount of as character. The next to last term in the Hamiltonian allows for the direct effect of any externally applied field, H_a, and the last is the nuclear quadrupole interaction with the electric field gradient (EFG). The quantity $\langle r^{-3} \rangle_{\mathrm{effm}}$ depends on the radial distribution of the unpaired electrons, and has significant value only for wave functions located on the iron. In low spin ferric complexes the orbital, dipolar, and contact parts of the Hamiltonian are often of comparable size. The electric quadrupole interaction, $\mathscr{H}_Q$, usually is dominated by the charge distribution of the iron d electrons.

If the ground electronic state is a Kramers doublet it is convenient to treat the Mössbauer problem in the spin Hamiltonian formulation, using the following $S = \frac{1}{2}$ spin Hamiltonian.

$$\mathscr{H} = \beta H_a \cdot \underline{g} \cdot S + I \cdot \underline{A} \cdot S + \frac{1}{4} Q V_{zz} \left[I_z^2 - I \frac{(I+1)}{3} + \frac{\eta}{3} (I_x^2 - I_y^2) \right]$$

$$- g_N^* \beta_N I \cdot H_a \tag{2}$$

This is made equivalent to (1) by insertion of appropriate quantities $\underline{g}$, $\underline{A}$, V_{zz} and η. These of course depend on the detailed nature of the electronic ground doublet. Because (2) contains both nuclear and electronic operators, the basis wave functions must contain nuclear and electronic parts, and the problem is one of spin coupling. In the presence of an external field, however, a simplification is possible. If the applied field is strong compared to the magnetic hyperfine interaction, i.e. for $H_a \gg 10$ gauss, the first term dominates (2) and we may solve it to find the electronic eigenstates and their value of $\langle S \rangle$. This may then be used to replace the operator S in the remainder of the Hamiltonian to determine the nuclear eigenstates and the Mössbauer spectrum. The spin is thus equivalent to an internal magnetic field $H_i = -\underline{A} \cdot \langle S \rangle / g_N \beta_N$. The members of the ground doublet have opposite spin values, and produce Mössbauer spectra which are distinguishable only when the direct effect of the applied field on the nucleus is significant.

II   Electron in an isolated orbital state

We now compare Mössbauer and ESR spectroscopy in the case of an unpaired electron which lies in an orbital which is isolated on the energy scale. This isolation prevents orbital admixture and results in complete quenching of the orbital angular momentum. The g value is then just the free spin value, and the ESR spectrum does not distinguish between σ (e.g., $3z^2 - r^2$) and π (e.g., xy) types of orbitals. The Mössbauer spectrum does sense the difference through the $\underline{A}$ tensor, as shown diagrammatically in Fig. 1. The essential result is that the hyperfine A tensor has one large component for a π orbital,

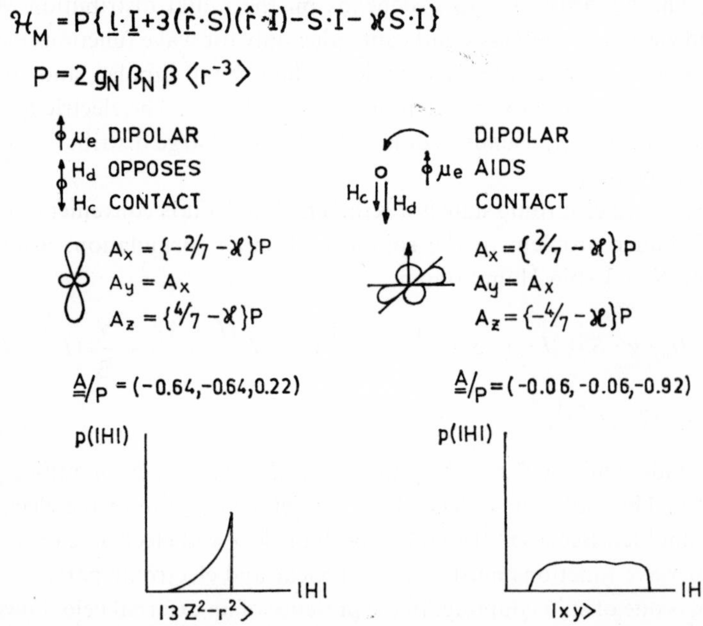

$$\mathcal{H}_M = P\{\underline{L}\cdot\underline{I} + 3(\hat{r}\cdot S)(\hat{r}\cdot I) - S\cdot I - \varkappa S\cdot I\}$$

$$P = 2\,g_N\,\beta_N\,\beta\,\langle r^{-3}\rangle$$

$\oint \mu_e$ DIPOLAR
$\uparrow H_d$ OPPOSES
$\downarrow H_c$ CONTACT

$A_x = \{-2/7 - \varkappa\}P$
$A_y = A_x$
$A_z = \{4/7 - \varkappa\}P$

$\underline{A}/_P = (-0.64, -0.64, 0.22)$

DIPOLAR
$\oint \mu_e$ AIDS
$H_c \updownarrow H_d$ CONTACT

$A_x = \{2/7 - \varkappa\}P$
$A_y = A_x$
$A_z = \{-4/7 - \varkappa\}P$

$\underline{A}/_P = (-0.06, -0.06, -0.92)$

$p(|H|)$ ... $|H|$ $|3Z^2-r^2\rangle$

$p(|H|)$ ... $|H|$ $|xy\rangle$

Figure 1   Diagram explaining qualitative features of Mössbauer spectra caused by unpaired electron spins in isolated orbitals. The sigma orbital is elongated (left side), and when the electron moment μ_e lies parallel to it the opposition of dipolar and contact interactions makes the net magnetic intearction with the nucleus small. When μ_e is transverse to the length of the orbital the two terms add. Thus the $\underline{A}$ tensor has one small and two large components. When an applied field forces μ_e to assume all possible directions in a polycrystalline sample the statistical preference for transverse directions means that at most nuclei the effective field will be near its maximum value of $\frac{1}{2}A_1/g_N\beta_N$, and a spectrum with well defined peaks will result. For the pi orbital $\underline{A}$ will be large when μ_e lies perpendicular to the plane of the orbital. Since H_{eff} (approximately) and the statistical weight both go as $\sin\theta$, there is an almost uniform distribution of H_{eff} values between zero and the maximum. The result will be a spectrum without well defined lines but two large components for a σ orbital. Thus when a polycrystalline specimen is put in a moderate external field, the π electron will give rise to a broad range of internal fields, while for the σ electron the field values will tend to cluster near the maximum. In the latter case we expect a resolved Mössbauer line spectrum, while the former should show a blur. Figure 2 (from Ooster-

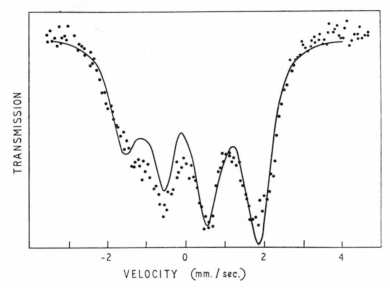

Figure 2 Mössbauer spectrum of hemoglobin nitric oxide at 4.2°K in an applied field of 500 G transverse to the gamma ray beam. The solid curve is calculated, using $A_\perp(Fe^{57m})$ = +1.33 mm/sec, $A_\parallel(Fe^{57m})$ = −0.46 mm/sec, values characteristic of a sigma orbital. A positive axial EFG is assumed of such strength that a quadrupole splitting of 1.5 mm/sec would be expected in the absence of magnetic interactions. A Lorentzian line of width 0.6 mm/sec has been folded in, and an isomer shift of 0.25 mm/sec is assumed. The velocity scale is relative to metallic Fe. (Oosterhuis 1969b)

huis 1969b) shows the theoretical and experimental spectra of haemoglobin nitric oxide, whose $g \simeq 2$ (Sancier 1962) implies that the unpaired spin is in an isolated orbital. The interpretation shows that it is in a σ bond. In order to get a reasonable theoretical spectrum an amount of EFG contribution from the paired electrons was assumed which was just enough the reverse the sign of that from the unpaired one. Nonetheless the conclusion of σ bonding is strong, for no combination of EFG with the magnetic interactions of an unpaired π electron could produce the observed spectrum. Figure 3 shows a molecular orbital scheme which would explain the results. The lowering of the σ bonding level below the π antibonding levels is the essential result. If

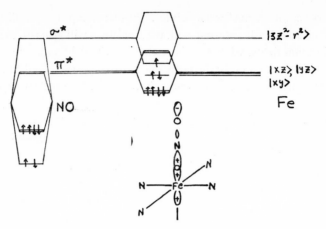

Figure 3 Energy level scheme for haemoglobin nitric oxide. Mössbauer results indicate that unpaired spin is in a sigma orbital, hence that the levels are ordered as shown. If the pi bonding orbitals shown have equal amplitudes of iron t_{2g} and nitric oxide pi*, the EFG produced is just enough to reverse the sign of that produced by the unpaired electron. These are the assumptions made in the calculation of the previous figure

$|xz\rangle$ and $|yz\rangle$ were shared equally between Fe and NO, the net EFG of the lowest six electrons shown on the diagram would be just sufficient to reverse the sign of the EFG caused by the unpaired electron (assumed equally shared with NO).

III   Typical low spin case

As a second example we consider the case of cytochrome c (Lang 1968) a fairly typical low spin haem protein. For this material the model of Griffith has been found to give good agreement with ESR (Salmeen 1968) as well as Mössbauer results. The cubic field is assumed very large, effectively preventing admixture of e_g states into the low lying t_{2g}. The latter are assumed to be split by a rhombic field and we take axes such that $|xy\rangle$, $|xz\rangle$, $|yz\rangle$ are in ascending order in energy. The six available states formed by combining the t_{2g} with spinors are occupied by five electrons, and the problem is equivalent to the problem of a single hole. Spin-orbit coupling distributes the hole over the three orbitals, and the character of the ground state is determined by the orbital splittings. The splittings thus determine the g values and the Mössbauer spectrum, neglecting the small direct lattice effects upon the latter. We may therefore regard ESR and Mössbauer spectroscopy as means of determinming the splitting and examine their relative accuracy.

Figures 4 and 5 show experimental Mössbauer spectra of cytochrome *c*, together with calculated results. In each case the solid curve is calculated assuming orbital splittings determined from the measured *g* values*, (1.24, 2.24, 3.06), which correspond to orbital energies $E_{yz} = 3.30\zeta$, $E_{xz} = 1.83\zeta$,

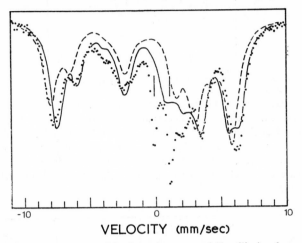

VELOCITY (mm/sec)

Figure 4   Mössbauer spectrum of ferricytochrome c of T. utilis in absence of applied magnetic field at 4.2°K. The solid curve is the calculated spectrum based on g values (1.24, 2.25, 3.07) which are in essential agreement with those of horse heart ferricytochrome c (Salmeen 1968). The dashed curve corresponds to g = (1.0, 2.4, 3.0) and provides an indication of the relative sensitivity of Mössbauer spectroscopy. A line width of 1.2 mm/sec has been assumed in the calculations to take rough account of relaxation effects, transferred hyperfine interaction with ligand nuclei, and distortion of the frozen protein. Data are from (Lang 1968)

relative to E_{xy}. (Here ζ is the spin-orbit coupling constant.) The dashed curves are for $g = (1.0, 2.4, 3.0)$ which in turn corresponds to $E_{yz} = 2.30\zeta$, $E_{xz} = 0.9\zeta$. Since promotion of the hole from $|yz\rangle$ to $|xz\rangle$ is the main effect of the spin-orbit coupling, it is perhaps more meaningful to compare the energy difference between these orbitals, and we see that these are 1.47ζ and 1.40ζ for the two cases. Because of the high energy involved, promotion of the hole to $|xy\rangle$ is not very important in determining the *g* values or the Mössbauer spectra, and hence the measurements do not determine this orbital level well. Since the dashed curves of Figs. 4 and 5 are only slightly less satisfactory than the full curves, we can say roughly that the accuracy of Möss-

---

* The ESR results refer to horse heart cytochrome (Salmeen 1968) while Mössbauer spectra are for T. utilis (Lang 1968). Usually however there is little species dependence in the ESR of haem proteins.

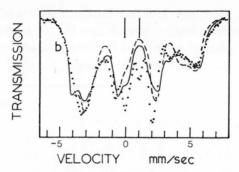

Figure 5 Mössbauer spectrum of ferricytochrome c of T. utilis in a transverse magnetic field of 500 gauss at 4.2°K. The solid curve is the calculated spectrum based on $g = (1.24, 2.25, 3.07)$, near that of horse heart ferricytochrome (Salmeen 1968). The dashed curve corresponds to $g = (1.0, 2.4, 3.0)$, and provides an indication of the relative sensitivity of Mössbauer spectroscopy. A line width of 0.6 mm/sec has been assumed in the calculations to take rough account of relaxation effects and distortion of the frozen protein. Data are from (Lang 1968)

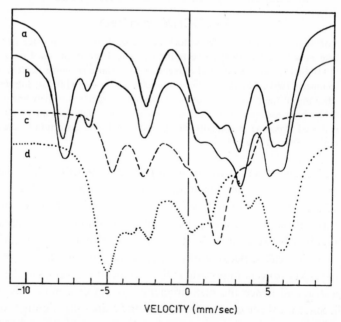

Figure 6 Zero field calculated Mössbauer spectra of cytochrome c. Curve (a) is a repeat of the solid curve of Fig. 4 and corresponds to coincident cubic and rhombic field axes. Curves (b), (c), and (d) are computed on the assumption of a rhombic field related to the above by rotations of 45 degrees about the x, y, and z axes

bauer spectroscopy in this case corresponds to a 10 to 20 per cent uncertainty in the *g* values.

We have seen that the *g* values may be used to determine the rhombic field splittings of the t_{2g} orbitals. However, ESR measurements on polycrystalline or frozen solution samples do not determine the orientation of the rhombic field relative to the cubic field in the molecule. Although this relative orientation does not affect the principal *g* values, it does affect the size of the components of the *A* tensor as well as the orientation of the A and g tensors relative to the EFG tensor produced by the d electron charge. Thus the Mössbauer spectra, even for polycrystalline samples, are sensitive to the alignment of the rhombic field within the molecule. Examples of this may be seen in Figs. 6 and 7. These again refer to zero field and small field spectra of the

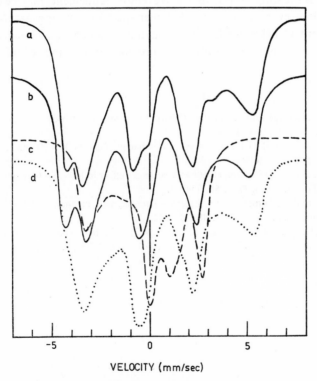

VELOCITY (mm/sec)

Figure 7   Calculated Mössbauer spectra of cytochrome c in a small transverse magnetic field. Curve (*a*) is a repeat of the solid curve of Fig. 5, and corresponds to coincident cubic and rhombic field axes. Curves (*b*), (*c*), and (*d*) are computed on the assumption of a rhombic field related to the above by rotations of 45 degrees about the *x*, *y*, and *z* axes respectively

cytochrome c. In each figure, curve (*a*) is a repeat of the spectrum determined from the measured *g*'s, assuming rhombic and cubic systems are coincident. Curves (*b*), (*c*) and (*d*) correspond to rotations of the rhombic field by 45 degrees about *x*, *y*, and *z* respectively. It is clear that the largest effects are produced by rotations normal to planes with widely different *g* values. The Mössbauer results on cytochrome *c* indicate that the *x* axis of the rhombic field lies near to a cubic field axis. Relative rotations about the *x* axis cannot, however, be ruled out.

In summary we see that paramagnetic Mössbauer spectroscopy appears to be a less precise tool than ESR, but can probe more diverse features of the electronic distribution. The lower precision is partly attributable to the lack of a sufficiently refined calculation of the Mössbauer spectra. The model used here is the simplest possible, and eventual refinements may be expected. Nevertheless, even at present Mössbauer spectroscopy appears useful as a complement to other methods of studying paramagnetic biological molecules, and it is unique in its ability to provide certain kinds of information.

References

Lang, G., D. Herbert, and T. Yonetani (1968) "Mössbauer spectroscopy of cytochrome c". *J. Chem. Phys.* **49**, 944–50.

Oosterhuis, W.T. and G. Lang (1969a) "Mössbauer effect in $K_3Fe(CN)_6$". *Phys. Rev.* **178**, 439–56.

Oosterhuis, W.T. and G. Lang (1969b) "Mössbauer effect in low-spin (d^7) complex molecules of Fe." *J. Chem. Phys.* **50**, 4381–7.

Salmeen, I. and G. Palmer (1968) "Electron paramagnetic resonance of beef-heart ferricytochrome c". *J. Chem. Phys.* **48**, 2049–52.

Sancier, K.H., G. Freeman, and J.S. Mills (1962) "Electron spin resonance of nitric-oxide–hemoglobin complexes in solution". *Science* **137**, 752–4.

Mössbauer effect studies of hyperfine interactions in the iron-sulphur proteins

C. E. JOHNSON

Materials Physics Division, AERE Harwell, Berks, England

Abstract

The application of measurements of the magnetic hyperfine interaction of Fe^{57} by the Mössbauer effect to the study of iron in biological molecules is described. Results on the iron sulphur proteins are reviewed.

Introduction

IN THIS PAPER I shall describe the application of the Mössbauer Effect to the study of the iron in some iron–sulphur proteins[1]. These are a group of molecules which are found in many diverse organisms; in plants, animals and bacteria. They are involved in oxidative electron transfer processes, i.e. they have large negative redox potentials. Their crystal structure is not known, but they are characterized by being diamagnetic in the oxidized state and paramagnetic with a rather unusual EPR signal with $g \sim 1.94$ when reduced.

The Mössbauer Effect (or NGR—nuclear gamma-ray resonance) is the newest of the resonance techniques to be applied to the study of biological molecules, and it is worth discussing first its relation to the other well-established resonance methods (NMR and EPR) which are the subject of this conference.

The Mössbauer Effect as a spectroscopic probe in biomolecules

In the first place the Mössbaner Effect is not a *magnetic* resonance technique; it can be observed equally well in diamagnetic as in paramagnetic materials. This in itself can be a great advantage, for example in the iron–sulphur proteins both the oxidized and reduced molecules may be studied.

The use of the Mössbauer Effect in biochemistry is rather different from that of NMR, which is mainly applied to detailed structural problems and to determining the conformation of the protein molecule as a whole. The Mössbauer Effect is essentially a local probe, and hence the information obtained is more similar to that obtained from EPR which is used (a) to study the chemical state and bonding of the resonant atoms and (b) to obtain qualitative data on the local structure and symmetry in their neighbourhood. Whereas EPR is sensitive only to ions (e.g. Fe^{3+}, Co^{2+}) or free radicals which have unpaired electron spins, the Mössbauer Effect is specific to certain nuclei (e.g. Fe^{57}, Sn^{119}) which have low-energy γ-rays (a few tens of keV) decaying to the ground state.

In fact for paramagnetic ions the Mössbauer and EPR spectra give complementary information. In both hyperfine splitting may be observed, but whereas the Mössbauer Effect measures the effective magnetic field produced at the nuclei by the electrons, EPR measures the magnetic field produced at the electrons by the nuclei. The condition for observing hyperfine interaction by the two methods is the same, i.e. that the relaxation time τ of the electron spins should be longer than $\hbar/A$, where A is the strength of the hyperfine interaction. This means essentially that the electronic magnetic moment must remain stationary for a time long compared with the precession period of the nuclei in the field of the electrons (or the electrons in the field of the nuclei). Otherwise if $\tau \sim \hbar/A$ both the EPR and Mössbauer lines broaden and overlap and the magnetic hyperfine interaction cannot be resolved. If the electron spins flip fast compared with $\hbar/A$ no magnetic effects are seen in the Mössbauer spectrum, which then looks similar to that of a diamagnetic substance. Finally when $\tau \ll \hbar/g\beta H$ no EPR signal may be obtained, although the Mössbauer spectrum is still observable provided the sample is still in the solid state. So the conditions for observing hyperfine effects in EPR and Mössbauer spectra are that the relaxation times should be long. This may be achieved by cooling the specimen to low temperatures, which increases the spin–lattice relaxation time T_1, and by using magnetically dilute samples, which increases the spin–spin relaxation time T_2. Biological materials are almost by definition magnetically dilute, so that hyperfine interactions are observable in their Fe^{57} Mössbauer spectra at low temperatures.

When magnetic effects are not seen in the Mössbauer spectrum, there is still useful information to be obtained from the chemical isomer shift and the electric quadrupole splitting. The centre of the spectrum is sensitive to the chemical state of the iron atom (whether it is ferrous or ferric, and high or low spin) and on the degree of convalency. The quadrupole splitting is a probe of the local stereochemistry, and measures qualitatively the distortion from cubic symmetry and may also help in assigning the chemical state of the iron.

Effect of a magnetic field on the Mössbauer spectrum from a slowly relaxing paramagnetic ion

A difficulty is encountered in interpreting the Fe^{57} Mössbauer spectra of biological molecules, because for strongly covalent bonding the chemical shift and quadrupole splitting cannot easily distinguish between high-spin Fe^{3+}, low-spin Fe^{3+} and low-spin Fe^{2+}. Hence it is important to understand the magnetic hyperfine spectra which are observed at low temperatures. These of course cannot be observed for low-spin Fe^{2+}, which is diamagnetic. For slowly relaxing paramagnetic ions the application of an external magnetic field usually produces a simpler Mössbauer spectrum, which is easier to interpret than the zero–field spectrum. To illustrate this we shall consider the simple example of an ion with spin $\frac{1}{2}$ with isotropic g-value and hyperfine coupling. This seems to describe approximately the behaviour of some of the iron–sulphur proteins which we have studied, and it will also serve to demonstrate the relation between the EPR and Mössbauer spectra and to introduce the more complicated cases where the ion is anisotropic.

The energy levels of the ion are given by the spin–hamiltonian

$$\mathscr{H} = AS \cdot I + g\beta H \cdot S$$

and are shown in Fig. 1 as a function of the magnetic field H for nuclei with spin $I_g = \frac{1}{2}$ (coupling constant A_g) and $I_g = \frac{3}{2}$ (coupling constant $A_e = 0.562A_g$). The subscripts g and e refer to the ground and excited states of Fe^{57}. The Mössbauer Effect involves transitions between these states with the selection rules $\Delta m = 0, \pm 1, \Delta M = 0$. The EPR spectrum involves transitions within the ground state only, and with selection rules $\Delta m = 0, \Delta M = \pm 1$; m and M are the projections of the nuclear and electron spins respectively on the axis of quantization.

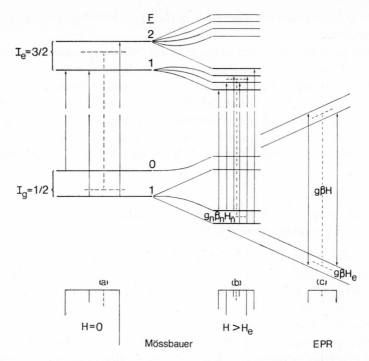

Figure 1   Energy levels of a paramagnetic ion with spin-Hamiltonian $\mathcal{H} = AS \cdot I + g\beta H \cdot S$ for $S = \frac{1}{2}$ and $I = \frac{1}{2}$ and 0. The Mössbauer transitions are shown (a) for $H = 0$ and (b) for $H > H_e$ (note that there is an identical set of lines between the $S_z = +\frac{1}{2}$ levels). The EPR transitions are shown in (c)

In zero field the hyperfine interaction is not equivalent to an effective magnetic field, but couples S and I to give a total angular momentum $F = S + I$, rather like the situation in a free atom. For slow electron-spin relaxation the Mössbauer spectrum shows the asymmetrical three-line pattern of Fig. 1(a).

When the magnetic field is much larger than $A/g\beta$, the fluctuating components $A/2\,(S_+I_- + S_-I_+)$ of the hyperfine interaction tensor are suppressed, and the splitting of the hyperfine levels is $A/2$. The Mössbauer spectrum becomes symmetrical (apart from quadrupole interactions) and shows a six-line Zeeman pattern similar to that found in ferromagnets (Figure 1 (b)). The effective field at the nucleus is $H_n = A/2g_n\beta_n$. In the EPR spectrum the line is split by $2H_e$, where $H_e = A/2g\beta$ is the field at the electrons produced by the nuclei (Fig. 1(c)). For Fe^{57} $g_n = 0.18$ and since $\beta/\beta_n = 1840$, H_n is

$1840/0.18 \sim 10^4$ times H_e. Typical values for $\Delta H = 2H_e$ in the iron-sulphur proteins are about 20 G, and for H_n about 200 kG.

For anisotropic g- and A-values the splittings become a function of the angle which the field makes with the axes of the ion, and in a specimen where these axes are randomly oriented the lines will become broadened.

The iron–sulphur proteins

The structure of the iron–sulphur proteins (sometimes known as the non-haem iron proteins) has not been determined. They have been characterized by EPR spectroscopy, largely due to the work of Beinert and his collaborators. They are diamagnetic in the oxidized state, but on reduction they become paramagnetic and show an unusual EPR spectrum with average g-value less than 2. By observing the growth of the EPR signal as the proteins are titrated with a reducing agent as well as by more conventional techniques, the number of electrons transferred per molecule on reduction has been measured. These data and the g-values are given for a number of iron-sulphur proteins in Table I. For spinach ferredoxin and some other proteins one electron is transferred per two iron atoms, and it is of great importance to know how this electron is distributed in the molecule. By enriching some of these proteins in Fe^{57} and S^{33} hyperfine splitting has been observed, [2-7]

Table I Properties of the iron-sulphur proteins

	No. of Fe/ molecule	e/mole transferred on reduction	g_x	g_y	g_z
Spinach ferredoxin	2	1	1.893	1.962	2.048
Euglena ferredoxin	2	1	1.90	1.96	2.06
Putidaredoxin	2	1	1.94	1.94	2.01
Adrenodoxin	2	1	1.94	1.94	2.01
Azotobacter	2	1	1.93	1.94	2.00
Xanthine oxidase	8	8	1.899	1.935	2.022
Chromatium ferredoxin	4	2	1.891	1.944	2.059
Clostridium ferredoxin	7–8	1–2	1.892	1.960	2.005

establishing that the unpaired electron is localized in the region of the iron and labile sulphur atoms. Two of them (putidaredoxin[4] and adrenodoxin[5]) showed hyperfine splitting due to interaction with a pair of iron nuclei, showing that the electron interacts equally with both iron atoms.

Proteins at the bottom of the table are more complex. For example xanthine oxidase contains molybdenum and flavin, which also accept electrons on reduction. Nevertheless, a $g = 1.94$ signal is observed[8] similar to that of the simpler molecules.

Mössbauer Effect data

Mössbauer spectra have been obtained of Fe^{57} in plant ferredoxin from spinach[9,10] and *Euglena*[11] (with Dr D. O. Hall of King's College, London) and in xanthine oxidase[12] from milk (with Dr R. C. Bray of the Chester Beatty Research Institute, London). The *Euglena* was grown on enriched Fe^{57} (90%), the other proteins contained natural iron (2% Fe^{57}). The ferredoxins were in frozen solution, and the xanthine oxidase was precipitated in alcohol and concentrated by centrifuging. Reduction was by sodium dithionite.

In the oxidized state all the proteins showed a chemical shift of ~0.2 mm/ sec relative to iron and a quadrupole splitting of 0.6 mm/sec. On applying a field of 30 kG to *Euglena* ferredoxin at 4.2 °K no hyperfine splitting was observed, and from the splitting due to the direct effect of the field on the nucleus the sign of the electric field gradient was found to be positive. This is potentially a useful piece of structural information; if the number and symmetry of the ligand environment were known, it would enable us to say whether the distortion from cubic symmetry was a compression or an elongation along the symmetry axis.

The Mössbauer spectra of reduced *Euglena* ferredoxin at 4.2 °K are shown in Fig. 2. The effect of applying a field of a few hundred gauss is similar to that described for the case of $S = \frac{1}{2}$ and isotropic hyperfine coupling and shown in Fig. 1; the spectrum is asymmetrical in zero field (Fig. 2(a)) and becomes a symmetrical Zeeman pattern in the field (Fig. 2(b)). The agreement is not perfect—in zero field the lines are broad and many extra lines are observed, but in the field there seems to be an almost constant effective field at the iron nuclei of about 170 kG. The same behaviour is found for reduced spinach ferredoxin and for xanthine oxidase. In Fig. 3 the spectra of the three proteins at 4.2 °K in a small field are compared. (The field was parallel to the γ-rays, so the intensity of the two $\Delta m = 0$ transitions is small). It is seen that the effective field is closely the same in all three proteins, although the intensities of the Mössbauer absorption are different owing to the different amounts of Fe^{57} in the samples.

At high temperatures the electron-spin relaxation times become short and

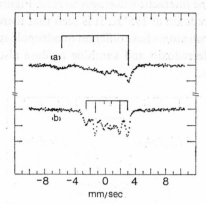

Figure 2 Mössbauer spectra of reduced *Euglena* ferredoxin at 4.2 °K (a) in $H = 0$, (b) in $H = 500$ G

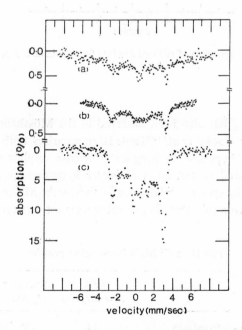

Figure 3 Mössbauer spectra of reduced (a) xanthine oxidase, (b) spinach ferredoxin and (c) *Euglena* ferredoxin at 4.2 °K and in a field of 100 G parallel to the γ-rays

the magnetic hyperfine interaction averages to zero. Figure 4 shows the spectrum of *Euglena* ferredoxin at 195°K. It is seen that there are four lines, so apparently the two iron atoms have different quadrupole splittings and chemical shifts. Spinach ferredoxin and xanthine oxidase also behave similarly at high temperatures.

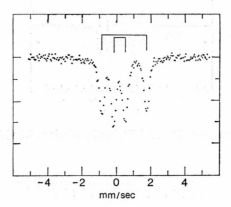

Figure 4   Mössbauer spectrum of reduced *Euglena* ferredoxin at 195°K

Conclusions

Hyperfine splitting has also been observed in the Mössbauer spectrum of putidaredoxin by Cooke *et al.*[13]Table II summarises both the EPR and Mössbauer Effect hyperfine splitting data for iron sulphur proteins which have been measured so far. In *Azotobacter* hyperfine splitting has been observed in the ESR spectrum[2] but not (so far) by the Mössbauer Effect.[14] However, apart from this, there is general agreement between the two kinds of measurement.

Table II Fe^{57} hfs in iron-sulphur proteins

	EPR $A = 2H_e$ (G)	Mössbauer H_n (kG)
Spinach ferredoxin	14	185
Euglena ferredoxin	–	170
Putidaredoxin	14	200
Adrenodoxin	14	–
Azotobacter vinlandii	22	–
Xanthine oxidase	–	180

The similarity of the Mössbauer spectra of these iron–sulphur proteins suggest that the iron is in the same state in all of them, in spite of their different origins, functions and complexity of their molecular structures. Since xanthine oxidase, which has eight iron atoms per molecule, behaves similarly to the simpler ferredoxins in which one electron is shared between two iron atoms upon reduction, it would seem probable that the iron atoms occur in pairs and that each pair accepts one electron when the molecule is reduced. Studies with enriched Fe^{57} would be of great interest to see whether there are differences in detail between the iron in the proteins. Observations on hyperfine interactions in reduced bacterial ferredoxins would also be interesting, since Mössbauer spectra of the oxidized proteins[15][16] indicate that there is more than one kind of iron atom present in these molecules.

References

1 D.O.Hall and H.C.W.Evans, *Nature* **223**, 1342 (1969).
2 Y.I.Shethna, P.W.Wilson, R.E.Hansen and H.Beinert, *Proc. Nat. Acad. Sci.* **52**, 1263 (1964).
3 G.Palmer, *Biochem. Biophys. Res. Comm.* **27**, 315 (1967).
4 J.C.M.Tsibris, R.L.Tsai, I.C.Gunsalus, W.H.Orme-Johnson, R.E.Hansen and H.Beinert, *Proc. Nat. Acad. Sci.* **59**, 959 (1968).
5 H.Beinert and W.H.Orme-Johnson, *Ann. New York Acad. Sci.* **158**, 336 (1969).
6 W.H.Orme-Johnson, R.E.Hansen, H.Beinert, J.C.M.Tsibris, R.C.Bartholomaus and I.C.Gunsalus, *Proc. Nat. Acad. Sci.* **60**, 368 (1968).
7 D.V.DerVartanian, W.H.Orme-Johnson, R.E.Hansen, H.Beinert, R.L.Tsai, J.C.M.Tsibris, R.C.Bartholomaus and I.C.Gunsalus, *Biochem. Biophys. Res. Comm.* **26**, 569 (1967).
8 G.Palmer, R.C.Bray and H.Beinert, *J. Biol. Chem.* **239**, 2657 (1964).
9 C.E.Johnson and D.O.Hall, *Nature* **217**, 446 (1968).
10 C.E.Johnson, R.C.Bray, R.Cammack and D.O.Hall, *Proc. Nat. Acad. Sci.*, in press.
11 C.E.Johnson, E.Elstner, J.F.Gibson, G.Benfield, M.C.W.Evans and D.O.Hall, *Nature* **220**, 1291 (1968).
12 C.E.Johnson, P.F.Knowles and R.C.Bray, *Biochem. J.* **103**, 10C (1967).
13 R.Cooke, J.C.M.Tsibris, P.G.Debrunner, R.Tsai, I.C.Gunsalus and H.Frauenfelder, *Proc. Nat. Acad. Sci.* **59**, 1045 (1968).
14 G.V.Novikov, L.A.Syrtsova, G.I.Likhtenshtein, V.A.Trukhtanov, V.F.Rachek and V.I.Gol'danskii, *Doklady Akad. Nauk SSSR* **181**, 1170 (1968).
15 D.C.Blomstrom, E.Knight, N.D.Phillips and J.F.Weiher, *Proc. Nat. Acad. Sci.* **51**, 1085 (1964).
16 T.H.Moss, A.J.Bearden, R.G.Bartsch, M.A.Cusanovich and A.San Pietro, *Biochemistry* **7**, 1951 (1968).

ESR study of iron-containing complexes in animal tissues

YA. I. AXHIPA and L. P. KAYUSHIN

Institute of High Nerve Activity and Neurophysiology, Acad. Sci., Moscow, USSR
Institute of Biological Physics, Acad. Sci., Pushchino (Moscow region), USSR

NOWADAYS A GREAT DEAL of information, concerning the ESR absorption of different animal tissues is accumulated. In some cases in lyophilized and frozen samples of certain organs and blood an ESR signal of composite structure can be observed (1–10), the nature of which is far from being clear in every case.

In previous articles (6, 7) we reported that ESR spectra of tissues, taken after the animal had been affected by $NaNO_2$, were identical to those of metabolizing yeast and of rat liver cancerous tissue. Here we present our experiments in detail.

The rats were injected intravenously with a 1% $NaNO_2$ solution on the basis of 1 ml per 100 g of weight. It resulted in methaemoglobinaemia. The ESR spectra of lyophylized tissues were recorded at 77°K (Fig. 1). The ESR signals of all tissues analyzed such as different parts of the nervous system, liver, heart, kidney, spleen and blood exhibited a composite structure. The triplet ESR spectrum had the g-factor value equal to 2.01.* On this spectrum one more ESR signal with $g = 2.00$ was superimposed, which is due to a free radical, usually found in tissues.

Similar results were obtained both *in vivo* and *in vitro*, when blood and isolated tissues were treated by $NaNO_2$ and gaseous NO. Nevertheless we

---

* There was a misprint in the previous paper: see ref. (6): the g-factor value shown in the spectra of Figs. 1 and 2 should read 2.01 and not 2.03.

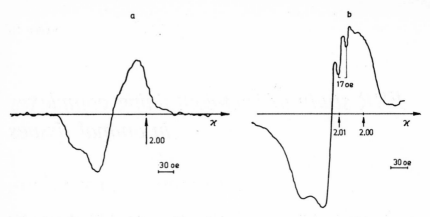

Figure 1 ESR spectrum of spleen after the animal had been treated with $NaNO_2$; a) room temperature, computer averages of 16 spectra. b) 77°K, modulation amplitude – 16 Oe.

failed to observe the triplet signal in adrenal tissue. In this case the ordinary ESR spectrum with $g = 2.00$ was obtained.

It is well known that some nitrites when injected into the blood interact with haemoglobin, binding to the ferrous ion, oxidizing it to the ferric form and forming a stable compound, unable to combine with oxygen. The result is a blocking of the transport function of the erytrocytes.

The structure of the ESR signal of hemoglobin, placed in NO medium, as described by Gordy and Raxroad (11), ls very similar to that of the ESR signals of blood and tissues in our experiments.

Therefore we might suggest that the ESR signal fine structure appears as a result of haemoglobin–NO complex formation with the unpaired electron spin located mainly on the NO nitrogen atom and that the signal asymmetry is due to *g*-factor anisotropy.

Since a triplet ESR signal was also detected in blood samples and its intensity was greater than that of nervous and other tissue ESR signals, the last ones could be considered as being due to haemoglobin–NO complex formation in the blood of tissue capillaries. The signal could appear on NO binding to myoglobin (metmyoglobin) of smooth muscles of blood-vessels, which could not be removed when a whole tissue is analyzed.

However Gordy and Raxroad showed that the characteristic triplet signal also occurred when cytochrome C was placed in a NO medium. This allows us to suppose that the ESR signal which we obtained is related not only to methaemoglobin, but also to an NO interaction with other heme-containing compounds, in particular with cytochromes and heme-containing exzymes, which are very abundant in mitochondria. To examine this possibility, ESR absorption of liver mitochondria of normal rats treated with $NaNO_2$ in the above-mentioned way was studied.

Rat liver mitochondria of control animals show a singlet ESR signal with a *g*-factor of 2.00 and halfwidth of about 15 Oe. Mitochondrial ESR spectra of $NaNO_2$ treated animals have a width of 45 Oe, and a *g*-factor of 2.035.

The fine structure is unclear due to the strong anisotropy of the *g*-factor, which is probably of the same origin as the tissue ESR spectra of these animals. In Fig. 2 the ESR signal with a *g*-factor 2.00 can be seen too, which is

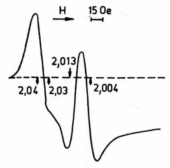

Figure 2 ESR spectrum of lyophilized samples of mitochondria after the animal had been treated with $NaNO_2$

characteristic for normal tissues. If the fine structure signal is supposed to be caused by a *g*-factor anisotropy, one may calculate the values of *g*-perpendicular and *g*-parallel for this spectrum. They equal respectively 2.04 and 2.013. The results of these investigations indicate that $NaNO_2$ introduced into the organism damages not only haemoglobin but heme-containing respiratory enzymes too. The latter conclusion is confirmed by experiments which demonstrate nitrosyl-ferricytochrome C formation on $NaNO_2$ addition to skeletal muscle mitochondria, kept under anaerobiosis (12).

Under conditions of NO excess concentration (i.e. O_2 deficiency), the nitrosyl groups can take part in respiratory metabolism. In this case, the ESR spectra of heme-containing complexes may be due to NO interaction both with cytochromes and heme-enzymes (12).

In this connection one can suppose that the damaging of tissue respiration induced by nitrites, is caused not only by the formation of inactive haemoglobin derivatives and disordering of the transport function of blood, but that oxygen starvation is due to a reduced consumption of oxygen by the tissues as well.

The death of animals, treated with nitrites in such doses as to cause met-haemoglobin formation, confirms the hypothesis that nitrites affect not only haemoglobin, but also other heme-containing compounds and possibly non-heme iron too, which take part in the biological redox processes.

The stability of the ESR signal shape of adrenal tissue after $NaNO_2$ administration may well be explained by the high ascorbic acid metabolism in this organ, the ascorbic acid being a strong reducing agent of the nitrosyl-hemoglobin complex.

KCN also possesses a similar ability to change the ESR absorption of blood and erytrocytes. The injection of this compound results in the appearence of a triplet ESR signal with a splitting of 14 Oe and a *g*-factor value of 2.01 in the samples of blood and erythrocytes (Fig. 3). The nature of this

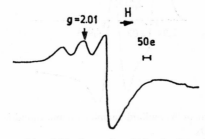

Figure 3 ESR spectrum of lyophilized samples of blood taken after the animal had been treated with KCN

signal is analogous to that of triplet ESR signal described above. The ESR spectra of all the tissues of animals treated with KCN were studied. However, in every tissue except blood, singlet ESR signals are observed, similar to those of lyophilized samples of the control animals. However, the intensity of the signals changed.

So the change of intensity and structure of tissue ESR spectra of animals, treated by $NaNO_2$ and KCN can be associated both with terminal oxidation damaging and with the formation of free radical complexes of iron with compounds involved in these processes.

The triplet ESR signal discovered by us in the blood and erythrocytes after KCN intravenous injection first of all confirms the specificity of the KCN action in the organism, in contrast with the results relative to $NaNO_2$, when triplet ESR signals were found in every treated tissue.

It may be supposed that the triplet ESR signal of blood, arising after KCN treatment of the animal, is due to cyanmethemoglobin formed on KCN binding to methemoglobin, that is always present in blood of normal animals and readily combines with cyanide compounds.

It would seem that KCN binding to cytochrome C should result in the appearance of the ESR signal for this complex. The absence of such a signal in mitochondria and whole tissue preparations after KCN administration leads to the conclusion that our knowledge concerning the mechanism of inhibitory effect of cyanide compounds on tissue respiration has to be completed.

A signal identical to those that we discovered in animal tissues after $NaNO_2$ injection has been observed by Commoner and coworkers in malignant rabbit liver tissue. Nevertheless the authors have not explained the origin of these ESR signals (4).

In Vanin and Nalbandjan's paper published in 1965 (5), ESR spectra of metabolizing yeast have been described. These ESR spectra were in tissues of $NaNO_2$–treated animals. These authors called these signals "sulphuric" since they considered them to be due to an unpaired electron localization on sulphur atoms. In a later paper (8) the authors suggested that the ESR spectra of lyophylized animal tissues treated by NO are due to nitrosyl-iron complexes. When comparing ESR spectra which have a *g*-factor value equal to 2.035 with the ESR signals of nitrosyl–ferrous–cysteine complex and nitrosyl–nonheme iron complex, a constituent part of Green's respiratory complex, and with the ESR signals of biological materials exhibiting *g*-factor values of 2.03, these authors (8) became convinced that the latter signal was not of "sulphuric" origin.

Woolum, Tiezzi and Commoner in their article published in 1968 (13) on the ESR study of the NO–iron complexes with amino acid, demonstrated an effect of protein imidazol groups on the ESR spectra of these complexes.

The comparison of our data with those of McDonald *et al.* (14) allows us to conclude that ESR spectra of complex shape, appear on complex formation between NO and heme and non-heme iron.

The ESR spectra, arising in animal tissues treated with methemoglobin-forming agents, seem therefore to be due to iron complexing with nitrogen atoms of protein active groups. This conclusion follows from the above mentioned data as well as from a great deal of different model experiments (14).

It is interesting to observe that there is an absolute analogy between the ESR spectra observed in methaemoglobinaemia (6, 7, 10) and those of animal tissues under certain kinds of cancerogenesis and also of some metabolizing microorganisms (4, 5, 9).

It is possible that the ESR spectra coincidence was caused by the fact that in both cancerogenesis and the course of metabolic processes of microorganisms and in some other cases the conditions were favourable to the appearance of nitrogen compounds able to from paramagnetic complexes with iron-containing substances. In particular the appearance in animal tissues of the ESR signal with a g-factor of 2.035 may be associated with a nitrosyl group taking part in respiratory processes under conditions of insufficient O_2 supply to tissues.

References

1   Hashimoto, Yamana T., and Mason, H.S., *J. Biol. Chem.* **237**, PC 3843, 1962.
2   Nebert, W.W., and Mason, H.S., *Cancer Res.* **23**, 823, 1963.
3   Mallard, J.R., and Kent, M., *Nature* **204**, 1192, 1964.
4   Vithayathil, A.J., Ternberg, J., and Commoner, B., *Nature* **307**, 1246, 1965.
5   Vanin, A.F., and Nalbandjan, P.M., *Biophysics* **10**, 167, 1965.
6   Azhipa, Ya.I., Kayushin, L.P., and Nikishkin, E.I., *Biophysics* **10**, 167, 1965.
7   Azhipa, Ya.I., Kayushin, L.P., and Nikishkin, E.I., *Second International Biophysics Congress Abstracts* A 14(a)/02 3.4. Vienna, Austria, September 5–9, 1966.
8   Vanin, A.F., Blumenfeld, L.A., and Chetverikov, A.G., *Biophysics* **12**, 829, 1967.
9   Saprin, A.N., Shabalkin, B.A., and Koslova, N.M., *Dokl. AN USSR*, **181**, 1520, 1968.
10 Azhipa, Ya.I., Kayushin, L.P., and Nikishkin, E.I., *Biophysics*, 14, 852, 1969.
11 Gerdy, W., and Rexroad, H.N., in *Free Radicals in Biological Systems*, M.S.Blois, H.W.Brown, Jr., R.M.Lemmon, R.O.Lindblom, and M.Weissbluth, Ed., Academic Press Inc., New York, N.Y., 1961, pp. 268–273.
12 Walters, C.L., and Casselden Taylor, A.M., *Biochem. Biophys. Acta* **143**, 310, 1967.
13 Woolum, G.C., Tiezzi, E., and Commoner, B., *Biochem. Biophys. Acta* **160**, 311–320, 1968.
14 McDonald, C.C., Phillips, W.P., and Mower, F., *J. Am. Chem. Soc.* **87**, 3319, 1965.

Study of quinones in photosynthetic systems

D. H. KOHL, P. M. WOOD, M. WEISSMAN and J. R. WRIGHT

Washington University, St. Louis, Missouri 63130, U.S.A.

TWO ESR SIGNALS are observed when O_2-evolving photosynthetic systems are illuminated. The data which were used to establish and characterize these two signals are shown in Fig. 1 (1). Signal I decays rapidly in the dark, has a half-width of about 9 gauss and no observable hyperfine splitting. Signal II is light induced, persists for tens of minutes in the dark, has a half-width of ca. 19 g and reveals partially resolved hyperfine splitting. The hyperfine structure of Signal II is due to interaction with protons as is demonstrated by the structure of the signal generated by totally deuterated organisms (Fig. 2).

It has been alleged that the free radical, plastosemiquinone, is the source of Signal II observed in photosynthetic materials. Much indirect and some direct evidence links Signal II with plastoquinone, an important photosynthetic electron carrier. We have recently provided direct evidence of this relationship in experiments in which the lipids, including plastoquinone, are extracted from chloroplasts with heptane and then plastoquinone is added back (2). When protonated plastoquinone is added back, a signal is observed which is indistinguishable from Signal II. On the other hand, when deuterated plastoquinone is added back, a narrow signal is observed which is characteristic of the signal observed in totally deuterated organisms. While these experiments demonstrate that Signal II is indeed due to a free radical derived from plastoquinone, one may not infer, *a priori*, that the free radical involved is plastosemiquinone. In fact there is one experimental result which argues strongly against identifying Signal II with plastosemiquinone; namely, an authentic semiquinone of plastoquinone produces a symmetrical spectrum even when immobilized in a random glass while Signal II is distinctly asym-

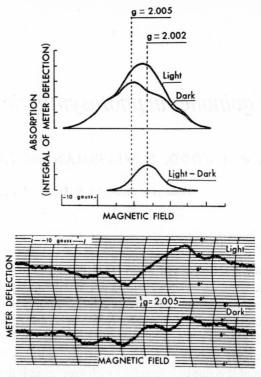

Figure 1 ESR signals from spinach chloroplast fragments washed twice with 0.5 M sucrose solution. The lower curves are from the actual spectrometer recording; the upper curves, the integrals derived from them. $T \sim 35\,°C$ (from Ref. 1)

metrical. This is seen in Figs. 3, 4, 5. Signal II is seen in Fig. 3 with its characteristic asymmetric structure and ca. 5 gauss splittings. Two examples of Signal II from widely different biological sources and two laboratories are presented to emphasize that the observed asymmetric shape is an intrinsic property of Signal II. The semiquinone of plastoquinone is seen in Fig. 4, here generated in ethanolic KOH. Higher concentrations of the semiquinone may be obtained with borohydride reduction of plastoquinone. The overall line width is 17.7 gauss and splitting constants which are deduced from measurements of high resolution spectra of the two extreme groups of lines provide an excellent fit over the entire spectrum. Of course one would not expect this spectrum to resemble Signal II even if Signal II were due to the semiquinone since organic extraction data make it clear that at least part of the plastoquinone in the chloroplast is tightly bound and thus the anisotropic contribution to the hyperfine structure would not average out as it does here.

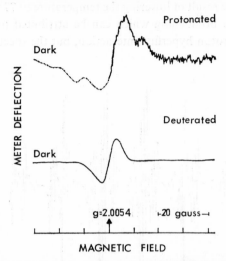

Scenedesmus obliquus

Protonated

Dark

Deuterated

Dark

METER DEFLECTION

g=2.0054   ⊢20 gauss⊣

MAGNETIC FIELD

Figure 2 The influence of growth in totally deuterated medium on Signal II for the green alga, *Scenedesmus obliquius*. The organisms which produced the signal recorded in the top trace were grown on a normal protonated medium. The bottom trace was generated by organisms in which all hydrogen nuclei were deuterons. Mod. amp. = 1 gauss (from Ref. 5)

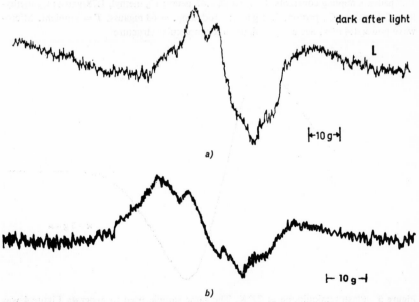

dark after light

L

⊢10 g⊣

a)

⊢ 10 g ⊣

b)

Figure 3 Typical Signal II. a) As seen in tobacco chloroplast fragments which had first been lyophilized and then resuspended in water. The apparent variable base line is due to a signal from Mn^{+2}. Mod. amp. = 3.0 gauss. b) As seen in living cultures of Chlamydomonas (after Ref. 3)

Figure 5 shows the result of lowering the temperature to 77°K. The resultant spectrum shows a broadening which can be attributed to the anisotropic contribution of proton hyperfine interaction, but the spectrum is decidedly symmetric.

Figure 4 High resolution spectrum of authentic semiquinone of plastoquinone. Plastoquinone in ethanol treated with KOH. The distance between the extreme lines a–d is 17.7 gauss. Coupling constants: C_2 methyl, 1.87 gauss; C_3 methyl, 1.78 gauss; C_5 methylene, 2.37 gauss; C_6 proton, 2.01 gauss; Mod. amp. = 63 mgauss, T = ambient. Microwave power 1.4 mW. See entry 2, figure 6 for molecular structure

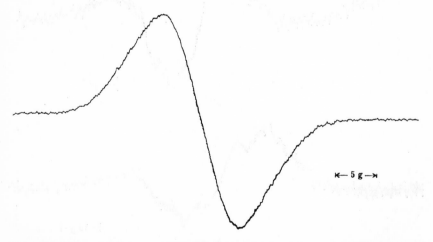

Figure 5 Plastosemiquinone at 77°K. The same sample used to generate Figure 4 was placed in liquid N_2 and the above spectrum recorded. Mod. amp. = 1.0 gauss. Microwave power = 1.4 mW

At this point it became clear to us that we were in for a long set of experiments and since it is tedious to isolate plastoquinone, we shifted to studies involving α-tocopherol and α-tocopherolquinone (see Fig. 6 for molecular structures). Note that the tocopherols are in the fused ring chroman configuration rather than in the single ring quinol. We preferred α-tocopherol to the more closely related γ-tocopherol since methyl substitution at carbon-six on the ring makes free radicals derived from α-tocopherol less reactive than those derived from γ-tocopherol. Results entirely analogous with those observed with plastoquinone are obtained from α-tocopherol. The α-tocopherolsemiquinone ambient spectrum generated in ethanolic-KOH is fit by a computer generated spectrum using coupling constants which are very close to those obtained by Fritsch, Tatwawadi and Adams (4) for the same radical generated electrolytically. When the temperature of this sample was reduced to 77°K, the line broadened in the same manner as did plastosemiquinone seen in Fig. 5 although the spectrum is not quite as broad as the corresponding spectrum for plastosemiquinone (half-width about 7 gauss vs. 9 gauss, overall width about 25 gauss vs. 30 gauss).

A most interesting thing is observed as the microwave power is increased substantially when α-tocopherol in ethanolic-KOH is in the cavity. The spectrum attributed to α-tocopherolsemiquinone saturates at low power and the spectrum observed in Fig. 7 emerges as the microwave power is increased. We attribute this spectrum to α-tocopheroxyl free radical. The name is meant to describe the alkoxyl substituted phenoxy free radical, whose molecular structure is depicted in Fig. 6. Splitting constants consistent with those observed in substituted phenoxy radicals give a good fit in computer simulated spectra. Recall that the spectrum of plastochromanoxyl would be expected to differ from the spectrum of α-tocopheroxyl only by the absence of the contribution from the methyl substituent at carbon-six in the latter.

A sample which generates the spectrum seen in Fig. 7 can be prepared substantially free of the semiquinone. When such a sample is brought to low temperature, the result seen in Fig. 8 is observed. Not only is this spectrum asymmetric, but it looks remarkably like Signal II from photosynthetic material. (Compare with Fig. 3.)

A similar result is observed if α-tocopherol or γ-tocopherol in ethanol are first taken to 77°K and then irradiated in the ultraviolet. In these experiments the temperature must be allowed to rise briefly after exposure to ultraviolet radiation in order to allow radicals formed by the solvent to decay.

The results of these studies with plastoquinone analogues present the possibility of resolving the apparent dilemma alluded to earlier—the dilemma

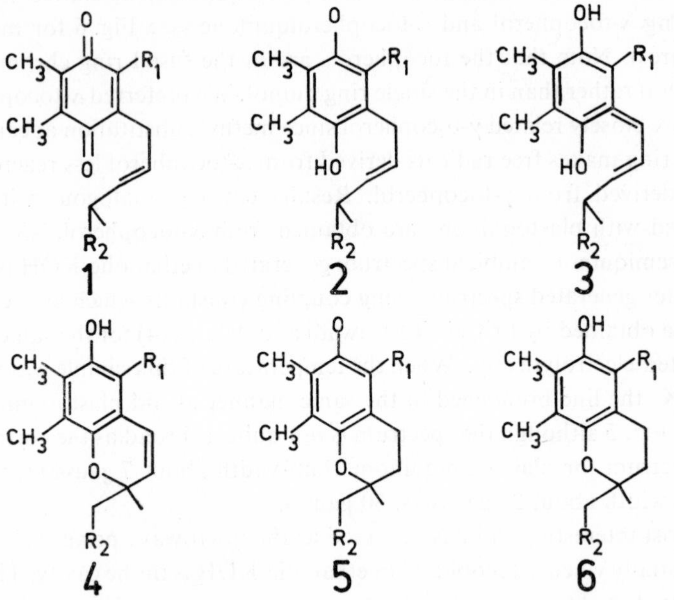

Figure 6 Molecular structures

	$R_1 = H$ $R_2 = $ eight isoprenoid units	$R_1 = CH_3$ $R_2 = $ phytyl	$R_1 = H$ $R_2 = $ phytyl
1	Plastoquinone A	α-tocoquinone*	γ-tocoquinone*
2	Plastosemiquinone A	α-tocopherolsemiquinone	γ-tocopherolsemiquinone
3	Plastoquinol A	α-tocoquinol	γ-tocoquinol
4	Plastochromenol-8 (solanochromene)	—	—
5	Plastochromanoxyl free radical	α-tocopheroxyl free radical	γ-tocopheroxyl free radical
6	Plastochromanol-8	α-tocopherol	γ-tocopherol

* Sometimes called tocopherolquinone, although some reserve the latter name for
γ-OH tocoquinone.

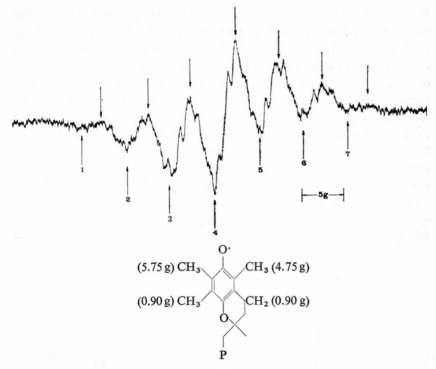

Figure 7 α-tocopheroxyl free radical. This spectrum was observed to grow out of a room temperature spectrum of α-tocopherol semiquinone in ethanol as the microwave power was increased. This spectrum itself shows signs of microwave power saturation. Microwave power = 320 mW.

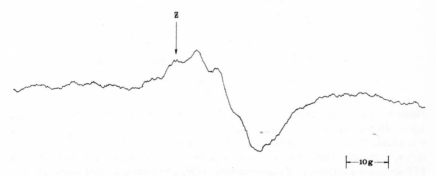

Figure 8 α-tocopheroxyl at 77°K. Modulation amplitude = 3.2 gauss. This high modulation amplitude was used since most photosynthetic samples, such as that seen in Figure 3a, have been recorded at that value. Microwave power = 2.0 mW

being that it has been directly demonstrated that the free radical responsible for producing Signal II in photosynthetic material is directly derived from plastoquinone while the symmetric line shape of immobilized plastosemiquinone suggests that this radical is not the source of Signal II. On the basis of data presented here in part, we propose that plastochromanoxyl free radical is the main contributor to Signal II and that it plays a role in the reduction of plastoquinone to plastoquinol observed in the light although there is no adequate explanation for the dark persistence of Signal II. Participation of this radical is consistent with all of the data since it may be directly derived from plastoquinone and its line shape (more precisely the line shape of an analogue) when immobilized closely mimics Signal II.

One approach toward testing this conclusion is to look for changes in the optical spectrum upon illumination of O_2-evolving photosynthetic material which changes correspond with changes in the concentration of either plasto-

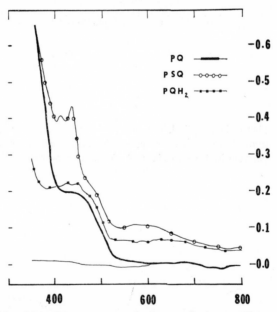

Figure 9   Visible absorption spectrum of plastosemiquinone. The spectrum of ca. 1.3 μmoles plastoquinone (PQ) per ml. ethanol was recorded and then a chip of KBH_4 was added. The absorption between 410–440 mμ and in the region around 600 mμ grew for ca. 25 minutes. These changes were correlated with the changes in the intensity of the ESR spectrum attributed to plastosemiquinone (PSQ). Reduction to plastoquinol (PQH_2) was virtually complete after 100 minutes. The difference in the absorption at 800 mμ probably is due to scattering by the particles of KBH_4. The spectra were drawn by a recording spectrophotometer. The symbols serve to identify the curves rather than being data points

chromanoxyl or plastosemiquinone. This requires, as a first step, that one be able to identify the optical lines associated with these intermediates. It would also be useful in this regard, and in general, to have optical extinction coefficients for these intermediate compounds. In practice it has proved easier to record the optical spectrum of the semiquinones than of the chromanoxyls. The optical spectrum of plastosemiquinone is presented in Fig. 9. By comparing the intensity of the ESR signal which this preparation generated with the ESR spectrum of diphenypicrylhydrazyl (DPPH) and using the optical spectrum of DPPH as the primary standard, it has been possible to estimate the extinction coefficient of plastosemiquinone in ethanol; $\Delta OD_{437} = 10,000$ liters mole^{-1} cm^{-1}.

While we consider this work on the optical properties of free radicals derived from plant quinones to be very promising, and we intend to pursue it for its intrinsic interest, there is a more direct test of our hypothesis that it is the chromanoxyl which is predominantly responsible for Signal II. Professor A. Ehrenberg has pointed out that ENDOR experiments should allow one to observe the magnitude of the isotropic splittings which are hidden in Signal II by the anisotropic contribution to the hyperfine structure. The system ideally lends itself to such an experimental approach since the difference between the isotropic splittings in the semiquinones and the chromanoxyls is large (ca. 2 gauss vs. ca. 5 gauss). We eagerly await the result of this experiment from Ehrenberg's laboratory.

References

1   B. Commoner, J. J. Heise, B. B. Lippincott, R. E. Norberg, J. V. Passonneau and J. Townsend, *Science*, **126** (1957) 57.
2   D. H. Kohl and P. M. Wood, *Plant Physiology*, **44** (1969) 1439–1445.
3   E. C. Weaver, *Arch. Biochem. Biophys.*, **99** (1962) 193.
4   J. M. Fritsch, S. V. Tatwawadi and R. N. Adams, *J. Phys. Chem.*, **71** (1967) 338.
5   D. H. Kohl, J. Townsend, B. Commoner, H. L. Crespi, R. C. Dougherty and J. J. Katz, *Nature*, **206** (1965) 1105–1110.

Electron paramagnetic resonance of free radicals in hydrazine involving processes*

L. BURLAMACCHI and E. TIEZZI

Institute of Physical Chemistry, University of Florence, Florence, Italy

Introduction

CERTAIN RECENT STUDIES have emphasized that free radicals and other paramagnetic species may be involved in carcinogenetic processes[1]. In particular, paramagnetic iron-nitrosyl complexes have been found to play a probable role in biological processes of this type[2]. The carcinogenetic activity of hydrazine has also been outlined by Juhasz and co-workers[3]. The involvement of hydrazine in free radical reactions has already been reported by Adams and Thomas[4]. They detected the electron paramagnetic resonance (EPR) spectrum of a transient, intermediate species, identified as the radical cation $[N_2H_4]^+$, by means of EPR continuous flow technique.

A previous study of paramagnetic complexes between Fe^{II}, NO_2^-, amino-acids or proteins and L-ascorbic acid[2] has been extended to systems containing Fe^{II}, NO_2^-, vitamin C, and several nitrogen containing bases[5]. It was observed that hydrazine may react in aqueous solution both with the Fe–NO group and the 1-ascorbic acid, giving rise to paramagnetic species which show nitrogen hyperfine structure.

The importance of L-ascorbic acid in fundamental biological processes and the above considerations about hydrazine led us to undertake an EPR study

* This research was supported by the Italian National Research Council (C.N.R.).

of these kinds of free radicals and to extend our research to some substituted hydrazines. The presence of hydrazine and substituted hydrazines in several radical involving reactions was studied.

This work deals with four different aspects of the radicalic reactions of the hydrazines: (1) hydrazine as a ligand in iron-nitrosyl complexes, in aqueous solution. (2) the interaction between Vitamin C and hydrazine (or substituted hydrazines) giving rise to free radicals. (3) the oxidation mechanism of hydrazine and substituted hydrazines. (4) hydrazine as the producing agent of nitroaromatic free radicals.

Experimental

EPR spectra were recorded by means of a Varian V-4502 X-band spectrometer with a field modulation of 100 Kc/sec. Nuclear hyperfine splittings, g-factors, and linewidth values were obtained by comparison with peroxylamine disulphonate (Fremy salt) as a reference standard, and with the aid of a dual sample cavity. Flat sample—cells (0.3 mm thick) were used for aqueous solutions. Transient radicals were prepared using a two-stream continuous flow system. The magnetic field was measured using a Varian F-8 nuclear fluxmeter with an external frequency counter. The accuracy was ± 0.04 G.

Aldrich Chem. L-ascorbic acid, Eastman Kodak 95% hydrazine and monomethyl hydrazine, Fluka asymmetric dimethyl hydrazine were used without further purification. $Na^{15}NO_2$ was obtained from Tracerlab, Waltham, Mass. (U.S.A.)

Hydrazine–Fe–NO paramagnetic complexes

The complexes were prepared in aqueous solution following a procedure already described[2]. The two spectra in Fig. 1 refer to Fe-^{14}NO-hydrazine and to Fe-^{15}NO-hydrazine complexes respectively. The variation from three to two lines when the isotope ^{15}N is used ($I = \frac{1}{2}$), unequivocally shows that the coupling constant is due to the nitrogen nucleus of the NO group. The values of the nitrogen splitting constants $a^{14}NO = 14.8$ G and $a^{15}NO = 21.1$ G are in agreement with the nuclear g factors of the two nitrogen isotopes.

On the basis of g-values and hyperfine coupling constants, octahedral or pyramidal structures which correspond to species I and II found in reference (6) must be attributed to the complex. By analogy with NH_3,[7] the octa-

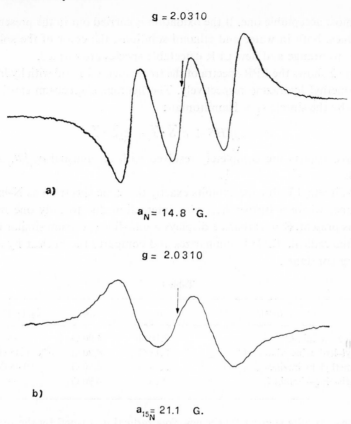

g = 2.0310

a)

a_N = 14.8 G.

g = 2.0310

b)

a_{15_N} = 21.1 G.

Figure 1 a) EPR spectrum of Fe–^{14}NO–hydrazine complex; b) EPR spectrum of Fe–^{15}NO–hydrazine complex

hedral structure should most probably correspond to a high spin d^7 complex. Due to its very short electron spin relaxation time, the complex should be EPR undetectable. Pyramidal complexes with structure II are generally four and five-membered chelate rings. In this case the three-membered hydrazine ring seems unlikely. Thus if we assume the pyramidal structure, a rectangular-based pyramid with four hydrazine molecules and one NO group at the apex can be suggested.

Vitamin C—hydrazine free radicals

The free radical anion formed by oxidation of vitamin C has already been studied by Lagercrantz[8]. Several structures have been considered by Russell *et al.*[9] They have proposed a furanic structure, i.e. semiquinone type radical, as the most acceptable one. If the oxidation is carried out in the presence of hydrazines, both in water and ethanol solutions, the color of the solution changes to orange and new EPR detectable species are formed.

Figure 2 shows the EPR spectra of the free radicals formed with hydrazine and N-methyl hydrazine respectively. The hydrazine spectrum can be described by the simple spin Hamiltonian:

$$\mathcal{H} = g\beta HS + a_1 \vec{S} \cdot \vec{I} + a_2 \vec{S} \cdot \vec{I}$$

The three triplets are completely resolved with an unusual a_{N_2}/a_{N_1} ratio: $= 2.24$.

N-N-dimethyl hydrazine exhibits exactly the same spectrum as N-methyl hydrazine, where a further hyperfine interaction due to only one methyl group is present. Semicarbazide displays a nine-line spectrum similar to the hydrazine radical. Table I summarizes and compares the nuclear hyperfine coupling constants.

Table I

Free radicals	a_{N_1}	a_{N_2}	$a_H (-CH_3)$
Hydrazine–vitamin C	2.18 G	4.90 G	–
N-methyl–hydrazine–vitamin C	2.18 G	4.90 G	0.45 G
N-N-dimethyl–hydrazine–vitamin C	2.18 G	4.90 G	0.45 G
Semi-carbazide–vitamin C	2.2 G	4.90 G	–

The above results suggest that a new free radical is formed by the reaction between vitamin C and hydrazine. The main feature of this radical is the presence of some unpaired electron spin density at the hydrazine nitrogens. The unusual difference between the two splitting constants of N_1 and N_2 strongly suggests that the attack is only on one side of hydrazine. This is supported by the fact that the three other asymmetrically substituted hydrazines give the same kind of free radical.

We can only suggest probable structures for the free radicals under study. By analogy with the vitamin C free radical, a semiquinonic structure can be proposed:

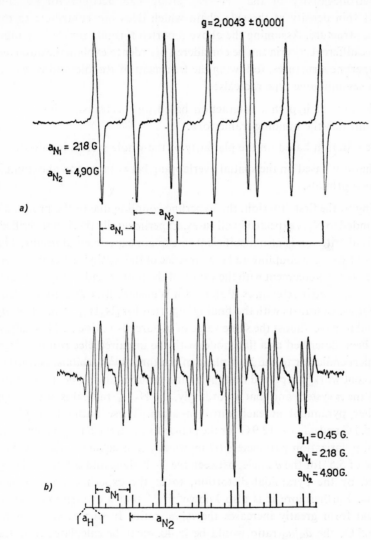

Figure 2 EPR spectra of: a) hydrazine–Vitamin C free radical; b) N–methyl–hydrazine–Vitamin C free radical

The electronegativity of the $\rangle$N$-$N$\langle$ group can account for the low unpaired spin density at the side chain which does not contribute to the hyperfine structure. Assuming the above reported formula for the free radicals, three different criteria can be considered in order to explain the observed EPR hyperfine structures, following the treatment of Rabold and co-workers[10] on semiquinone type radicals:

(a) mechanism of hyperfine interaction based on the transmission of spin density through the bonding framework.

(b) steric criterion based on the planarity of the whole system involved.

(c) mechanism based on the spatial overlapping between proton and quinoidal oxygen orbitals.

According to the first criterion, the hyperfine coupling due to the proton directly bonded to N_1 is expected in all cases. Experimentally this is not verified. The second criterion is also insufficient to explain experimental results. The absence of hydrogen coupling and the presence of the methyl hyperfine interaction are not in agreement with the values of the proton and methyl-splitting constants reported in references 10 and 11 for Wunster's Blue and p-phenylene diamine ions. Contrasts with the finding reported for $[N_2H_4]^+$,[4] in which a_H was found to have almost the same value of a_N, are also evident. This radical ion has been described as a flat model with the unpaired electron in a $2p_z\pi$ orbital perpendicular to the $X-Y$ plane of the three $sp^2\sigma$ orbitals. Nevertheless, a recent work[12] reported that the amino group is always bent out ot the plane of the π system and that accordingly, the $N_2H_4^+$ radical is not planar, but rather pyramidal at each nitrogen atom. These authors calculated $a_N = +6.4$ G and $a_H = -12.9$ G for the planar form on the basis of computed sigma–pi polarization parameters (Q matrices). Consequently, they derived the value of 5.5° for the ϑ angle, between the $X-Y$ plane and the new orbitals produced by the pyramidal distortion, to fit the experimental hyperfine constants. Furthermore it is also known[4,12,13] that a distortion toward a pyramidal form greatly increases the a_N/a_H ratio. In our case, assuming $a_H \leq 0.05$ G, the a_N/a_H ratio would be ≥ 40. ϑ can be calculated from the coupling constants by using the equation[13]:

$$a_N = a_N (2s) [\varphi^2 \, \varrho_\pi + (1 - \varphi^2) \varrho_\sigma] \quad \text{where} \quad \varphi^2 = 2 \tan^2 \vartheta$$

The second term in parenthesis may be assumed to be negligible because a_H was not detected. The unpaired spin density ϱ_π in N_1 may be roughly evaluated[14,15] to be $0.09 - 0.03$ giving a value of ϑ ranging between 8° and 14°. These conclusions enable us to suggest a rather pyramidal hybridization for

$N_{(1)}$ in the $-N_{(2)}-N_{(1)}\!\!\begin{smallmatrix} \diagup R \\ \diagdown R' \end{smallmatrix}$ fragment. In order to explain the methyl protons hyperfine coupling, a long range interaction via spatial overlapping between the quinoidal oxygen and the methyl group can be taken into account as suggested by the third criterion. Finally the interaction with only one methyl group in the N—N-dimethyl hydrazine radical may be correlated with a steric hindrance of the side chain on the $-N(CH_3)_2$ group.

An alternative, although less probable, structure can be proposed:

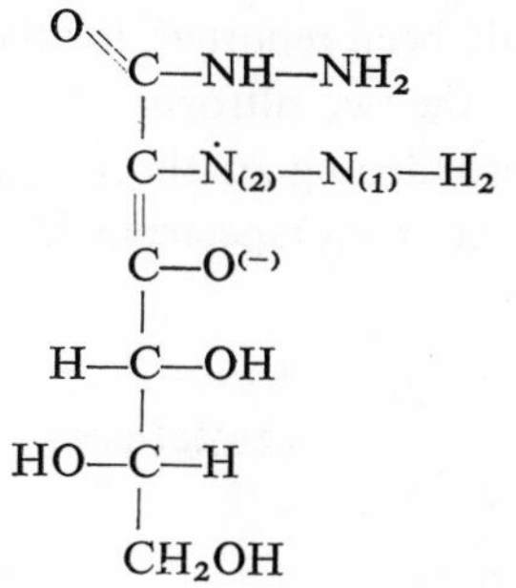

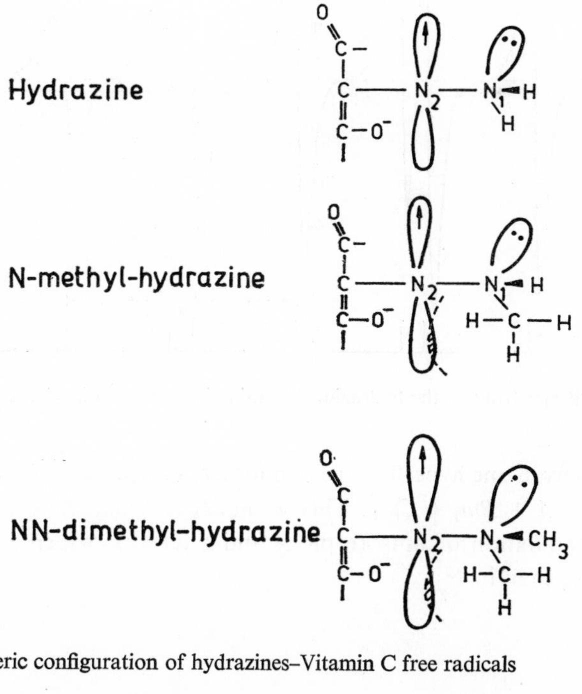

$$
\begin{array}{l}
\overset{O\diagdown}{}C-NH-NH_2 \\
| \\
C-\dot{N}_{(2)}-N_{(1)}-H_2 \\
\| \\
C-O^{(-)} \\
| \\
H-C-OH \\
| \\
HO-C-H \\
| \\
CH_2OH
\end{array}
$$

Hydrazine

N-methyl-hydrazine

NN-dimethyl-hydrazine

Figure 3 Steric configuration of hydrazines–Vitamin C free radicals

This may be obtained by the reaction of the first hydrazine molecule on the ring of ascorbic acid, followed by a nucleophylic attack of the second hydrazine molecule via water elimination. The maximum yield of the free radical, at pH = 8.6, is in good agreement with this mechanism. In the above form a somewhat pyramidal structure for N_1 is still assumed, with the unpaired electron localized in a p orbital on N_2. Under these conditions the steric configurations of Fig. 3 can be drawn. Here we suggest an interaction between the unpaired electron in the $N_2 - p$ orbital and the methyl protons, lying in the same plane, via orbital overlapping. The unusual difference between a_{N_1} and a_{N_2}, which has already been reported, fits the suggested scheme well.

The non-equivalence of the two nitrogen atoms is also shown by the different anisotropic line broadening of the two nitrogen triplets. Figure 4 represents the hydrazine–vitamin C spectrum in a viscous solution. The line-

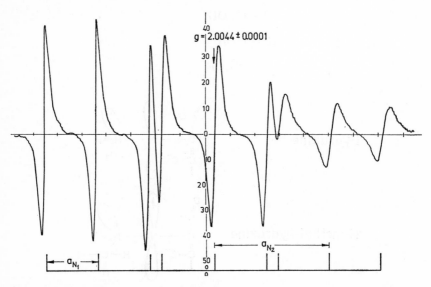

Figure 4 EPR spectrum of the hydrazine–Vitamin C free radical in viscous solution

width varies from one hyperfine line to another, according to the known relation: $1/T_2 = A + Bm_1 + Cm_1^2$. This asymmetric broadening is related to tumbling modulation of anisotropic A and g tensors, which appear to be greater in N_2 than in N_1.

Free radicals in hydrazines oxidation mechanism

Oxidation involving processes lead to the production of radical cations of the general form $[N_2R_4]^+$ using inorganic oxidants such as Ce^{+4} and Fe^{+3}. In these reactions pure hydrazine produces detectable free radical concentration[4]. Under the same experimental conditions (flow technique) monomethyl and dimethyl hydrazines give rise to an almost undetectable concentration of free radicals. However, we observed free radicals by oxidizing the N-methyl and N-N-dimethyl hydrazines with ceric ammonium sulphate in acid-aqueous solution. The related EPR spectra show a very complicated nuclear hyperfine structure because of the presence of the methyl protons and because of the different conditions of the two nitrogen atoms. At least five different hyperfine coupling constants are expected for the monomethyl-hydrazine radical ion: a_{N_1}, a_{N_2}, a_{CH_3}, one for the single proton, and another for the amine-protons. Further experiments are in progress with isotopic species in order to clarify the EPR hyperfine structure.

Hydrazine as radical source

The formation of free radicals from substituted hydrazines ought to be easier than from pure hydrazine because of the lower dissociation energy of the N–N bonds. The value of $H_2N–NH_2$ dissociation energy is approximately 60 kcal, which is little more than the 54 kcal found for HO–OH dissociation energy in hydrogen peroxide. The bond dissociation energies in substituted hydrazines vary in the range of 30–60 kcal, thus suggesting that substituted hydrazines should behave as radical sources analogous to the peroxides. However, little study has been done from this point of view and only few hydrazines have been used as radical producing agents in chain reactions. Furthermore, pure hydrazine is a better radical source than the substituted ones. In fact, we were able to form free radicals, treating nitro-aromatic compounds with pure hydrazine in water-ethanol solution. We didn't arrive at the same result using substituted hydrazines.

P-chloronitrobenzene, dinitrobenzene, O-nitrobenzoic acid, p-nitrobenzoic acid, and other similar compounds react with hydrazine giving rise to the corresponding well-known free radical anions. The related EPR spectra are widely reported in literature and it is only noteworthy to comment that our data completely agree with the described EPR parameters. As an example,

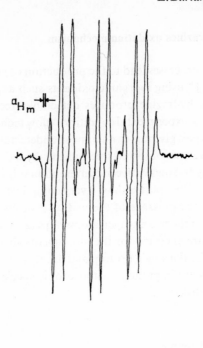

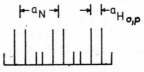

Figure 5 EPR spectrum of meta-nitro-benzaldehyde radical anion

we have the EPR spectrum of the radical anion derived from meta-nitro-benzaldehyde as shown in Fig. 5.

At this stage of our research it is only possible to suggest a probable scheme for the formation of free radicals. We propose a one-electron transfer process

$$\bigcirc\!\!-NO_2 + N_2H_4 \rightarrow \left[\bigcirc\!\!-\dot{N}O_2\right]^- + N_2H_4^+$$

Due to its high instability, the $N_2H_4^+$ radical ion is not detected during the recording of the more intense nitroradical spectrum.

Acknowledgments

The authors are very grateful to Dr. J.Woolum, Department of Biology, Washington University, St. Louis (Mo.) (U.S.A.) and to Dr. G.Pedulli, Institute of Industrial Chemistry, University of Bologna, Bologna (Italy) for fruitful discussions and precious suggestions on this work.

References

1   A.J.Vithayathil, J.L.Ternberg and B.Commoner, *Nature*, **207**, 1246 (1965).
2   J.C.Woolum, E.Tiezzi and B.Commoner, *Bioch. Biophys. Acta*, **160**, (3), 311 (1968).
3   J.Juhasz, J.Balo and B.Szende, *Nature*, **210**, 1377 (1966).
4   J.Q.Adams and J.R.Thomas, *J. Chem. Phys.*, **39**, 1904 (1963).
5   L.Burlamacchi, G.Martini and E.Tiezzi, to be published.
6   L.Burlamacchi, G.Martini and E.Tiezzi, this book, p. 137.
7   W.P.Griffith, J.Lewis and G.Wilkinson, *J. Chem. Soc.*, 3993 (1958).
8   C.Lagercrantz, *Acta Chem. Scand.*, **18**, 562 (1964) and references reported therein.
9   G.A.Russell *et al.*, *Rec. Chem. Prog.*, **27**, 3 (1966).
10 G.P.Rabold *et al.*, *J. Chem. Phys.*, **46**, (3), 1161 (1967).
11 J.R.Bolton, A.Carrington and J. Dos Santos-Verga, *Mol. Phys.*, **5**, 615 (1962).
12 R.Poupko and B.L.Silver, in press.
13 T.Cole, *J. Chem. Phys.*, **35**, (4), 1169 (1961).
14 C.L.Talcott and R.J.Myers, *Mol. Phys.*, **12**, (6), 549 (1967).
15 R.R.Falle, *Canad. J. Chem.*, **46**, 1703 (1968).

ESR studies of radicals derived from glycine and alanine

R. POUPKO, A. LOEWENSTEIN and B. L. SILVER

Technion—Israel Institute of Technology, Haifa, Israel

RADICALS WERE PRODUCED from glycine and alanine in solution by reaction in a flow system with Ti^{3+}/H_2O_2. Some previous studies of this system have been reported[1,2,3,4]. The radical species formed depends on the pH, and all the present results were obtained in basic media. In order to avoid precipitation of titanium at these pH's it was complexed with EDTA. It was shown that EDTA does *not* affect the nature of the radicals formed.

The results are summarized in Table I and the spectra shown in Figs. 1 and 2.

Table I Hyperfine parameters

Compound	Relative intensities	Hyperfine splitting (G)	Assumed radical
$NH_2CH_2COO^-$	(1:1)	$a_{CH}^H = 13.8$	
	(1:1:1)	$a_{NH_2}^N = 6.1$	NH_2CHCOO^-
	(1:1)	$a_{NH_2}^H = 2.9\text{–}3.6^a$	
	(1:1)	$\phantom{a_{NH_2}^H =} 2.3\text{–}3.2^a$	
$NH_2CH(CH_3)COO^-$	(1:3:3:1)	$a_{CH_3}^H = 13.9$	
	(1:1:1)	$a_{NH_2}^H = 5.1$	$NH_2C(CH_3)COO^-$
	(1:1)	$a_{NH_2}^H = 1.7\text{–}2.1^a$	
	(1:1)	$\phantom{a_{NH_2}^H =} 0\text{–}0.5^a$	

^{a} See Fig. 3 for the temperature variation of the splitting constants.

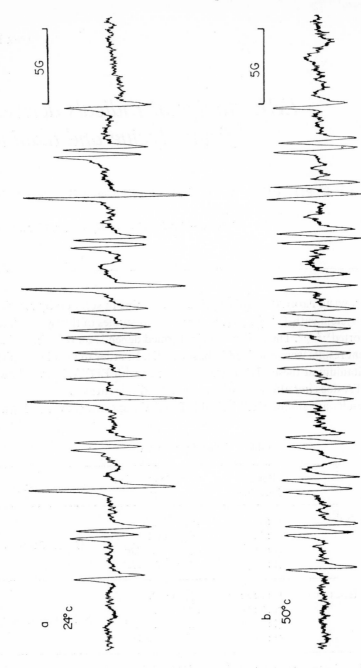

Figure 1 First derivative spectra of the radical derived from glycine in basic media

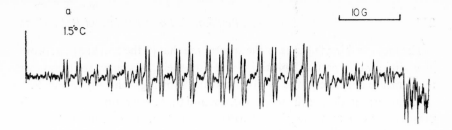

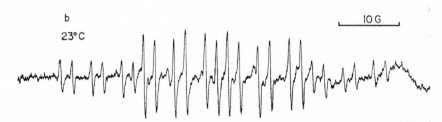

Figure 2 First derivative spectra of the radical derived from α alanine. The broad line at the right hand side at 23 °C is due to Ti-EDTA complex

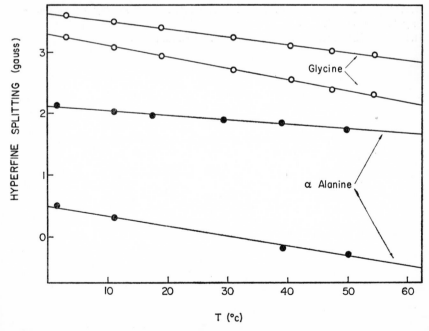

Figure 3 Hyperfine splittings of the amine hydrogens in radicals derived from glycine and alanine in basic media as a function of the temperature

The most striking features of the spectra are that (i) the two amine protons give different hyperfine splittings and (ii) their hyperfine splittings are temperature dependent, the behaviour being shown graphically in Fig. 3.

The interpretation of the inequivalence and the temperature dependence is presumably due to some kind of torsional motion. A difficulty is that the nitrogen hyperfine splitting is temperature independent.

The magnitudes of the nitrogen and amine proton splittings were evaluated from estimated spin densities and calculated polarization constants. The results for the glycine and alanine radicals are: $a_{NH}^{H_1} = 5.77$ G, $a_{NH}^{H_2} = 5.82$ G; $a_{NH_2}^{N} = 2.07$ G. It is obvious that the calculated splitting constants are in disagreement with the experimental results. However if it is assumed that the NH group is bent relative to the rest of the radical, the splitting constants calculated for angles of 6.5° and 5.5° for glycine and alanine respectively, are close to the experimental results.

References

1   R. Poupko, B. L. Silver and A. Loewenstein, *Chem. Comm.* 453 (1968).
2   H. Taniguchi, K. Fukui, S. Ohnishi, H. Hatano, H. Hasegawa and T. Maruyama, *J. Phys. Chem.* **72**, 1926 (1968).
3   R. F. Florin, F. Sicilio and L. A. Wall, *J. Res. Nat. Bur. Standards, Physics and Chemistry*, **72A**, 49 (1968).
4   H. Paul and H. Fischer, *Ber. Bunsenges. Phys. Chem.* (in press).

EPR of short lived free radicals in aqueous solution of biomolecules

C. NICOLAU* and H. DERTINGER

Institut für Strahlenbiologie, Kernforschungszentrum Karlsruhe, Germany

Summary

Free radicals produced by the reaction of OH˙ and NH_2^- with pyrimidine bases and nucleosides have been identified using the EPR-flow technique. Another series of experiments was carried out by means of a three-way flow system in order to elucidate the mechanism of chemical radioprotection by sulfhydryl compounds. It was shown that organic free radicals can be restituted via hydrogen donation from the SH-group of the protective molecules.

SHORT LIVED ORGANIC free radicals generated by the reaction of the OH radical with biomolecules characterize the early stage of radiation damage in aqueous solution. It is for this reason that these important intermediates are of increasing interest in radiation-biological research aiming at the detection of the primary reaction sites of biomolecules. In the paper of Poupko, Silver and Loewenstein given at this congress an investigation has been reported dealing with the reaction of amino acids and simple amines with OH˙ in a conventional EPR-flow system. The following contribution to the application of "magnetic resonances in biological research" is concerned with the reaction of OH˙ with another series of organic compounds of considerable biological interest: the constituent molecules of nucleic acids. In extension of these investigations the reaction of these molecules with the small radical

---

* Permanent address: Institute of Inframicrobiology "Stefan Nicolau", Bucharest, Rumania.

$NH_2^{\cdot}$ is studied, too, in order to obtain additional information about the sites of radical attack in these compounds. Finally, an investigation on the mechanism of chemical radioprotection, effected by SH-compounds will be reported.

The flow system used has been developed by Dixon and Norman (1) and since then applied successfully in the course of various other investigation of this kind (2, 3, 4). It consists mainly of two reservoirs of solutions, one containing $TiCl_3$ and the other H_2O_2 or NH_2OH, depending on whether $OH^{\cdot}$ or $NH_2^{\cdot}$ being generated. These two solutions of pH 2 which also contain the organic molecules are mixed in a four-jet mixing chamber. The mixture subsequently passes through a flat sample cell inside the microwave cavity thus allowing the detection of the organic free radical produced by $OH^{\cdot}$ or $NH_2^{\cdot}$ attack. Commercially available reagents were used in all experiments without further purification. The experiments were carried out by means of a conventional 3 cm-EPR-spectrometer (AEG); *g*-factors were calibrated with respect to the DPPH-value of 2.0036.

The site of attack in pyrimidine bases and nucleosides

The parameters of the EPR-spectra obtained from the reaction of $OH^{\cdot}$ and $NH_2^{\cdot}$ with the pyrimidine bases and the corresponding nucleosides are summarized in Table 1. In several cases more than one spectrum is observed,

Table 1   EPR-parameter of the $OH^{\cdot}$-adducts (2, 4)

Molecule	*g*-Factor	Hyperfine coupling (Gauss)	
Thymine	$\begin{cases} 2.0029^b \\ 2.0032^c \end{cases}$	$a(\text{Met}) = 22.4,$	$a(\text{H}) = 18.7$ $a(\text{H}) = 15.1$
Cytosine[a]	2.0027^b	$a(\text{H}_\text{I}) = a(\text{H}_\text{II}) = 18.3$	
Uracil[a]	2.0029^b	$\begin{cases} a(\text{H}_\text{I}) = 21.2, & a(\text{H}_\text{II}) = 18.1 \\ a(\text{N}) = a(\text{H}_\text{III}) = a(\text{H}_\text{IV}) = 0.8 \end{cases}$	
Thymidine	$\begin{cases} 2.0029^b \\ 2.0032^c \end{cases}$	$a(\text{Met}) = 23.6,$	$a(\text{H}) = 18.0$ $a(\text{H}) = 10.8$

[a] Same values as nucleosides; [b] $C_{(5)}$-adduct; [c] $C_{(6)}$-adduct.

indicating that different reaction sites exist in the molecules exposed. However, as can be easily derived from the hyperfine coupling data of Table 1, the reaction of $OH^{\cdot}$ and $NH_2^{\cdot}$ only affects the $C_{(5)} = C_{(6)}$-bond of the pyrimidine ring, leading to the formation of either $C_{(5)}$- or $C_{(6)}$-adducts (Fig. 1). The results are thus analogous to those obtained from EPR-investigations of the irradiated dry compounds (5, 6, 7).

The attachment of sugar to the pyrimidine base apparently does not lead to the production of detectable EPR-signals, that could be attributed to this compound. However, alterations in the spectra of the bases were observed in

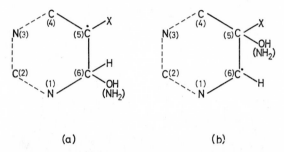

(a)   (b)

Figure 1   (a) $C_{(6)}$-adducts, and (b) $C_{(5)}$-adducts of the pyrimidines, produced by the reaction with OH˙ and $NH_2^{•}$. (Thymine: X = CH_3; cytosine and uracil: X = H)

the case of thymidine and uridine. The coupling constants of thymidine after OH˙ attack differ in some respects from those of the free base as can be seen from Table 1. The most significant change is a considerable decrease in the proton splitting at the $C_{(6)}$-position of the $C_{(6)}$-adduct. Moreover, the relative yield of the two adducts is different in thymine and thymidine (4). An analogous observation is made in the case of uridine after $NH_2^{•}$ attack (Fig. 2). Both adducts are formed in a nearly equal amount, whereas only the $C_{(5)}$-adduct is found in uracil (2). The fact that the two adducts are simultaneously formed in uridine, as well as the observation that they differ in the hyperfine coupling constants (Table 2), although they are indistinguishable with respect to the

Table 2   EPR-parameter of the $NH_2^{•}$-adducts (2, 4)

Molecule	g-Factor	Hyperfine coupling (Gauss)		
Thymine[a]	2.0027[b]	a(H) = 16.7,	a(N) = 5.05	
	2.0031[c]	a(Met) = 23.1,	a(H) = 16.8,	a(N) = 7.2
Cytosine[a]	2.0028[b]	$a(H_I)$ = 11.8,	$a(H_{II})$ = 18.6,	a(N) = 11.8
Uracil	2.0029[b]	$a(H_I)$ = 18.0,	$a(H_{II})$ = 34.2,	a(N) = 6.4
Uridine	2.0029[b]	$a(H_I)$ = 18.0,	$a(H_{II})$ = 34.2,	a(N) = 6.4
	2.0029[c]	$a(H_I)$ = 18.0,	$a(H_{II})$ = 37.0,	a(N) = 4.5

[a] Same values as nucleosides;   [b] $C_{(5)}$-adduct;   [c] $C_{(6)}$-adduct.

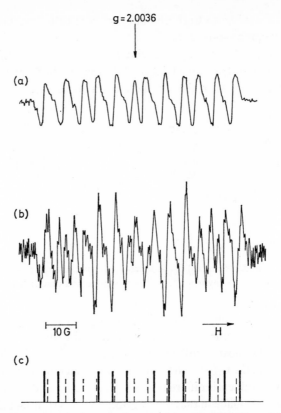

Figure 2 (a) EPR-spectrum of uracil after reaction with NH_2^- ($C_{(5)}$-adduct). (b) EPR-spectrum of uridine after reaction with NH_2 (c) Stick diagram of the uridine signal; solid lines: $C_{(6)}$-adduct; dashed lines: $C_{(5)}$-adduct

number of interacting nuclei, is attributed to structural effects caused by the attachment of sugar (4). Also the observation, mentioned above, that both adducts are formed in different amounts in two of the bases and the corresponding nucleosides, is consistent with this assumption.

Radical site transfer and radioprotection by sulfhydryl compounds

The flow system does not only allow investigations on radical formation, but also permits transfer processes of a radical site to be studied. It is well known from radiation-biological investigations that SH-containing substances such as thiols and mercaptanes are effective radioprotectors which differ significantly in this respect from ordinary radical scavengers (8). It has been suggested that this protective capacity is due to the ability of sulfhydryls to eliminate free radicals by a restitution mechanism (9, 10). An investigation was, therefore, undertaken to present evidence for such a mechanism. For this purpose a modified mixing chamber was used which allowed admixture of a third, cysteine-containing solution shortly after the formation of the organic free radicals (three-way flow system). Upon introduction of this solution the radical spectrum disappeared and a triplet signal centered at $g = 2.0105$ was observed (Fig. 3), which is caused by a radical of the type

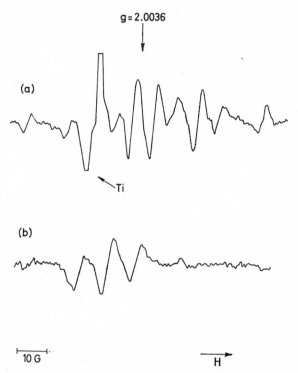

Figure 3 (a) EPR-spectrum of thymine after OH· attack; the strong line termed "Ti" belongs to a titanium-complex (12). (b) Signal of the radical RS· obtained after admixture of a cysteine solution in a three-way mixing chamber; same spectrometer settings in (a) and (b)

$-CH_2-\dot{S}$ (11). This phenomenon is best explained by assuming that the free radical (M·) produced by the reaction of OH· or NH_2^- is removed by hydrogen donation from the SH-group:

$$M· + RSH \rightarrow MH + RS·$$

Such a process is energetically favoured by the low SH bonding energy of 3.7 eV as compared with 4.1 to 4.4 eV for CH or NH (8). It could be established in control experiments, as well as by an investigation of the concentration dependence of the RS˙-formation that the observed effect is not due to the direct reaction of abundant OH and NH_2 radicals with cysteine (3).

This process of transfer via hydrogen donation of the original radical site to the sulfhydryl compound is not the only way in which restitution and therefore protection against radiation damage can be achieved. It could be shown that disulphide compounds exhibit a similar effect. However, in this case an electron donation from the disulphide to the free radical seems to account for the restitution process (3). Thus, the flow system in combination with EPR-technique proves to be a powerful method for studying processes relevant for radiation biology.

References

1   W.T.Dixon and R.O.C.Norman, *J. Chem. Soc.* 3119 (1963).
2   Cl.Nicolau, M.McMillan and R.O.C.Norman, *Biochim. Biophys. Acta* **174**, 413 (1969).
3   Cl.Nicolau and H.Dertinger, *Radiat. Res.*, in the press.
4   H.Dertinger and Cl.Nicolau, *Biochim. Biophys. Acta*, in the press.
5   J.B.Cook, J.P.Elliott and S.J.Wyard, *Molec. Phys.* **13**, 49 (1967).
6   B.L.Pruden, W.Snipes and W.Gordy, *Proc. Natl. Acad. Sci. U.S.* **53**, 917 (1965).
7   A.Müller, in: *Progress in Biophysics and Molecular Biology*, Vol. 17, p. 99, Pergamon Press, Oxford, New York (1967).
8   H.Dertinger and H.Jung, *Molekulare Strahlenbiologie*, Springer-Verlag, Berlin, Heidelberg, New York (1969).
9   P.Alexander and A.Charlesby, in: *Radiobiology Symposium* (Z.M.Bacq and P.Alexander, eds.), p. 49, Butterworth, London (1955).
10 P.Howard-Flanders, *Nature* **186**, 485 (1961).
11 W.A.Armstrong and W.G.Humphreys, *Canad. J. Chem.* **45**, 2589 (1967).
12 H.Fischer, *Ber. Bunsenges. Phys. Chem.* **71**, 685 (1967).

A distribution scheme
for radiation damage in DNA

SILVANO GREGOLI, MARTIAL OLAST and
ALBERT BERTINCHAMPS

*Laboratoire de Biophysique et Radiobiology, Faculté des Sciences,
Université libre de Bruxelles, 67 rue des Chevaux, Rhode St.Genèse, Belgium*

Introduction

IN ORDER TO STUDY the direct effect of radiations on their main biologica target, a large amount of work has been performed, especially using ESR spectroscopy, on the separate constituents of the nucleic acids (1–5).

The final damage to the molecule of DNA is, however, very different from the simple addition of the damage to the constituents, for in a large number of systems thymine and cytosine appear to be the main site of radiation injury in DNA (6–10). These two bases taken separately are, however, not markedly more radiosensitive than the others. What could therefore be taken as a weak point of the DNA molecule is, in fact, a point of convergence of the radiation damage. This convergence must be due either to an energy transfer, or more probably to a radical transfer from one nucleic acid constituent to the other.

We have therefore undertaken a study of the effect of radiation, not on the separate constituents of nucleic acids but on a series of complexes which they form together. The results of this investigation enabled us to propose a mechanism for the internal redistribution of the radiation damage in these complexes. A similar mechanism could probably account for the preferential destruction of the pyrimidine bases in DNA.

221

Single nucleotides

The ESR spectra of the four DNA nucleotides, γ-irradiated and observed at 25°C, are shown in Fig. 1. The radicals giving rise to the above spectra are not completely known. The structure of the dTMP spectrum results from the superimposition of two (or more) distinct patterns: the octet of the H-addi-

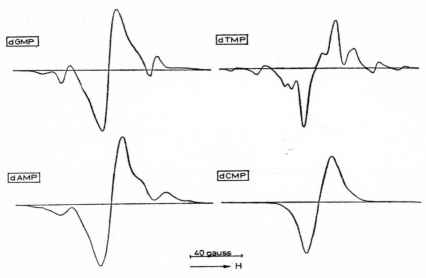

Figure 1   First derivative spectra of γ-irradiated deoxyguanilic acid (dGMP), deoxyade-nilic acid (dAMP), deoxythymidilic acid (dTMP) and deoxycytidilic acid (dCMP). All the nucleotides are in the form of Na_2 salts. Irradiations and observations at 25°C. Dose 3 Mrad

tion radical described by Gordy (11) and another signal, whose correspond-ing radical (s) is unknown, overlapping the octet in the central region. For the other nucleotides it has also been suggested that, except possibly for the dCMP, the radicals are formed by the addition of a proton to the base ring (12).

Purine–pyrimidine nucleotide complexes

In the DNA molecule, the four bases are submitted to essentially two types of interactions: hydrogen bonding between the complementary bases and stacking between the vertically overlapping bases. In aqueous solutions, on the other hand, the nucleic acid derivatives form complexes whose stability is exclusively provided by the stacking forces, the number of hydrogen bonds being negligible (13–16). In an effort to gain more detailed information on the mechanism by which the radiation damage may migrate within the DNA molecule towards some preferential sites, we have undertaken a systematic study by ESR of the spin distribution in these complexes.

They were obtained by freeze-drying aqueous solutions of equimolar amounts of two (or more) nucleotides. Irradiations, (Co^{60}), were performed under vacuum to a dose of 3 Mrad.

Figure 2 shows the spectra of the irradiated complexes between two complementary nucleotides. We compared in each case the spectrum of the complex with the spectrum of a mechanical mixture of the same compounds

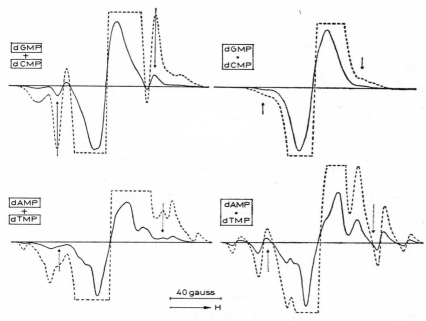

Figure 2 ESR spectra of the mechanical mixtures (left) and of the complexes (right) obtained from the complementary nucleotides. The spectra obtained by increasing the sensitivity are superimposed as a broken line. The solid arrows indicate the position of the dGMP satellites. The dotted arrows indicate the position of the dAMP satellites. Irradiations and observations at 25 °C. Dose 3 Mrad

in the same equimolar ratio. In both cases, one may observe in the spectrum of the complex the disappearance of the purine nucleotide pattern and the strengthening of the pyrimidine one. Since quantitative measurements showed that the total number of spins were unchanged, one may infer the existence of a spin transfer from the purine to the pyrimidine nucleotides.

Figure 3 shows that this type of transfer occurs even between non-complementary nucleotides.

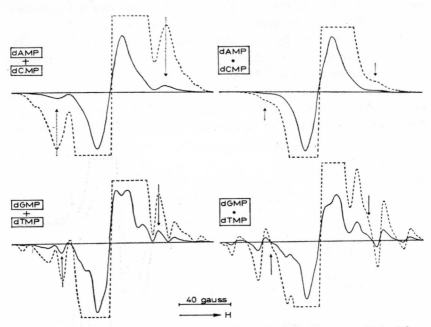

Figure 3   ESR spectra of the mechanical mixtures and of the complexes obtained from the non-complementary nucleotides. Irradiations and observations at 25 °C. Dose 3 Mrad

Figure 4 shows the spectra of the complexes between two purine nucleotides and one pyrimidine nucleotide.

The same measurements performed on the RNA nucleotides, the nucleosides and the bases indicate the same general tendency of the spins to migrate from the purine to the pyrimidine derivatives.

Considering the way the molecules are associated in the complexes, the above results (and in particular those obtained on the non-complementary nucleotides) strongly suggest that the stacking forces rather than the H bonds are instrumental in the migration of the damage.

For the transfer to occur it appears necessary for the donor nucleotide to be close enough to the acceptor nucleotide. Otherwise the spins would find themselves in the material impossibility of being transferred. It may happen that this condition is not completely fulfilled. For example, if, among several

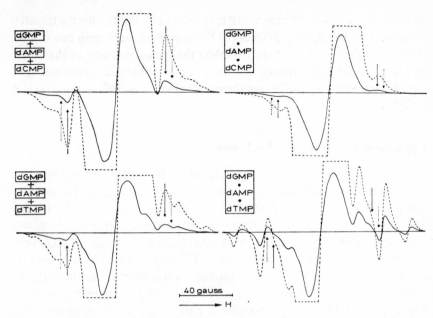

Figure 4 ESR spectra of the mechanical mixtures and of the complexes containing two purine and one pyrimidine nucleotide in the molar ratio 1 : 1 : 1. Irradiations and observations at 25 °C. Dose 3 Mrad

nucleotides, some have a greater tendency to self-associate than to co-associate with the others, one may find large stacks of pure nucleotides which decrease the number of stacks containing different nucleotides. This appears to be the case for the complex between GMP and CMP. The constant of self-association of G being larger than the constant of co-association of G-C (17),

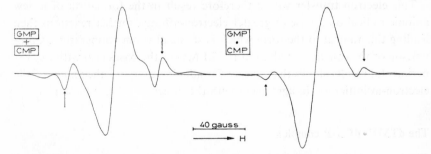

Figure 5 ESR spectra of the mechanical mixture and of the complex between guanilic acid (GMP) and cytidilic acid (CMP). The solid arrows indicate the GMP satellites. Irradiations and observations at 25 °C. Dose 3 Mrad

the formation of stacks of pure GMP reduces the possibilities for the transfer to occur. As a result, (Fig. 5) traces of the purine donor pattern can still be observed in the spectrum of the complex: the typical satellites of the GMP spectrum are strongly reduced but have not completely disappeared. One may estimate that only the 70 % of the spins have been transferred from GMP to CMP.

A hypothesis for the transfer mechanism

Alongside the path followed by the transfer, we have tried to investigate the physical reasons which determine the migration of the damage. The observation that the four DNA nucleotides irradiated and observed at 77°K give singlets very similar to one another and probably due to trapped electrons, prompts us to extend to the case of the other bases the hypothesis of Lenherr and Ormerod (18) concerning the mode of formation of the H-addition radical on the thymine. These authors produced evidence that this radical was formed by a two-step reaction, the first step leading to an anionic radical, Thy^-, stable at 77°K and responsible for the singlet observed at this temperature. A rise in temperature would increase the probability of the anion reacting with a neutral molecule, RH, to give finally the H-addition radical.

If in the same way we attribute the singlets which can be observed at 77°K for the four nucleotides to the resonance of anionic radicals, a simple interpretation of our results could be given. The primary radioproducts would consist of anionic radicals distributed at random along the stacks. Since the base stacking produces an overlapping of the π-electron clouds, the added electrons would in a second time be redistributed in a more stable way on the bases having the lowest empty molecular orbital of lowest energy.

This electron transfer would therefore result in the formation of a new anionic radical on a base of greater electron-affinity, further reactions then leading this radical to the form which is stable at room temperature. Comparison of the calculated values of the LEMO of the bases (19) with our experimental results shows that, except for adenine-thymine, the values of the electron-affinities are in agreement with the direction of the transfer.

The dTMP • dCMP complex

The two pyrimidine nucleotides both act as spin acceptors, so it is of interest to know whether a transfer can occur between them and whether in a complex containing the two pyrimidine nucleotides, one of them constitutes a preferential site for the final radiation damage. Since the theoretical values of the energy orbitals of dTMP and dCMP indicate a greater electron-affinity for this latter, a spin transfer in its direction would be expected.

Figure 6 shows the spectrum of the complex between dTMP and dCMP, and of the corresponding mechanical mixture. The interpretation of this result is somewhat complicated by the fact that, as mentioned above, the

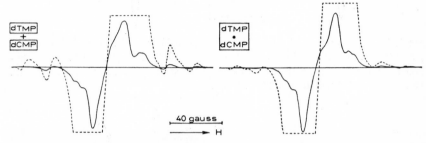

Figure 6 ESR spectra of the mechanical mixture and of the complex between the two pyrimidine nucleotides: dTMP and dCMP. Irradiations and observations at 25°C. Dose 3 Mrad

radiation produces on the dTMP two distinct radical species: the H-addition radical (τ_1) and another radical, unknown (τ_2). By comparing the two above spectra with the spectra of the single nucleotide components, one can see that the spectrum of the complex does not reproduce the singlet of the radical of pure dCMP as expected in a case of total transfer. The outermost satellites typical of the H-addition radical (τ_1) are drastically reduced but in the central region other lines of the dTMP spectrum do persist. This suggests that in the complex, only the formation of the H-addition radical (τ_1) has been prevented. The spectrum of the complex should then consist mainly of the superimposition of the second radical of the thymine (τ_2) on the broad singlet of the dCMP.

A confirmation of this point could, of course, be obtained by the graphical superimposition of the unknown τ_2 pattern on the dCMP singlet. For this purpose, and knowing that the H-addition radical is formed through an anionic phase, we have tried to isolate the τ_2 pattern by irradiating the pure dTMP in the presence of electron scavengers (iodoacetamide, I_2, N_2O). As may be seen in Fig. 7 the scavenger has hindered the formation of the H-addition radical (the outermost satellites have disappeared), leaving in the center a pattern which can be attributed to the τ_2 radical. Figure 8 shows that the graphical superimposition of this pattern on the dCMP singlet re-

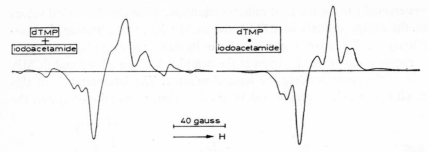

Figure 7 ESR spectrum of the dTMP γ-irradiated in the presence of iodoacetamide (10%) (right). The left-hand spectrum shows that a mechanical mixture of the two compounds in the same proportions gives exactly the same spectrum as the pure dTMP. Irradiations and observations at 25 °C. Dose 3 Mrad

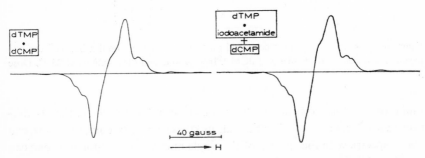

Figure 8 Left-hand side: spectrum of the complex between the two pyrimidine nucleotides. The spectrum on the right is obtained by the graphical superimposition of the dCMP singlet to the pattern of the dTMP treated with iodoacetamide. Irradiations and observations at 25 °C. Dose 3 Mrad

produces exactly the pattern of the dTMP · dCMP complex. From these results we may therefore conclude that, of the thymine radicals, only the τ_1 radical goes across an anionic physe and is thus the only which is able to transfer its electron to the dCMP of greater electron-affinity.

Four nucleotide complex

Figure 9 shows the spectra of the equimolar complex formed with the four DNA nucleotides, together with the corresponding mechanical mixture. As expected, no traces of the two purine nucleotides are detectable, and the damage is localized on the pyrimidine moieties. The four nucleotide spectrum does not, however, exactly reproduce the spectrum of the dTMP · dCMP complex: the lateral structure corresponding to the τ_1 radical is much more apparent in this case. This is possibly due to the formation, in the complex, of stacks containing only one of the two pyrimidine nucleotides. In these conditions, the $T \rightarrow C$ transfer becomes materially impossible and the Thy^- anion can follow its evolution towards the H-addition radical.

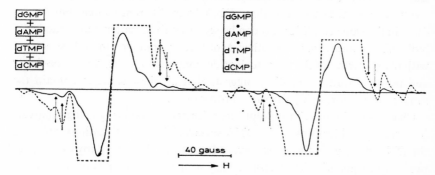

Figure 9 ESR spectrum of the mechanical mixture and of the complex obtained from the four DNA nucleotides in the same molar ratio. Irradiations and observations at 25°C. Dose 3 Mrad

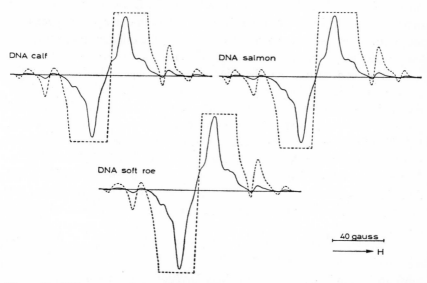

Figure 10 ESR spectra of three different DNA (Sigma) respectively from calf thymus, salmon sperm and soft roe. The three samples were dissolved in bidistilled water, rapidly frozen in liquid nitrogen and then freeze-dried. Irradiations and observations at 25°C. Dose 3 Mrad

The DNA spectrum

There have been many attempts to interpret the DNA spectrum in terms of its constitutent spectra, but since the literature on this subject continues to be somewhat contradictory (5–9, 20–23), we have tried to analyze this spectrum by making use of the results obtained on the nucleotide complexes. Figure 9 shows the spectra of three DNA from different sources but freeze-dried in the same conditions. It can be seen that these patterns are practically indistinguishable from one another. Moreover, a comparison of these spectra with the spectrum of the four nucleotide complex, together with time-decay and power saturation studies, suggests that all these samples contain the same radicals in roughly the same proportions.

Changes in the pH of the solutions prior to freeze-drying bring relatively important modifications to the DNA spectra. These modifications, affecting the DNA samples in the same way, did not bring, however, any evidence of purine-like signals in the spectrum. They seem to correspond only to changes in the ratio of the two thymine and cytosine components of the spectra. The same changes in the spectrum shapes were observed by freeze-drying different buffer solutions at the same pH, but always with the same features for the three DNA samples. These results suggest that the DNA spectrum does not depend on the DNA source, that the final radiation damage is in each case localized on the two pyrimidine bases, and that the different preparation of the samples may influence in some way the efficiency of the $T \rightarrow C$ transfer.

It is thus conceivable that the migration of damage in DNA follows the mechanism which we have proposed for the complexes. The primary anionic population randomly distributed along the DNA chain may in a second phase be re-arranged by the electron transfers towards the pyrimidine bases. It is likely that this transfer occurs vertically along the stacked bases, the overlapping of the π-electron orbitals along the double helix leading to the formation of conductive channels allowing a certain electron mobility (24, 25). The range in which the electron can move before reaching the pyrimidine acceptor, could in this case be long enough to overcome a certain number of bases.

This work was carried out under Association contract Euratom–Université libre de Bruxelles No. 007-61-10 BIAC.

Contribution No. 510 of the Biology Division of Euratom.

References

1 Alexander, C., Jr., and Gordy, W., *Proc. Natl. Acad. Sci.*, **58**, 1277 (1967).
2 Bernhard, W., and Snipes, W., *Physics*, **59**, 1038 (1968).
3 Herak, J.N., and Gordy, W., *Proc. Natl. Acad. Sci.*, **56**, 7 (1966).
4 Holmes, D.E., Ingalls, R.B., and Myers, L.S., Jr., *Int. J. Radiat. Biol.*, **12**, 415 (1967).
5 Müller, A., *Progress in Biophysics and Mol. Biol.*, **17**, 101 (1967).
6 Pershan, P.S., Schulman, R.G., Wyluda, B.J., and Eisinger, J., *Physics*, **1**, 163 (1964).
7 Cook, J.B., and Wyard, S.J., *Int. J. Radiat. Biol.*, **11**, 357 (1966).
8 Ehrenberg, A., Rupprecht, A., and Strom, G., *Science*, **157**, 1317 (1967).
9 Ehrenberg, A., Ehrenberg, L., and Löfroth, G., *Nature*, **200**, 376 (1963).
10 Setlow, R.B., and Carrier, W.L., *J. Mol. Biol.*, **17**, 237 (1966).
11 Pruden, B., Snipes, W., and Gordy, W., *Proc. Natl. Acad. Sci.*, **53**, 917 (1965).
12 Holmes, D., Ingalls, R., and Myers, L.S., Jr., *Int. J. Radiat. Biol.*, **13**, 225 (1967).
13 Broom, A.D., Schweizer, M.P., and Ts'o, P.O.P., *J. Amer. Chem. Soc.*, **89**, 3612 (1967).
14 Schweizer, M.P., Broom, A.D., Ts'o, P.O.P., and Hollis, D.P., *J. Amer. Chem. Soc.*, **90**, 1042 (1968).
15 Ts'o, P.O.P., in *Molecular Association in Biology*, Pullman edit., 1968, New York Academic Press, p. 39.
16 Solie, T.N., and Schellman, J.A., *J. Mol. Biol.*, **33**, 61 (1968).
17 Gellert, M., Lipset, M.N., and Davies, D.R., *Proc. Natl. Acad. Sci.*, **48**, 2013 (1962).
18 Lenherr, A.D., and Ormerod, M.G., *Biochim. Biophys. Acta*, **166**, 298 (1968).
19 Pullman, B., and Pullman, A., in *Quantum Biochemistry*, 1963, New York Interscience Publishers, p. 215.
20 Ormerod, M.G., *Int. J. Radiat. Biol.*, **15**, 220 (1966).
21 Gordy, W., Pruden, B., and Snipes, W., *Proc. Natl. Acad. Sci.*, **53**, 751 (1965).
22 Herak, J.N., and Gordy, W., *Proc. Natl. Acad. Sci.*, **55**, 698 (1966).
23 Van de Vorst, A., *Int. J. Radiat. Biol.*, **12**, 153 (1967).
24 Duchesne, J., Depireux, J., Bertinchamps, A., Cornet, N., and Van der Kaa, J.M., *Nature*, **188**, 405 (1960).
25 Hedén, C.G., and Rupprecht, A., *Acta Chem. Scand.*, **20**, 583 (1966).

Amine oxidases: a new class of copper oxidases

B. MONDOVÌ, G. ROTILIO, A. FINAZZI AGRÒ
and E. ANTONINI

Institute of Biological Chemistry, University of Rome,
Regina Elena Institute for Cancer Research, Rome, and
Center for Molecular Biology of the National Research Council, Rome, Italy

1 Introduction

SEVERAL, IF NOT ALL, of the copper proteins so far known are oxidases: no other catalytic activity has been discovered in copper proteins, and this fact itself indicates, that copper has such properties as to give it a precise functional role in biological systems. This suggests that copper, in copper oxidases, fulfils a specific function which, besides the individual differences, might be basically the same in all copper enzymes. The discovery and the study of new classes of copper enzymes can bring about further evidence and details in this sense.

In the assignment of function to copper in copper containing enzymes, EPR techniques have played an outstanding role. In general, EPR can give information on the possible function of a metal in enzymes (maintenance of a particular protein conformation; binding of substrate—either O_2 or the other substrate; direct involvement in the electron transfer during the catalytic process), and tell what kind of electronic environment the metal possesses in the protein. This method has been widely used in the study of copper oxidases and the data obtained, together with those derived from other techniques, mainly optical spectroscopy, have provided a detailed picture of the interactions of the metal in these systems. At the present time only a rough classi-

233

fication of copper oxidases can be attempted, on the basis of various criteria. The most homogeneous class of these proteins appear to be that characterized by the following parameters:

a) high extinction coefficients in the visible;
b) high redox potential;
c) unusually small hyperfine splitting in EPR spectra;
d) relatively low g values.

All these features denote a high degree of distortion around the cupric ion, which has been reported to be highly favourable, either thermodynamically or kinetically, to a cyclic Cu^{++}–Cu^+ conversion during the catalytic activity.

This group includes the "blue" copper oxidases (laccase, ascorbate oxidase, etc.) and cytochrome oxidase, which all produce H_2O in the reaction with oxygen.

The rest of the copper oxidases can be grouped together mainly owing to the lack of the properties indicated above. Some of them (amine oxidases, uricase, galactose oxidase) produce H_2O_2. Others (tyrosinase, dopamine-β-hydroxylase) are oxygenases.

Here we shall report the results of some recent studies on amine oxidases, which threw light on the role of copper in these enzymes and indicate that the behaviour of this metal in this oxidative system might not be qualitatively different from that in "blue" copper oxidases.

2 The amine oxidases

Amine oxidases are a wide class of enzymes which catalyze the oxidative deamination of amines according to the following reactions:

$$RCH_2NH_2 + O_2 + H_2O \rightarrow RCHO + NH_3 + H_2O_2$$

In order to avoid confusion, we would like to classify these enzymes into two classes, according to Blaschko (1).

a) Amine oxidases inhibited by carbonyl reagents ("diamine oxidase" type). They are soluble enzymes which contain copper and pyridoxal phosphate as prosthetic groups. They include animal, plant and bacterial enzymes and act on primary amino groups of short aliphatic diamines, histamine, benzylamine, spermine, mescaline (2).

b) Amine oxidases not inhibited by carbonyl reagents ("monoamine oxidase" type). This class is much less known than the other one; some of them seem to contain copper (3) and a flavin (4) as prosthetic groups. In mitochondria of animals "monoamine oxidase" probably represents a complex system of enzymes with relatively narrow substrate specificities (5). They act on primary and secondary amines, but on the latter only if the substituent is a methyl group. Substrates for these enzymes are: catechol compounds as dopamine, norepinephrine, epinephrine, tryptamine derivatives, histamine. It should however be pointed out that Gorkin (6) obtained in vitro transformation of mitochondrial "monoamine oxidase" into a diamine oxidase-like enzyme.

No data are so far available on the quaternary structure of amine oxidases. The molecular weight determination on pig kidney diamine oxidase, made both by sedimentation–diffusion and by approach to equilibrium suggest that the enzyme may undergo reversible association–dissociation. The values obtained with this enzyme over a concentration range of 1 to 10 mg/ml are not multiples of the minimum molecular weight obtained by copper analysis i.e. 90.000 (7).

The specific activity of pig kidney diamine oxidase appears to depend on enzyme concentration. At 38 °C in 0.1 M phosphate buffer pH 7.4 the turnover number, calculated on minimum molecular weight, is about 100, when the enzyme concentration is about 0.01 mg/ml, while at 2.5 mg/ml the turnover number decreases to about 30. Amine oxidases, on storage or after treatments such as freezing and thawing, undergo changes of various kind, which often involve modification of the state and reactivity of the metal and of pyridoxal phosphate (8, 9, 10, 11, 12).

The visible spectra of amine oxidases show a broad absorption band around 500 nm. No bands are present at longer wavelengths. Optical rotatory dispersion curves of pig kidney diamine oxidase, show no Cotton effects in the visible range. The helical content, measured by the intensity of the rotation at 233 nm appear to be very low (13), as in other copper protein.

Very few data are available on the structure and properties of the protein part of amine oxidase. Wang *et al.* (14) were able to titrate, after a very long incubation of plasma amine oxidase with urea, one SH group per subunit. Pig kidney diamine oxidase exposed to guanidine has a molecular weight corresponding to the monomer (i.e. to the minimal molecular weight calculated on the copper content). A further dissociation into fragments of about 45,000 molecular weight is produced by treating this enzyme with mercaptoethanol. Mercaptoethanol does not have this effect in the absence of guanidine (13).

3   Copper in amine oxidases

All amine oxidases sensitive to carbonyl reagents contain copper. This metal was found to be present also in a mitochondrial amine oxidase (3). Flavins are present in a mitochondrial enzyme, while in the carbonyl-reagents sensitive enzymes they are absent (2, 7). The average copper content of amine oxidases is 0.07–0.1 %.

Copper is firmly bound to these enzymes. Dialysis against Tris and EDTA is without effect (15). Most of the copper is removed easily by treatment of amine oxidases with diethyldithiocarbamate. In pig kidney diamine oxidase about 30 % of the copper does not react with diethyldithiocarbamate (7) and does so only after exposure of the enzyme to urea (13). The "copper free" enzymes are completely inactive (7) and the activity is restored after the addition of suitable amount of copper. Other metals are inactive (7, 21).

The first report of an EPR spectrum of an amine oxidase is that of Yamada *et al.* (16) for beef plasma amine oxidase. Other soluble amine oxidases were then investigated (7, 17, 18), whereas for mitochondrial amine oxidase, only a preliminary communication by Nara *et al.* is available (3). The EPR bonding parameters of the enzymes so far studied are shown in Table I together with those of some other copper enzymes and copper complexes.

The parameters are quite similar in all amine oxidases and indicate that the cupric ions are present in these enzymes in a complex of tetragonal symmetry and in an equivalent environment. The type of bonding appear to be strikingly different from that of the "blue" copper oxidases. According to the current opinions in copper bonding (19, 20), one can speak, in the case of amine oxidases, of a complex of symmetry only slightly distorted from the square planar, and of considerably ionic character, in the sense of a low degree of delocalization of the unpaired hole of the cupric ions.

As to the possible copper ligands, for most of the amine oxidases indirect evidence exists for the presence of nitrogen ligands: in fact the g values and bonding parameters are comparable to those of complexes in which copper is bonded to nitrogen atoms (see Table I). In the case of pig kidney diamine oxidase, however, at high resolution, several superhyperfine lines are detected in the region of $g_\perp$, with a splitting of about 14 gauss which suggests the involvement of nitrogen ligands (7). A particularly controversial point is the amount of Cu^{++} detectable by EPR (see Table II). It is well known that the fact that the EPR detectable copper is less than that determined by chemical assay would mean either that cuprous copper is also present or that two or more cupric atoms are coupled with one another. How-

Table I EPR parameters of amine oxidases and other copper proteins and copper complexes

Ligand	Reference	Coordinating atoms	g_m	$g_{\parallel}$	Hps (gauss)	$A_{\parallel}$ (cm⁻¹)	$\frac{4}{7}a^2 + \varkappa$
Beef plasma amine oxidase	16		2.053	2.266	155	0.015	0.75[a]
Pig plasma amine oxidase	18		2.060	2.294		0.016	0.77[a]
Pig kidney diamine oxidase	7		2.063	2.25	149	0.018	0.81
Pig kidney diamine oxidase	17		2.05	2.17			
Cytochrome oxidase	34		2.03	2.197		0.009	0.48
Laccase	19		2.048	2.262	169	0.018	
Laccase Cu++ Type 2	31		2.036	2.265		0.016	0.75
Erythrocuprein	19		2.063	2.28		0.019	
Galactose oxidase	32		2.04	2.282		0.017	
Dopamine β-hydroxylase	30		2.056	2.23		0.018	0.78
Histidine	19	4N	2.063	2.316		0.018	0.84
Oxalate	19	4O	2.078	2.337		0.017	0.80
EDTA	19	2N, 4O	2.09	2.13		0.015	0.70[a]
Cysteine	33	2N, 2S	2.03			0.019[b]	

[a] Calculated according to ref. 19.   [b] In the original paper the value is expressed in Mc/sec.

Table II   EPR detectable copper in some amine oxidases

	Percentage with respect to chemically determined copper
Beef plasma amine oxidase (16)	73%
Pig kidney diamine oxidase	80%
Pig plasma amine oxidase (18)	82%

ever, Yamada and Yasunobu (21) found no cuprous copper by a chemical method in plasma amine oxidase. If a dipolar interaction seems not probable because of the presence of well defined and relatively sharp EPR spectra, an exchange interaction could not be excluded.

In contrast with the results obtained with soluble amine oxidases Nara et al. (3) reported that the EPR spectrum of the beef liver mitochondrial amine oxidase resembles that of cytochrome oxidase. This would indicate that the copper is present in a complex of the "covalent" type, like that of the "blue" oxidases.

4   EPR studies of the catalytic reaction of amine oxidases

The catalytic mechanism of diamine oxidases is envisaged to consist of two distinct phases: one anaerobic, formally equivalent to a transamination, which involves the binding of the substrate to the pyridoxal moiety of the enzyme and the formation of the aldehyde from the amine: the other which consist in the oxidation of the pyridoxamine enzyme by molecular oxygen (22).

The relevant experimental facts in this context are: a) the results obtained by Buffoni (23) and Kumagai et al. (24) who isolated the pyridoxal-histamine compound from pig plasma amine oxidase and pig kidney diamine oxidase respectively after reduction with borohydride in the presence of ^{14}C histamine; b) the experiments of Reed and Swindell (25) and Finazzi Agrò et al. (26) who demonstrated the formation of the aldehydes products in anaerobiosis respectively for plasma amine oxidase and pig kidney diamine oxidase. In the case of the pig kidney enzyme Finazzi Agrò et al. (26) also provided kinetic evidence for the formation of two binary complexes (i.e. enzyme–substrate and reduced enzyme–O_2) during the catalytic activity of diamine oxidase.

When substrate is added in anaerobic conditions to the enzyme, the absorption at 500 nm shows a 10% decrease and the change is reversed on oxygenation. A peak at about 350 nm becomes evident in the presence of cadaverine added anaerobically; this absorption disappears after exposure to air.

Spectropolarimetric analysis of the reaction between pig kidney diamine oxidase and substrate indicated that after anaerobic addition of substrate an extrinsic Cotton effect appears around 350 nm (13). Circular dichroic spectra of pig kidney diamine oxidase (28) show a positive peak at 330 nm and a negative peak at 420 nm. On addition of the substrate the peaks are slightly lowered and a new positive peak appears at 470 nm.

The pre-steady state kinetics of the enzyme was studied in a rapid mixing apparatus by following the optical density changes at 500 nm on mixing the enzyme with cadaverine and oxygen (26).

Figure 1 shows the time course of the optical density change at 500 nm after mixing the enzyme with cadaverine in the absence of oxygen. Two distinct phases are present in the progress curves: a fast one with half times in the range of tenths of a second, followed by a plateau lasting seconds and by an even slower further optical density decrease. There is an obvious de-

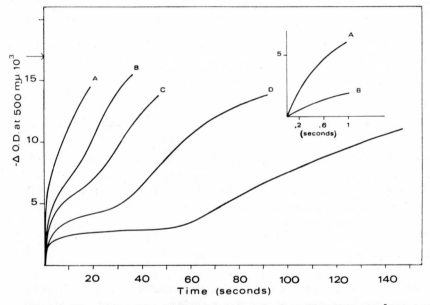

Figure 1 Time course of the reaction of pig kidney diamine oxidase (1.83×10^{-5} M in 0.1 M potassium phosphate buffer pH 7.4) with various amounts of cadaverine under anaerobic conditions. Concentration of cadaverine: $A = 5 \times 10^{-4}$ M; $B = 1.25 \times 10^{-4}$ M; $C = 6.25 \times 10^{-5}$ M; $D = 3.13 \times 10^{-5}$ M; $E = 1.57 \times 10^{-5}$ M. The arrow indicates the asymptotes for curves A, B, C and D. The asymptote for curve E corresponds to an optical density of 13×10^{-3}. Insert shows details of the fast phase of the curve A and B. Molar concentrations of the enzyme are expressed on the basis of a minimum molecular weight of 90,000 calculated from the copper content

pendence of various phases of the reaction on substrate concentration, other conditions being equal. These also depend on oxygen concentration (Fig. 2). As shown in the figure, when the enzyme is mixed simultaneously with oxygen and substrate the plateau region lasts longer. At a high enough oxygen concentration the slow phase tends to disappear and is replaced by a slow change

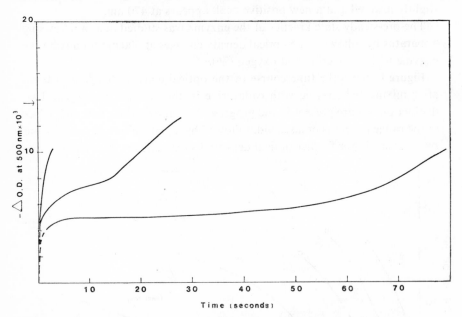

Figure 2 Time course of the reaction of pig kidney diamine oxidase (1.83×10^{-5} M in 0.1 M potassium phosphate buffer pH 7.4) after mixing with 5×10^{-3} M cadaverine and various oxygen concentrations. From the left to the right: no oxygen; 1.35×10^{-4} M oxygen; 6.75×10^{-4} M oxygen. The arrow indicates the asymptotes.

in the opposite direction which restores the normal oxidized spectrum of the enzyme. The reduced enzyme reacts with the oxygen with a velocity comparable to the fast phase of reduction.

EPR studies of the reaction between amine oxidase and substrate have been performed in the case of beef plasma oxidase (16) pig plasma benzylamine oxidase (18) and pig kidney diamine oxidase (7, 17). Neither Yasunobu *et al.* (16) nor Buffoni *et al.* (18) observed any change in intensity of the copper signal after anaerobic addition of substrate. Furthermore Yamada and Yasunobu (21) have determined the valence state of the copper during the anaerobic incubation with substrate by a chemical method involving the cuprous copper specific reagent neocuproine and detected no change of cupric copper. It should be pointed out in this regard that 0.3 mM neocuproine produces a 33% inhibition of the mitochondrial beef liver amine oxidase (3).

Goryachenkova *et al.* (17) observed that the anaerobic addition of substrate to pig kidney diamine oxidase slightly decreases the intensity of the EPR signal, while *g* values and $A_{\parallel}$ did not change. Detailed experiments have been carried out by the Rome group in attempts to correlate all the data available about the catalytic activity of the enzyme, either from steady state or from pre-steady state kinetic experiments, with the state of the copper. As a result of this study the following picture can be presented at this moment. There are two types of change in the EPR signal following addition of excess substrate to the enzyme under anaerobic conditions; the two types of change follow each other in time and may be correlated to the various phases in the pre-steady state kinetics. The first one is observed when the enzyme solulion is frozen for the EPR spectrum in a short time after mixing with substrate, corresponding approximately to the rapid phase of the stopped flow experiments. An increase of about 20–30% of the EPR signal of the copper is observed with a clear change in line shape (Fig. 3). These facts should indicate unmasking of EPR undetectable copper, whatever is the explanation for this. Furthermore in the $g_{\perp}$ region of the spectrum an enhancement of the superhyperfine pattern is evident. It might be tempting to relate this to the binding of the amine substrate to copper; this possibility, however, has been ruled out by the fact that the superhyperfine structure is the same in the presence either of ^{14}N-putrescine (^{14}N spin = 1) or of ^{15}N-putrescine (^{15}N spin = $\frac{1}{2}$). This result is a clear evidence that the amine does not bind to the copper and is in agreement with the current view of the formation of a pyridoxal phosphate–amine complex as the first intermediate in the diamine oxidase reaction. However, the change in the superhyperfine structure on $g_{\perp}$ are different for each substrate tested, indicating that the formation of the enzyme substrate complex is reflected also in minor modifications in the environment of the copper. As an additional evidence for the conclusion drawn from the isotopes experiment it should be pointed out that the spin relaxation of copper does not change in the presence of substrate.

The second type of EPR change is detected in a range of minutes (2–15 min) after the anaerobic addition of substrate. The times parallel fairly well those of the second phase, in the kinetic experiments. At this stage a 25–30% reduction of this copper signal intensity is observed, without any change in the signal shape (Fig. 4). After exposure of the incubation mixture to air, the original shape and intensity are restored.

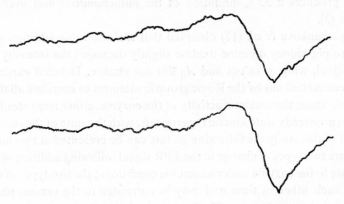

Figure 3 EPR spectra of pig kidney diamine oxidase in the absence and presence of cadaverine. Enzyme, 2%, was dissolved in 0.3 ml of 0.1 M phosphate buffer, pH 7.4, in an anaerobic cell. Upper curve, before addition of cadaverine. Lower curve, 1 min after addition of 0.003 ml of 0.02 M neutralized cadaverine hydrochloride. Microwave frequency was 9.16 GHz; microwave power 6.2 milliwatts, modulation amplitude 5 gauss, temperature 133 °K. The $g_{||}$ value is 2.29

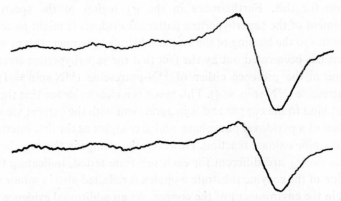

Figure 4 EPR spectra of pig kidney diamine oxidase, 7%, in 0.3 ml of 0.1 M phosphate buffer, pH 7.4 at various times after the anaerobic addition of 0.003 ml of 1 M neutralized cadaverine hydrochloride. Upper curve, 5 min. Lower curve, 60 min. For other details, see Fig. 3, except for microwave power, which was 20 milliwatts

The results of the rapid mixing experiments and of the EPR measurements have been incorporated into a minimum scheme which has been analyzed by analogue computer methods. The scheme tested is shown below.

$$1)\ Cu^{++}E + S \quad \overset{K_1}{\underset{K_2}{\rightleftharpoons}} \quad Cu^{++}ES$$

$$2)\ Cu^{++}ES \quad \overset{K_3}{\dashrightarrow} \quad Cu^{++}E^{oo} + P$$

$$3)\ Cu^{++}E^{oo} \quad \overset{K_4}{\underset{K_5}{\rightleftharpoons}} \quad Cu^+E^o$$

$$4)\ Cu^+E^o + O_2 \quad \overset{K_6}{\dashrightarrow} \quad Cu^{++}E + O_2^{--}$$

In this scheme $Cu^{++}E$ indicates the fully oxidized enzyme with absorption at 500 nm: S is the substrate (in this case the cadaverine). $Cu^{++}ES$ is the enzyme–substrate complex (i.e. the Schiff base between substrate and the enzyme-bound pyridoxal phosphate). $Cu^{++}E^{oo}$ indicates the enzyme reduced in a site other than copper (for instance the enzyme with bound pyridoxamine phosphate). The step $Cu^{++}ES \dashrightarrow Cu^{++}E^{oo}$ was assumed to be irreversible because with cadaverine as substrate the aldehyde obtained cyclizes rapidly yielding Δ^1-piperideine. Cu^+E^o indicates the enzyme with cuprous copper which then reacts with oxygen and regenerates the oxidized enzyme. Because in the EPR experiments it was impossible to obtain the reduction of all the copper present in the enzyme also with a large excess of substrate, it is assumed that the equilibrium in reaction 3 lies on the left. It was assumed that all the enzyme forms had distinct extinction coefficients at 500 nm. The computed curves are consistent with the qualitative aspects of the experiments including the dependence of various phases on substrate and oxygen concentrations; the possibilities of the scheme to account quantitatively for the observed time course of the reaction are being explored at present.

5 Conclusions

The findings exposed above outline distinct features of amine oxidase as a new class of copper containing oxidases. First of all, in addition to copper they contain another prosthetic group, which may be pyridoxal phosphate or flavin. This feature is shared with cytochrome oxidase. However, in the latter enzyme the succession of events on the two prosthetic groups is an unsettled point, in spite of a great deal of investigations. In amine oxidases, it appears reasonable that pyridoxal phosphate or flavin are involved in the first stage of the catalytic mechanism, the anaerobic one, whereas copper, which does not bind substrate, is involved in the second stage where oxygen is reduced by the enzyme.

The features of copper in these enzymes appear well established. The EPR spectrum speaks against an unusual deviation from square planar symmetry. The g values and hyperfine structure indicate nitrogen and oxygen ligands (29) rather than sulphur one; the presence of nitrogen atoms is also supported by the superhyperfine pattern in pig kidney diamine oxidase. A common finding appears to be a nearly constant (25–30%) amount of EPR undetectable copper, the nature of which is not clear.

EPR spectroscopy during the reaction of pig kidney diamine oxidase with substrate provides direct evidence for a change of valence, at least of part of the copper in the enzyme. It is not the first case of partial reduction of cupric copper in copper oxidase. Also dopamine-β-hydroxylase, an enzyme very similar, as to EPR parameters to amine oxidase, shows a partial reduction of copper during catalytic activity (29, 30). The "non blue" Cu^{++} of laccase too (31) show a partial and slow reduction in the presence of substrate: also in this case the EPR parameters are similar to those of amine oxidases. The partial reduction of copper is not inconsistent with a mechanism in which the metal is the pathway for electron transfer as is evident from inspection of the proposed scheme. However the possibility that the non-reducible copper might be an inactive one, due to a large fraction of inactive enzyme, is not ruled out. With regard to this, it should be pointed out that superhyperfine structure—a typical feature of pig kidney diamine oxidase—has so far not been detected in "native" copper enzymes while it is evident in "denatured" cytochrome oxidase.

If the picture outlined in this paper is correct, the role of copper in amine oxidases can be considered equivalent to that fulfilled in "blue" copper oxidases. The structural differences indicated by the EPR and optical spectra of the two classes of enzymes would explain the different kinetic efficiency of the two types of copper oxidases, rather than support the view of the absence of a change of valence of copper during the catalytic activity.

References

1 H. Blaschko, P. J. Friedman, R. Hawes and K. Nilsson, *J. Physiol.*, **145**, 384 (1959).
2 F. Buffoni, *Pharmacol. Reviews*, **18**, 1163 (1966).
3 S. Nara, B. Gomes and K. T. Yasunobu, *J. Biol. Chem.*, **241**, 2774 (1966).
4 I. Igaue, B. Gomes and K. T. Yasunobu, *Biochem. Biophys. Res. Comm.*, **29**, 562 (1967).

5 V.Z.Gorkin, N.V.Komisarova, M.I.Lerman and C.V.Veryovkina, *Biochem. Biophys. Res. Comm.*, **15**, 383 (1964).
6 V.Z.Gorkin, L.V.Tat'Yavenko and T.A.Moskvitina, *Biokhimiya*, **33**, 393 (1968).
7 B.Mondovì, G.Rotilio, M.T.Costa, A.Finazzi Agrò, E.Chiancone, R.E.Hansen and H.Beinert, *J. Biol. Chem.*, **242**, 1160 (1967).
8 K.T.Yasunobu, F.Achee, C.Chervenka and Tso Ming Wang, *Intern. symposium on pyridoxal enzymes*, K.Yamada, N.Katanuma and H.Wada ed., Maruzen Co. Ltd., Tokyo, p. 139 (1968).
9 B.Mondovì, M.T.Costa, A.Finazzi Agrò and G.Rotilio, *Pyridoxal Catalysis: Enzymes and Model Systems.*—E.E.Snell, A.E.Braunstein, E.S.Severin and Y.M.Torchinsky, ed., Interscience Publ., p. 403 (1968).
10 B.Mondovì, M.T.Costa, A.Finazzi Agrò and G.Rotilio, *Arch. Biochem. Biophys.*, **119**, 373 (1967).
11 F.Buffoni and F.R.S.Blaschko, *Proc. Royal Soc.*, *B*, **161**, 153 (1964).
12 A.Finazzi Agrò, G.Rotilio, M.T.Costa and B.Mondovì, *Boll. Soc. It. Biol. Sper.*, **20 bis**, 243 (1968).
13 B.Mondovì, M.T.Costa, A.Finazzi Agrò and G.Rotilio, *Intern. symposium on pyridoxal enzymes*, K.Yamada, N.Katanuma and H.Wada, ed., Maruzen Co. Ltd., Tokyo, p. 143 (1968).
14 Tso-Ming Wang, F.Achee and K.T.Yasunobu, *Arch. Biochem. Biophys.*, **128**, 106 (1968).
15 B.Mondovì, G.Rotilio and M.T.Costa, *Symp. on Chem. and Biol. Aspects of Pyridoxal Catalysis*, E.E.Snell, P.M.Fasella, A.Braunstein and A.Rossi Fanelli, ed., Pergamon Press Ltd. publ., Oxford, p. 415 (1963).
16 H.Yamada, K.T.Yasunobu, T.Yamano and H.S.Mason, *Nature*, **198**, 1092 (1963).
17 E.V.Goryachenkova, L.I.Shcherbatiuk and C.I.Zamaraev, in *Pyridoxal Catalysis: Enzymes and Model Systems*, E.E.Snell, A.E.Braunstein, E.S.Severin and Y.M.Torchinsky, ed., Interscience Publ., p. 391 (1968).
18 F.Buffoni, L.Della Corte and P.F.Knowles, *Biochem. J.*, **106**, 575 (1968).
19 B.G.Malmström and T.Vänngård, *J. Mol. Biol.*, **2**, 118 (1960).
20 W.E.Blumberg, *The Biochemistry of Copper*, J.Peisach, P.Aisen and W.E.Blumberg, ed., Academic Press, New York, p. 49 (1966).
21 H.Yamada and K.T.Yasunobu, *J. Biol. Chem.*, **237**, 3077 (1962).
22 C.M.McEwen, Jr., K.T.Cullen and A.J.Sober, *J. Biol. Chem.*, **241**, 4544 (1966).
23 F.Buffoni, in *Pyridoxal Catalysis: Enzymes and Model Systems*, E.E.Snell, A.E.Braunstein, E.S.Severin and Yu.M.Torchinsky, ed., Interscience Publ., p. 363 (1968).
24 H.Kumagai, T.Nagate, H.Yamada and H.Fukami, *Biochim. Biophys. Acta*, **185**, 242 (1969).
25 D.J.Reed and R.Swindell, *Feder. Proc.*, **28**, 891 (1969).
26 A.Finazzi Agrò, G.Rotilio, M.T.Costa and B.Mondovì, *FEBS Letters*, **4**, 31 (1969).
27 B.Mondovì, G.Rotilio, A.Finazzi Agrò, M.P.Vallogini, B.G.Malmström and E.Antonini, *FEBS Letters*, **2**, 182 (1969).
28 H.Yamada and Hamaguchi, *Intern. Symposium on Pyridoxal Enzymes*, K.Yamada, N.Katanuma and H.Nada, ed., Maruzen Co., Ltd., Tokyo, p. 151 (1968).
29 W.E.Blumberg, H.Goldstein, E.Lauber and J.Peisach, *Biochim. Biophys. Acta*, **99**, 187 (1965).
30 S.Friedman and S.Kaufman, *J. Biol. Chem.*, **241**, 2256 (1966).

32 B.G.Malmström, B.Reinhammar and T.Vänngård, *Biochim. Biophys. Acta*, **156**, 67 (1969).
32 W.E.Blumberg, B.L.Horecker, F.Kelly-Falcoz, and J.Peisach, *Biochim. Biophys. Acta*, **96** 336 (1965).
33 W.E.Blumberg and J.Peisach, *J. Chem. Phys.*, **49**, 1793 (1968).
34 H.Beinert, D.E.Griffiths, D.C.Wharton and R.H.Sands, *J. Biol. Chem.*, **237**, 2337 (1962).

EPR studies of the adrenocortical
11 β-hydroxylase system

HEINZ SCHLEYER, DAVID Y. COOPER and
OTTO ROSENTHAL

Harrison Department of Surgical Research and Johnson Research Foundation
University of Pennsylvania, Philadelphia, Pennsylvania, 19104

Abstract

The adrenocortical steroid 11β-hydroxylase, a mixed function oxidase of vertebrate tissue, has been resolved into (1) a reducing system with a reaction-specific flavoprotein, (2) an iron sulfur protein, and (3) a heme fraction containing P 450 in a solubilized, partially purified form. The enzymatic properties and redox functions of the active components are essentially maintained in these fractions. On recombination in appropriate proportions a fully reactive 11β-hydroxylase is reconstituted which is capable of hydroxylating deoxycorticosterone in the presence of TPNH and oxygen.

Spectroscopic, redox- and other properties of the components and of the reconstituted system have been investigated by EPR spectroscopy and suitable optical absorption methods. After a brief review of the presently available knowledge on mixed function oxidases some of these data are presented. The results are discussed in relation to the function of these constituents of the 11β-hydroxylase system and to the reaction mechanisms of mixed function oxidases in general.

Introduction

OUR INVESTIGATIONS are concerned with the detailed enzymatic mechanisms by which several hormones and a large variety of chemical compounds are metabolized in mammalian systems. The processes catalyzed by these enzyme

247

systems play an essential role in the biosynthesis and the degradation of hormone molecules and thus are involved in the regulatory pathways which maintain proper hormone levels *in vivo* in humans and mammals (1). In addition, these enzyme systems participate in the degradation and metabolism of numerous other molecules which are foreign to the organism such as many drugs and pharmaceuticals and some cancerogenic agents. These enzyme systems are often classified as "mixed function oxidases" (2) indicating that more than one function is performed by the atoms of the O_2 molecule in the overall reactions; occasionally the term "monooxygenases" (3) is used.

In a large number of these enzyme systems, including the specific case presented here, the overall process leads to a hydroxylation of a well defined organic molecule. In other closely related cases the processes of interest lead to a demethylation, the physiological breakdown of hemoglobin to biliverdin, and to the formation of bile acids. The mixed function oxidases are presumably systems which function at least in part as a concerted multienzyme system rather than a mixture of individual enzymes.

As widespread in scope as the types of reactions is the distribution of the multienzyme systems in the various organs. Where the intracellular localization is at all known, the enzyme activities are generally found in the microsomes, small organelles within the cells which form part of and can be isolated from the so-called endoplasmic reticulum. Liver tissue is an excellent source for microsomes—and much of the present knowledge is derived from investigations on these preparations. Since some of these multienzyme systems play an important role in the hormonal metabolism one might expect to find some of the enzyme activities enriched in endocrine tissues like ovaries, testes, hypophysis, and the adrenal glands. This has been demonstrated for a number of mixed function oxidases, including the one presented here. An interesting observation is that in endocrine tissues mixed function oxidase activities are found not only in the microsomes but also in mitochondrial preparations.

It has been shown—or is in other cases believed—that the overall enzymatic activities of many mixed function oxidases depend on the function of a common specific component with properties of a somewhat atypical heme protein. Demonstration of such a role stems mostly from studies of the CO-inhibition of the overall enzymatic activities, the light-dependent reversal of this inibition and, where measured, the detailed characteristics of the action spectrum for this de-inhibition which allows the identification of the responsible pigment in a manner identical to that used by Warburg in his classical studies on cytochrome oxidase.

The property of a light-reversible CO-inhibition of the C-21-hydroxylation of steroids was first observed by Ryan and Engel (4) in 1957 but a real interest in these multienzyme systems began to develop only a few years ago. Using the special spectrophotometric techniques developed for measurements on highly light-scattering preparations Garfinkel (5) and Klingenberg (6) found independently in 1958 the occurrence of a new optical absorption band in CO-difference spectra of liver microsome preparation. Cooper and his colleagues (7), in their subsequent extensive and detailed studies of the steroid-21-hydroxylase reaction, determined not only the overall stoichiometry of these reactions but established that the CO-combining pigment found in several of these multienzyme systems was to be considered the "terminal oxidase" for these reactions and was functionally involved in the mechanism of insertion of oxygen into the substrate molecules.

The CO-combining heme pigment, the common component thus of a variety of mixed function oxidases, was named "Cytochrome P-450" by Omura and Sato (8). This terminology is based on the difference spectroscopy (reduced pigment + CO − reduced pigment) of the CO compound which is formed after reduction. In reality this term is a misnomer since the pigment has an absorption maximum at 417 nm in the ferric form, a maximum at 419 nm in the ferrous form, and no absorption maximum at all at 450 nm (cf. Fig. 3). In most aspects this heme protein is not a typical cytochrome either but resembles in many of its properties more closely the hemoglobin–myoglobin series and in some of its reactions with the substrate molecules the compounds I and II, respectively, of peroxidase systems.

The steroid hydroxylase system

The adrenocortical 11β-steroid hydroxylase is not found in microsomes but is located exclusively in adrenal mitochondria. This multienzyme system is specific towards its substrate as well as the type of process it catalyzes: A hydroxyl group is stereospecifically inserted in the 11β-position of ring C of deoxycorticosterone. The reaction which occurs under physiological conditions in humans and mammals is shown in Fig. 1.

The overall stoichiometry has been determined. Both deoxycorticosterone and molecular oxygen are required for the reaction, and the main if not the only physiological electron donor is TPNH. The fate of the oxygen molecule has been studied by isotope methods in a number of related multienzyme systems containing P 450 as active component (9). It has been demonstrated that only one of the two atoms is inserted into the steroid molecule while the other one, not directly accounted for in these experiments, appears as water. Extensive studies have shown a rather high rate of substrate metabolism under physiological conditions. Typical turnover rates for deoxycortico-sterone are of the order of 9 moles of substrate stereospecifically hydroxylated per mole of P 450 per minute.

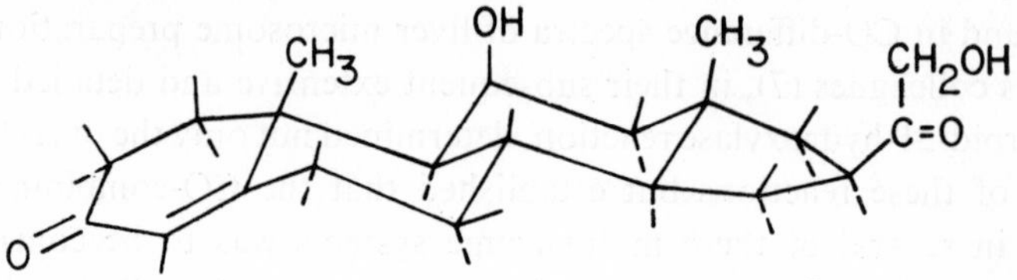

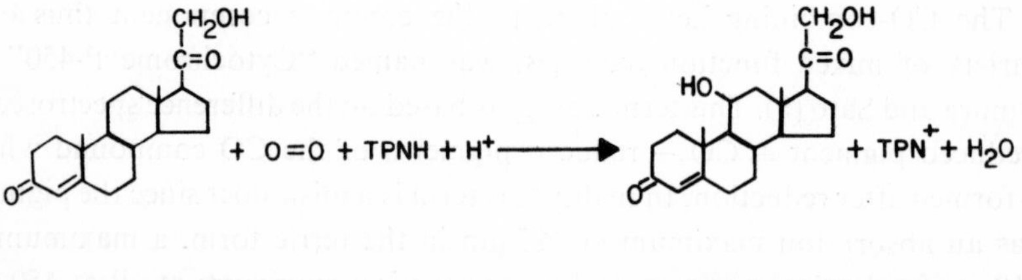

Deoxycorticosterone (DOC)   **Corticosterone (Cpd. B)**

Figure 1   The adrenocortical steroid 11β-hydroxylase reaction. Stoichiometry of the reaction and stereochemistry of the product

In previous studies the following electron transfer components of this multienzyme system have been recognized and suggested as representative of the pathway of electrons from the physiological electron donor to substrate and P 450:

$$\text{TPNH} \rightarrow \text{flavoprotein} \rightarrow \text{iron sulfur protein} \rightarrow \text{P 450}$$

In this electron transport chain the flavoprotein bestows the specificity to-wards TPNH under physiological conditions (DPNH can act, too, but only at higher concentrations and with lesser rates) and acts in turn as a specific reductase of the iron sulfur protein (non-heme iron protein). The systematic designation of this component of the multienzyme system should therefore be adrenocortical TPNH–iron sulfur protein reductase*. The heme protein

---

* The corresponding enzymes in microsomes are often called TPNH-cytochrome c reductase because of the use of cytochrome c as an artificial electron acceptor in the measurements; the natural acceptor in microsomes is still unknown.

P 450 is believed to be responsible for the activation of the oxygen molecule and might be involved in binding and, perhaps, even activation of the substrate molecule.

Sensitive optical techniques for absorption measurements on light-scattering materials made it possible to monitor the adrenocortical 11β-hydroxylase system in isolated adrenal mitochondria or tissues. Upon treatment with substrates, with a number of inhibitors of steroid metabolism, or with other chemical agents (for example nitrogeneous bases) optical absorption changes were detected by difference spectroscopy (10) but the effects are quite small and uncharacteristic, a fact which is not generally recognized in the literature. At present these difference spectra are of little help in the evaluation of the underlying chemical and enzymatic processes. Some may in fact represent changes in the environment of the P 450 molecule which are of the nature of "solvent effects" rather than the result of a direct interaction of the introduced agent with the heme group. There is a surprising lack of useful direct spectral information on the components of the electron transport chain which is understandable in view of the rather uncharacteristic spectral absorption properties of the heme protein P 450 in both its ferric and ferrous forms, and of the reduced and oxidized forms of the iron sulfur protein.

EPR spectroscopy is applicable to adrenal mitochondria or tissue homogenates under appropriate conditions and allows one to measure the heme protein P 450 and the iron sulfur protein of the adrenocortical 11β-hydroxylase system via their characteristic and well documented resonance absorptions. The half-reduced form of the flavoprotein should, at least in principle, be accessible by this technique although no results have been obtained as yet.

A major breakthrough in these studies was made when efforts in our laboratory led to the resolution of the adrenocortical 11β-hydroxylase system into its three main components of the electron transport chain by preparative enzymatic techniques (11). This enabled us to make full use of magnetic resonance spectroscopy in combination with other more convenient methods to study this and related multienzyme systems.

In the resolution procedure the following materials became available on a preparative scale:

(1) a reducing system with the reaction-specific flavin moiety of the TPNH–iron sulfur protein reductase,

(2) a fraction containing the iron sulfur protein component, and

(3) a heme fraction containing P 450 in a solubilized partially purified form.

Fractions (1) and (2) have been obtained in soluble form, the latter in highly purified and even crystallized form (12).

The adrenocortical 11β-hydroxylase is the only mixed function oxidase from mitochondria for which such a resolution has been achieved as yet. A somewhat related separation of a bacterial mixed function oxidase from Pseudomonas putida containing a soluble bacterial P 450 has been performed by Gunsalus and his colleagues (13) and Coon *et al.* (14) reported recently a partial separation of individual components of a microsomal mixed function oxidase system.

These resolution procedures made the separated components of the electron transport chain available for individual studies. It is of even greater importance that we were then able to reconstitute a fully reactive stereospecific steroid 11β-hydroxylase on recombination of these fractions in appropriate proportions which is capable of hydroxylating deoxycorticosterone in the presence of TPNH and O_2 (15). The maximal activities achieved in the reconstituted system are closely approaching those observed in solated adrenal mitochondria.

This proves that in the resolution experiments no severe damage or loss of functional activity has occurred. Detailed comparison of the EPR spectra of heme and iron sulfur protein from the resolved fractions and the reconstituted system with those of intact adrenal mitochondria provides additional evidence that no substantial alterations have resulted.

Redox studies on the resolved components

Availability of the electron transfer components in separate preparations made a detailed study of their redox properties possible, individually and in various combinations up to the complexity of the fully reconstituted multi-enzyme system. These redox measurements were further prompted by earlier findings (16) which suggested an unusually low redox potential for the oxygen-activating heme protein P 450. Techniques used in these studies were those of sensitive spectrophotometry, where applicable, and of EPR spectroscopy in combination with anaerobic titration techniques.

So far little effort has been put into the FAD-containing TPNH–iron sulfur protein reductase, mainly due to the difficulties of obtaining the enzyme in sufficient quantities. Results of a reductive titration experiment under strictly oxygen free conditions with TPNH as electron source are shown in Fig. 2; the spectra show a clean isosbestic point near 500 nm. Anaerobic titrations with dithionite yield a more complicated set of spectral data. A

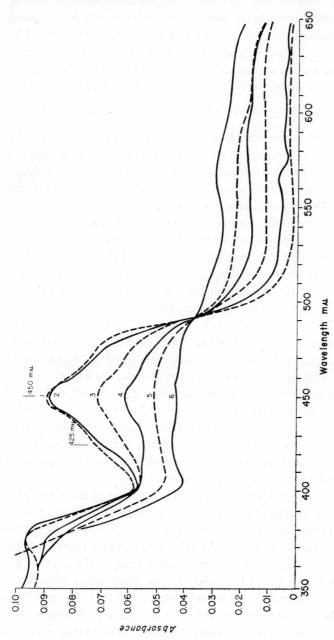

Figure 2 Reductive titration of TPNH–iron sulfur protein reductase. Flavoprotein concentration $8.0\,\mu\text{M}$; final volume 3.0 ml. Curve 1: oxidized enzyme in N_2 atmosphere; Curves 2–6: Spectra after consecutive additions of TPNH, 2.6 nmoles each, under strictly oxygen free conditions. Cary Model 14; pathlength 1.0 cm; temperature $+4°$.

preliminary analysis shows that the reduction of the flavin moiety occurs in two clearly discernible consecutive steps corresponding to the half-reduced and the fully reduced forms with proper one electron stoichiometry. The isosbestic point near 500 nm is lost as soon as the half reduction stage is reached indicating a more complex chemistry under these conditions. The following preliminary values for the redox potential have been measured:

$$E_0' (fp/fpH) = -250 \pm 10 \text{ mV}; \quad E_0' (fpH/fpH_2) = -345 \pm 5 \text{ mV}$$

Together with the redox potentials measured for the other components of the multienzyme system this would suggest that the reductase under physiological conditions is operative between the half reduced and the fully reduced form. This is reminiscent of the findings of Kamin and his colleagues (17) with the microsomal TPNH–cytochrome *c* reductase.

The iron sulfur protein (also termed "adrenodoxin") has been obtained in highly purified form by Suzuki and Kimura (12) and some of its basic properties have been measured. We have studied the redox behaviour of the iron sulfur protein preparation by following the bleaching of its optical absorption in the visible region of the spectrum and by the appearance of the EPR absorption characteristic for the reduced form of this enzyme. In a frozen solution it shows an anisotropic spectrum with nearly axial symmetry and the principal components $g_1 \approx g_2 = 1.94; g_3 = 2.01$ for the *g*-tensor in the spin Hamiltonian (cf. Fig. 7). Reductive titration with dithionite alone and in suitable mixtures with viologen dyes yields clear evidence for a reversible one electron transfer process with a measured value: $E_0' = -305 \pm 5$ mV at pH 7.0. This value differs markedly from recent values by Kimura (18) who lists adrenodoxin as a two electron transfer system with a redox potential of $+165$ mV. However, independent measurements by Suzuki, Estabrook and Cooper (19) have yielded values essentially identical with the ones reported here. The reason for this discrepancy is unknown.

The most fascinating but least understood component of the steroid 11β-hydroxylase and other related mixed function oxidases is the heme protein P 450. Its optical absorption characteristics are unique. The ferric form does not show any similarities to any of the other known heme proteins. Likewise, the spectral characteristics of the reduced form of the enzyme are atypical and not very different from the oxidized form with a red shift of only 2–3 nm and a slight decrease in absorption coefficient. Figure 3 illustrates this with the absolute spectra of the first mammalian P 450 preparation that maintained functional activity and a virtually unaltered chemistry. Almost all other information available at present is derived from difference

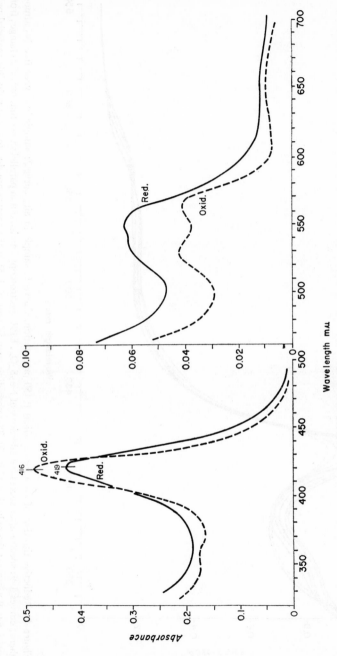

Figure 3 Absorption spectra of P 450. Heme concentration 3.2 μM, based on protoheme content. Dashed line: Ferric (Fe^{3+}) form; solid line: Ferrous (Fe^{2+}) form after reduction with Na$_2$S$_2$O$_4$

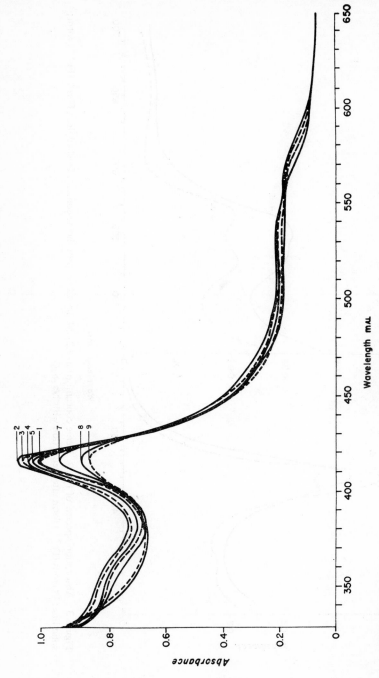

Figure 4 Reductive titration of an adrenocortical P 450 preparation. Curve 1: sample in air; curve 2: sample in oxygen free N_2 atmosphere; curves 3–9: spectra after successive additions of $Na_2S_2O_4$. EPR spectroscopy carried out in parallel shows that only the changes from curves 6 to 9 represent the reduction of P 450. The preceding changes (curves 1–5) in the spectra are due to other reducible material present in this particular preparation (cf. also Fig. 7, bottom trace, and Fig. 9)

spectroscopy and, in most cases, not of the free heme protein but of the CO-compound formed from the reduced heme.

Extensive redox studies with various reducing agents with and without added mediators as well as electrochemical reduction to determine half wave potentials were performed. An example of such an experiment is shown in Fig. 4 (cf. also the spectra in Figs. 7–9). Under conditions of strict oxygen exclusion—a factor which is crucial in these measurements (cf. Fig. 10)—we have obtained values of $E_0' = -400 \pm 10$ mV for the Fe^{3+}/Fe^{2+}-system of P 450 from the adrenocortical 11β-hydroxylase. This value agrees very well with our EPR results on adrenal mitochondria and the fully active reconstituted hydroxylase system (16). It should be mentioned that Waterman and Mason (20) found a rather low redox potential value for P 450 in a rabbit liver microsome preparation.

Our redox potential measurements on the isolated components of the resolved hydroxylase system are summarized in Fig. 5. They fall into two groups centered around -320 mV and -400 mV, respectively. The values

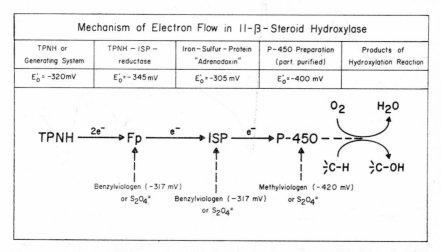

Figure 5 The electron transport system of adenocortical steroid-11β-hydroxylase

for the $TPN^+/TPNH$-couple, the reductance (fpH$/f$pH$_2$), and the iron sulfur protein (oxid./red.) are well matched. There is, however, a large gap in redox potentials between this group of components and the heme protein P 450. This explains why both flavoprotein and the iron sulfur protein are found maximally reduced in the presence of a large excess of TPNH while only a small fraction of the P 450 (20–30% under the conditions of anaerobic re-constitution experiments) is reduced. (A larger fraction can be reduced if an excess of CO is present in the anaerobic atmosphere.) As indicated electrons can be fed into the electron transport chain by artificial donors in efforts to reduce P 450 from its Fe^{3+}- to its Fe^{2+}-form.

EPR spectroscopy and the P 450 molecule

At present EPR spectroscopy provides the most valuable experimental approach to a study of the P 450 molecule. The availability of a partially purified active P 450 preparation makes these studies meaningful; it allows a much more detailed analysis of lineshape and other parameters of the EPR spectra than would be possible on the level of mitochondria, microsomes, or tissues. In all previous attempts to isolate and purify P 450 the molecule was invariably modified, leading most frequently to the formation of the so-

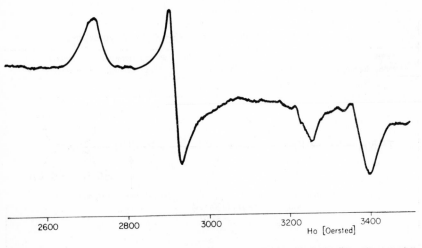

Figure 6 EPR spectrum of P 450. X-band spectrum at 77°K. Sample in 3 mm quartz tube (anaerobic cuvette), immersed in liquid nitrogen

called P 420 which has entirely different chemical properties (21) and is enzymatically inactive.

The only major paramagnetic species detected in a purified preparation (22) at magnetic fields up to 5000 Oersted in X-band is P 450 (Fig. 6). It has the characteristic properties of a low spin ($S = \frac{1}{2}$) heme protein with less than axial (i.e. rhombic) symmetry. The principal components of the g-tensor a 1.91, 2.24, and 2.42 for this preparation—values which are identical with those measured by us in isolated adrenal mitochondria and in the reconstituted hydroxylase system. The values are nearly identical to those found in microsomes by Mason *et al.* (23), who first reported this type of a spectrum as "Fe_x", and by many others. This supports the view that microsomal P 450 is indeed basically similar if not identical to the one studied by us in the steroid 11β-hydroxylase.

In a comparison with other available spectra of low spin heme proteins of our own and the literature these results show that P 450 exhibits by far the "smallest" anisotropy. All the available information, including chemical model studies and spectroscopic observations, are consistent with and suggest strongly that the ligands in the 5- and 6-position of ferric P 450 are nitrogen (like in histidine) and sulfur (as in a mercaptide), respectively. (A similar view has been expressed at this conference by Blumberg (24).) Modification products like P 420 do not have these properties and possess most likely a different ligand composition.

Additional information about the P 450 molecule can be obtained by EPR studies on preparations in which the purification has not been carried quite so far. This is best shown with a material in which a small amount of the iron sulfur protein is left behind (by omitting a chromatography step). In Fig. 7 the spectra of such a preparation under aerobic conditions and after nearly complete reduction with dithionite under oxygen free conditions are compared. The characteristic spectral features of the individual components present are given in the lower part; the comparison clearly illustrates the possibility of simultaneous quantitative measurements of P 450 and iron sulfur protein.

The lowest spectrum of Fig. 7 is that of a still unidentified species which is commonly seen in a variety of biological materials ranging from protozoa and bacteria to mammalian mitochondria. It has a much higher redox potential than any of the components of the 11β-hydroxylase system and can be readily reduced. No further paramagnetic species are detectable at either lower or higher fields. There is especially no evidence for the existence of any high spin heme protein with resonance absorption maxima near $g \sim 6$, a fact which has been claimed in various mixed function oxidases recently. There is also no detectable $g \sim 4.28$ resonance absorption due to rhombic Fe^{3+}.

Figure 8 shows the results of a quantitative reductive titration of the same P 450 preparation. As the titration proceeds the characteristic spectrum of the reduced iron sulfur protein develops and then the EPR absorption of

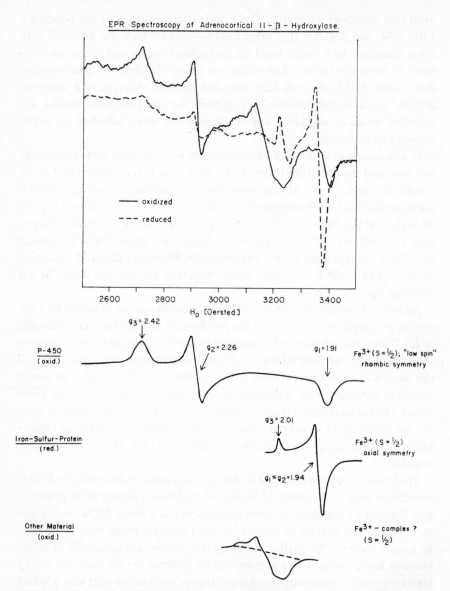

Figure 7   EPR spectroscopy of adrenocortical steroid 11β-hydroxylase. Partially purified P 450 preparation. X-band spectra at 77°K. The spectra of the detected components are shown separately, with their characteristic parameters, in the lower part of the figure

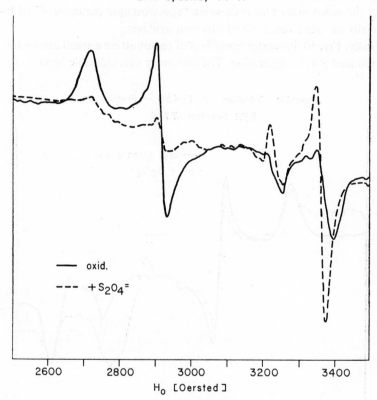

Anaerobic Titration of P-450 — Preparation.
EPR Spectra; 77°K

— oxid.

- - - +S₂O₄⁼

2600 2800 3000 3200 3400

H_0 [Oersted]

Figure 8 Reductive titration of a P 450 preparation. X-band spectra at 77°K. Heme concentration 107 μM. Reductant: $Na_2S_2O_4$. Solid line: Oxidized P 450; dashed line: after nearly complete reduction with $Na_2S_2O_4$. For descriptive details see text

P 450 disappears, with an overall stoichiometry of close to one electron equivalent per heme group.

A clean analytical separation of P 450 and the species responsible for the broad resonance absorption near $g \approx 2$ (cf. bottom trace of Fig. 7) can be readily achieved due to the large differences in redox potentials. In the presence of the oxidized form of this paramagnetic species the heme resonances are found to be broadened (cf. Fig. 9); this correlates with a corresponding decrease of the signal amplitudes in the first derivative spectra, but no loss of absolute integrated intensity of the P 450 absorption is observed. This observation is important in connection with the practice in biochemistry to determine concentrations on the basis of amplitude measurements. Effects such as these may well have led to some of the controversial statements about the redox state and even some "spectroscopic parameters" of P 450 especially in microsomal mixed function oxidases.

Finally, Fig. 10 illustrates the effects of re-admitting a small amount of air to a reduced P 450 preparation. The extremely autoxidizable heme is almost

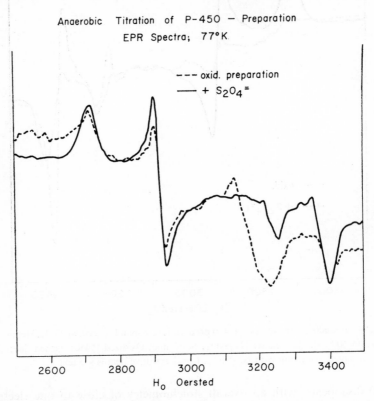

Figure 9 Reductive titration of a P 450 preparation. Conditions as in Fig. 8; for details see text. Dashed line: spectrum of the preparation in air; solid line: spectrum of the sample after addition of $Na_2S_2O_4$ equivalent to a concentration of 126 μM. Concentration of P 450 (still in Fe^{3+}-form) 107 μM

fully oxidized to the low spin ferric form while the iron sulfur protein is still maintained in the reduced EPR active form under the conditions of this experiment. No evidence for any intermediate state between ferrous and ferric heme P 450 has been found as yet under these conditions.

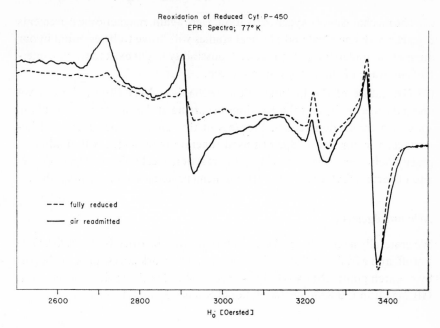

Figure 10 Effects of reoxidation on the EPR spectrum of a P 450 preparation. Conditions as in Fig. 8; for details see text

Outlook

Using the possibilities offered by the resolution and reconstitution of the steroid 11β-hydroxylase we have shown the present state of knowledge of the nature and reactions of mixed function oxidases. Application of magnetic resonance techniques to these complex systems presents a very challenging problem but at the same time one of the few promising approaches available. As pointed out, many questions are still left open and much further work will be needed before we can understand in detail the processes and mechanisms involved.

Three lines of further development in the field of mixed function oxidases are apparent—and in each of them the use of EPR spectroscopy will play an indispensible role:

(1) The nature of the P 450 molecule (and related compounds such as P 420). The success of this approach relies heavily on the availability of at least partially purified materials in place of the composite multienzyme systems, as described here.

(2) The mechanism of oxygen activation. Here again, magnetic measurements together with chemical and physical studies will prove to be essential in our attempts to understand the process. A possibility might exist to trap an inter-mediate stage of the reduction of oxygen.

(3) The nature of the hydroxylation mechanism. Very little is known yet about the mechanisms of the reactions occurring at the substrate molecule, in spite of a tremendous effort. Progress in this area will be of extreme im-portance for our knowledge of mixed function oxidases; it will in addition greatly help our understanding of the reaction mechanisms of peroxidases and of certain features of O_2-carrying hemeproteins such as hemoglobin.

Acknowledgement

We gratefully acknowledge the assistance of Mrs. Beatrice Novack, Dr. Olga Foroff, and Mr. Acie Slide in these studies. The work was supported in part by research grant AM 04484 of the U.S. Public Health Service and by grants GB 2451 and GB 8366 of the National Science Foundation.

References

1   D.Y.Cooper, H.Schleyer, R.W.Estabrook, and O.Rosenthal, *International Endo-crinological Congress*, Mexico City, July 1968 (in press).
2   H.S.Mason, *Advances in Enzymology* **19**, 79 (1957).
3   O.Hayashi, *Oxygenases* (ed. O.Hayashi), Academic Press, New York, 1962.
4   K.Ryan and L.Engel, *J. Biol. Chem.* **225**, 103 (1957).
5   D.Garfinkel, *Arch. Biochem. Biophys.* **77**, 493 (1958).
6   M.Klingenberg, *Arch. Biochem. Biophys.* **75**, 376 (1958).
7   D.Y.Cooper, R.W.Estabrook, and O.Rosenthal, *J. Biol. Chem.* **238**, 1320 (1963).
8   T.Omura and R.Sato, *J. Biol. Chem.* **237** PC 1375.
9   M.Hayano, M.C.Lindberg, R.I.Dorfman, J.E.H.Hancock, and W.E. von Doering, *Arch. Biochem. Biophys.* **59**, 529 (1955).
10 H.Remmer, J.Schenkman, R.W.Estabrook, H.Sasame, J.Gillette, S.Narasimhulu, D.Y.Cooper, and O.Rosenthal, *Mol. Pharmacol.* **2**, 187 (1966).
11 T.Omura, R.Sato, D.Y.Cooper, O.Rosenthal, and R.W.Estabrook, *Fed. Proc.* **24**, 1181 (1965).
12 K.Suzuki and T.Kimura, *Biochem. Biophys. Res. Comm.* **19**, 340 (1965).
13 I.C.Gunsalus, *Z. physiol. Chem.* **349**, 1610 (1968).
14 A.Y.H.Liu, K.W.Junk, and M.J.Coon, *J. Biol. Chem.* **244**, 3714 (1969).
15 D.Y.Cooper, S.Narasimhulu, O.Rosenthal, and R.W.Estabrook, *Advances in Chem-istry Series No. 77, Oxidation of Organic Compounds III*, p. 220 (1968).
16 D.Y.Cooper, H.Schleyer, and O.Rosenthal, *Z. physiol. Chem.* **349**, 1592 (1968).
17 H.Kamin, B.S.S.Masters, Q.H.Gibson, and C.H.Williams, jr., *Fed. Proc.* **24**, 1164 (1965).

18 T. Kimura, in *Structure and Bonding*, Vol. 5, Springer-Verlag, New York.
19 K. Suzuki, R. W. Estabrook, and D. Y. Cooper (in preparation).
20 M. Waterman and H. S. Mason, private communication.
21 T. Omura, and R. Sato, *J. Biol. Chem.* **239**, 2370, 2379 (1964).
22 D. Y. Cooper *et al.* (manuscript in preparation).
23 Y. Hashimoto, T. Yamano, and H. S. Mason, *J. Biol. Chem.* **237**, PC 3843 (1962).
24 W. E. Blumberg and J. Peisach, this book, p. 67.

Nuclear magnetic resonance studies of inhibitor and substrate association with lysozyme

M. A. RAFTERY and S. M. PARSONS

Gates and Crellin Laboratories of Chemistry*
and Church Laboratory of Chemical Biology
California Institute of Technology, Pasadena, California

THE THREE DIMENSIONAL STRUCTURE determination of lysozyme by X-ray analysis methods (1) followed by further structural studies of inhibitor complexes with the enzyme (2) has provided a detailed account of atomic interactions for association of various oligosaccharides with the crystalline enzyme. A groove extending down one side of the macromolecule has been shown to contain a series of six subsites which are capable of interacting with pyranoside rings. This is depicted in Fig. 1. Based on the X-ray analysis findings for association of NAG,[†] NAG—NAG, NAG—NAG—NAG, NAM, and NAG—NAM the scheme for relative binding modes shown in Fig. 1 has been proposed for these saccharides. In addition, a proposal for the location of the catalytic site of the enzyme, between sites D and E has been put forward. As a result of this, hydrolysis of larger oligomers of NAG or NAG—NAM would proceed as indicated in the figure.

Recent chemical evidence (3, 4) has substantiated the proposal for the location of the catalytic site and the mechanism for lysozyme catalysed hydro-

---

* Contribution No. 3997.

[†] NAG, N-acetyl-D-glucosamine; NAG—NAG, chitobiose; NAG—NAG—NAG, chitotriose; NAM, N-acetyl-D-muramic acid; NAG—NAM, N-acetyl-4-O-(2-acetamido-2-deoxy-β-D-glucopyranosyl)-D-muramic acid.

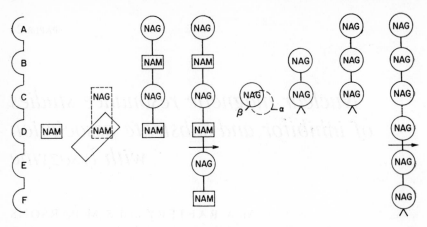

Figure 1 Scheme for the binding modes of several saccharides based on X-ray crystallographic analysis

lysis of a selected substrate has been shown to involve an enzyme bound carbonium ion. This essay describes the use of nuclear magnetic resonance to study the association with lysozyme of a variety of inhibitors and substrates, including a substrate whose hydrolysis by the enzyme has been shown to proceed through a carbonium ion intermediate (12).

The n.m.r. method which we have used to study the association of inhibitors and substrates with lysozyme has been detailed in several recent publications (5–9). Advantage is taken of the fact that the rates of formation and dissociation of enzyme complexes of the inhibitors and substrates are rapid. When nuclei in the associating small molecules experience a change in environment on becoming bound to the enzyme, chemical shift changes occur. In the fast exchange limit the observed spectrum of the small molecule represents a weighted average of enzyme bound and free species. It has been shown that for the equilibrium

$$E + I \rightleftharpoons EI$$

$$I_0 = E_0 \frac{\Delta}{\delta} - K_I \tag{1}$$

where I_0 and E_0 represent the total concentrations of inhibitor and enzyme, respectively, δ is the observed chemical shift of a nucleus in the associating small molecule (taking its chemical shift in the absence of enzyme as zero) and Δ represents the chemical shift of the same nucleus in the enzyme bound state. Thus the parameters K_I (the dissociation constant) and Δ can be evaluated. This gives a measure of the magnetic environment experienced by a nucleus of the small molecule when it is associated with the enzyme.

Previous studies, employing ultraviolet spectroscopic methods, on the association of β-(1-4)-linked oligosaccharides of 2-acetamido-2-deoxy-D-glucopyranose (20) have allowed calculation of the dissociation constants for the monosaccharide through the hexasaccharide in association with the enzyme. In that study it was shown that binding strength increased with increasing chain length up to the trisaccharide but that the tetra-, penta-, and hexasaccharides did not appear to bind any more strongly than did the trisaccharide. These results indicated that lysozyme contains three contiguous subsites to which acetamidopyranose rings bind strongly. This interpretation is in agreement with findings (2) employing X-ray analysis techniques to study the association of crystalline lysozyme with 2-acetamido-2-deoxy-D-glucopyranose, chitobiose, and, chiotriose. Although it could be shown by such an approach that 1:1 complexes were formed between the enzyme and the various monosaccharide inhibitors, it was not possible to relate this to association with only one of the three contiguous subsites rather than to multiple equilibria with all three sites. If this latter possibility were the case the dissociation constants and chemical shifts obtained would represent complex entities. The present study was undertaken in an attempt to resolve this question as well as that of the relative ways in which the mono-, di-, and trisaccharides of 2-acetamido-2-deoxy-D-glucopyranose and their methyl glycosides bind to the enzyme (Fig. 2).

Chemical shift parameters for the acetamido methyl protons of NAG and the β-(1-4)-linked di., tri-, and tetrasaccharides derived from it are shown in Table 1. In each case the resonance to highest field corresponds to the acetamido methyl group at the reducing end of the molecule. Similar parameters for the methyl-β-glycosides are given in the table and, in addition, the chemical shifts of the glycosidic methyl groups are shown.

In Table 2 the chemical shift changes obtained for various inhibitors upon complexation with lysozyme under various conditions are shown. It is evident that for all the saccharides with free reducing groups the acetamido methyl resonance at the reducing end of the molecule undergoes a large chemical shift to higher field upon binding to the enzyme. The methyl-β-glycosides upon complex formation are seen to display similar chemical shift changes in only the acetamido methyl resonances proximal to the glycosidic methyl group. In addition, the glycosidic methyl group resonances of all glycosides undergo a smaller chemical shift to lower field in the presence of the enzyme.

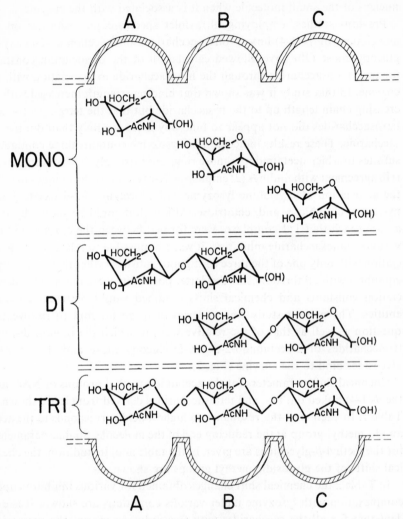

Figure 2 Possible binding modes of NAG oligomers to three contiguous subsites of lysozyme in solution

Table 1 Chemical shift data (apparent maxima) for methyl groups
in chitin oligosaccharides and their methyl glycosides

Inhibitor	Resonance[a]				
	CH_3—N_1[b]	CH_3—N_2[b]	CH_3—N_3[b]	CH_3—N_4[b]	OCH_3[c]
2-Acetamido-2-deoxy-D-glucopyranose	17.71				
Methyl 2-acetamido-2-deoxy-β-D-glucopyranoside	18.79				14.72
Chitobiose	18.35	15.63			
Methyl β-chitobioside	19.51	15.33			13.94
Chitotriose	18.32	16.28	15.83		
Methyl β-chitotrioside	19.53	16.28	15.90		13.33
Chitotetraose	18.35	16.54	16.54	15.98	

[a] The acetamido methyl groups are numbered 1, 2, 3, and 4 beginning at the reducing or glycosidic termini of the molecules. All chemical shifts are in Hz at 100 MHz. [b] Values relative to acetone; all chemical shifts to higher field. [c] Values relative to methanol; all chemical shifts to lower field.

This evidence shows that the reducing termini of all free acetamido sugars studied and the glycosidic termini of all the glycosides studied occupy the same magnetic environment on the enzyme. A binding scheme consistent with these results is shown in Fig. 3. All saccharides are shown to occupy the same subsite (subsite *C*) with other residues occupying subsites *A* and *B*. From the data in Table 2 it is possible to assign magnetic parameters to these three subsites. Acetamido methyl resonances in subsite *C* are shown to undergo chemical shifts of 0.5–0.8 ppm to higher field, while methoxyl resonances in the same subsite undergo down field shifts of approximately 0.2 ppm. Acetamido methyl groups in subsite *B* undergo no change in chemical shift while those occupying subsite *A* are shifted slightly to lower field (~ 0.08 ppm).

In addition to studies of binding of oligosaccharides and glycosides containing only NAG residues substrates which contain one NAG and one glucose residue have been studied. Compounds 14 and 15 in Fig. 3 are in this category. These compounds have been shown to be hydrolysed specifically at the glucosidic bond (11) by a carbonium ion mechanism (12, 13). In view of the suggested location of the catalytic site between subsites *D* and *E* (2) from X-ray analysis evidence and subsequent chemical modification studies which confirm this suggestion (3, 4) it was of interest to subject these compounds to study by the n.m.r. method. It was found that the acetamido methyl group underwent the anticipated chemical shift due to binding in

Table 2 Chemical shift data for inhibitors and substrates complexed with lysozyme at various pH and temperature values

Compound	Temp. (°C)	pH	Δ (ppm)[a]			
			CH$_3$—N$_1$[b]	CH$_3$—N$_2$[b]	CH$_3$N$_3$[b]	—OCH$_3$[c]
Methyl-β-NAG	31	4.9–5.4	0.54 ± 0.04	—	—	0.17 ± 0.03
	55	4.9–5.4	0.51 ± 0.03	—	—	0.16 ± 0.05
Chitobiose	45	4.9–5.4	0.57 ± 0.04	0	—	—
Methyl-β-chitobiose	35	4.9–5.4	0.60 ± 0.05	0	—	0.20 ± 0.05
Methyl-β-NAG	31	9.7	0.36	—	—	0.61 ± 0.02
Chitobiose	55	9.7	0.77 ± 0.04	0	—	—
Methyl-β-chitobioside	55	9.7	0.80 ± 0.04	0	—	0.61 ± 0.02
Chitotriose	65	9.7	0.61[d] ± 0.12	0	0.08	—
Methyl-β-chitotrioside	65	9.7	0.63[d]	0	0.08	0.19

[a] The acetamido methyl groups are numbered 1, 2, 3 beginning at the reducing or glycosidic termini of the inhibitor molecules.
[b] Values relative to acetone; all chemical shifts to higher field.
[c] Values relative to methanol; all chemical shifts to lower field.
[d] Not at fast exchange limit.

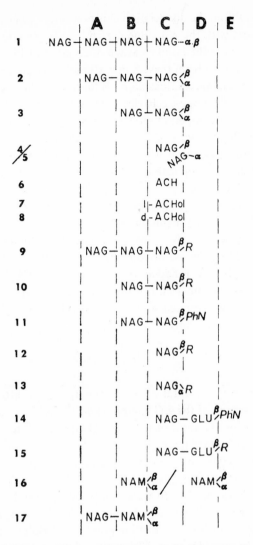

Figure 3 Scheme for relative modes of association with lysozyme of various saccharide inhibitors and substrates. Where α- and β-anomeric forms are indicated on a single line, no information on relative binding modes was obtained. Where α- and β-forms are depicted separately (as with α-NAG and β-NAG), different binding modes were elucidated. Where α- and β-forms are shown on the same molecule on two levels, both anomeric forms bind identically. Methyl groups are depicted by *R*, nitrophenyl groups by *PhN*

subsite C. In addition, the dissociation constant obtained ($K_S = 2 \times 10^{-2}$ M) was in agreement with the K_M value determined by kinetic studies of lysozyme catalysed glycosidic bond cleavage of compound 14.

The results obtained from these n.m.r. studies of complexation of the enzyme with a series of saccharides allow definition of their relative modes of association. Since the monosaccharide NAG was shown to bind only to subsite C, albeit in two competing modes depending on anomeric form (5, 6), it was of interest to attempt to define the specific features of the sugar which are necessary for association with this subsite. To that end it was found desirable to extend the applicability of the n.m.r. method on studying compounds 6, 7, and 8 in Fig 3.

Direct chemical shift measurements are not accurate for weakly associating enzyme inhibitors or for inhibitors experiencing only a slighly different magnetic environment on the enzyme surface. These limitations can sometimes be alleviated by the use of competition measurements. If an inhibitor competitive for subsite C is added to a solution of β-methyl-NAG and lysozyme the observed chemical shift of the NAG residue will decrease (5). This behavior is easily quantitated and in the simplest case gives, after good approximations to linear form,

$$C_0 = \frac{K_C}{K_I} E_0 \, \Delta_I \, \frac{1}{\delta_I} - K_C - \frac{K_C}{K_I} I_0 \qquad (2)$$

where C_0 represents the total added competitor concentration, K_C is the dissociation constant of the EC complex and the other symbols have been defined previously. The ratio of the slope of equation 2 to that of a similar plot of I_0 vs. $1/\delta_I$ obtained under identical conditions but without added competing inhibitor is equal to K_C/K_I. Two such experiments are shown in Fig. 4.

Varying concentrations of d_3-acetamidocyclohexane or pure l-trans-2-d_3-acetamidocyclohexanol (14, 15) were competed against NAG. The poorly binding l isomer was shown to be free of the much more strongly binding d isomer by the identities of its melting point and absolute rotation to those of demonstrably pure d isomer. The purity of the d isomer can be directly observed since lysozyme splits the resonances of the d and l forms, leaving the l resonance sharp and nearly unshifted. Thus Fig. 4 shows that acetamidocyclohexane and l-2-trans-acetamidocyclohexanol are equally poor competitors for subsite C binding and have dissociation constants of $12.6 \pm 2.7 \times 10^{-2}$ molar and $15.7 \pm 2.4 \times 10^{-2}$ molar, respectively.

Not determined thus far is Δ_I. A plot of I_0 vs. $1/\delta_I$ for inhibitors which undergo small poorly measured shifts will not determine Δ_I adequately. How-

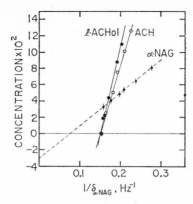

Figure 4 The concentration of d_3-acetamidocyclohexane (o) or l-*trans*-2-d_3-acetamido-cyclohexanol (●) *vs.* the inverse chemical shifts of the α anomer of 3.17×10^{-2} molar NAG in the presence of 3.2×10^{-3} molar lysozyme and 0.1 molar citrate pH 5.5, 31°. The dependence of I_0 *vs.* $1/\delta_I$ for αNAG (⧫) under the same conditions is also plotted. The β anomer is not shown

ever, by including the value of K_I obtained from the competition experiment as a data point on the I_0 axis, the slope, and thus Δ_I, is much better determined. Such a graph for pure *l-trans*-2-acetamidocyclohexanol, shown in Fig. 5, yields Δ_l equal to 0.22 ± 0.06 ppm. A similar plot for acetamido-cyclohexane gives Δ equal to 0.34 ± 0.07 ppm.

An analogous situation arises when direct measurements are made on the chemical shifts of two competitive molecules always held in equal concentration. Such situations are those of *dl* optical isomers and α and β anomers of some sugars. The equation governing this (6) may be expressed as

$$l_0 = \Delta_l E_0 \frac{K_{\overline{dl}}}{K_l} \frac{1}{\delta_l} - K_{\overline{dl}} - \frac{K_{dl}}{K_l} E_0 \qquad (3)$$

where d and l stand for the optical isomers and $K_{\overline{dl}}$ is a reduced dissociation constant for the *dl* mixture equal to $K_d K_l/(K_d + K_l)$. Provided that the last term in equation 3 is relatively small, a plot of the d and l inverse chemical shifts *vs.* their respective equal concentrations will have the same intercept in a manner identical with the α and β anomers of NAG (6).

For *dl-trans*-2-acetamidohexanol the intercept was well determined by the d isomer. After including it as a data point for the l isomer and utilizing the value of K_l found from the competition experiment, we calculate from equation 3 a Δ_l value of 0.23 ± 0.07 ppm (Fig. 5), in good agreement with the value determined from pure *l*. The *d* isomer, both in *dl* mixture and pure, gives a dissociation constant of $0.95 \pm 0.16 \times 10^{-2}$ molar and a Δ value of 0.91 ± 0.03 ppm.

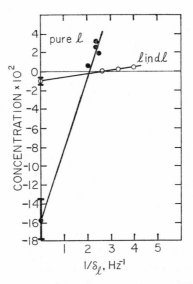

Figure 5 The concentration of *l-trans*-2-d$_3$-acetamidocyclohexanol *vs.* the inverse chemical shift of the *l* isomer in pure *l* (●) and in the *dl* mixture (○), 3.2×10^{-3} molar lysozyme and 0.1 molar citrate pH 5.5, 31°. The *d* isomer in the *dl* mixture is not shown. The ordinate data points with their errors were determined as described in the text

The similarity of acetamidocyclohexane and *l-trans*-2-acetamidocyclohexanol as well as the striking difference from *d-trans*-2-acetamidocyclohexanol in binding to subsite *C* is explained as follows. Providing that the ring conformations are similar, the hydroxyl group of one of the optical isomers will be related to lysozyme as the 1-OH in βNAG while that of the other isomer will correspond to the 3-OH of NAG. Thus we have a thermodynamically exact test of the relative importance of the β-1-OH and the 3-OH to NAG binding, subject to usual trepidations about models. Theoretical calculations (16) as well as empirical rotational correlations (17) indicate that the absolute configuration of the *d* optical isomer corresponds to the 3-OH of NAG. Our attempts to relate *d*-2-*trans*-acetamidocyclohexanol by oxidation to N-acetyl-α-aminoadipic acid of known absolute configuration were only partially successful, but the ORD curves of the product in the presence of cupric ion under mildly basic and acid conditions (18) indicated the same conclusion. Thus the 3-OH of NAG contributes a significant part of

Table 3 Data for binding of NAG various analogues to lysozyme

Inhibitor	$K_S(M)$	Δ(ppm)	
		$-NH-\overset{\overset{O}{\|\|}}{C}-CH_3$	OCH_3
β-NAG	$3.3\,(\pm0.2)\times10^{-2}$	0.51 ± 0.03	—
α-NAG	$1.6\,(\pm0.1)\times10^{-2}$	0.68 ± 0.03	—
β-Me-NAG	$3.3\,(\pm0.5)\times10^{-2}$	0.54 ± 0.04	0.17 ± 0.03
α-Me-NAG	$5.2\,(\pm0.4)\times10^{-2}$	0.55 ± 0.02	0
ACH	$12.6\,(\pm2.7)\times10^{-2}$	0.34 ± 0.07	—
AChol (l)	$15.7\,(\pm2.4)\times10^{-2}$	0.22 ± 0.06	—
AChol (d)	$0.55\,(\pm0.4)\times10^{-2}$	0.91 ± 0.03	—

subsite C binding energy while the β-1-OH contributes nothing, a conclusion in agreement with X-ray crystallographic evidence (2). Table 3 summarizes the results obtained with a series of NAG derivatives and related compounds.

In conclusion, the use of a simple application of n.m.r. to study the binding of substrates and inhibitors to lysozyme has shown that the binding properties of the enzyme in solution correspond in detail with these properties for crystalline enzyme as determined by X-ray analysis methods. The obvious conclusion is that since certain functional properties in both states are similar the structure of the enzyme should be similar in both states. In addition, the simplicity and rapidity of the n.m.r. technique allows study of enzyme substrate association (in cases where substrate hydrolysis is slow) and at the same time the method is sufficiently informative to yield a detailed analysis of the association process.

References

1 Blake, C.C.F., G.A.Mair, A.C.T.North, D.C.Phillips, and V.F.Sarma, *Proc. Roy. Soc. (London)* **B167**, 365 (1967).

2 Blake, C.C.F., L.N.Johnson, G.A.Mair, A.C.T.North, D.C.Phillips, and V.R.Sarma, *Proc. Roy. Soc. (London)* **B167**, 378 (1967).

3 Parsons, S.M., and M.A.Raftery, *Biochemistry* **8**, 4199 (1969).

4 Lin, T.Y., and D.E.Koshland, Jr., *J. Biol. Chem.* **244**, 505 (1969).

5 Raftery, M.A., F.W.Dahlquist, S.I.Chan, and S.M.Parsons, *J. Biol. Chem.* **243**, 4175 (1968).

6 Dahlquist, F.W., and M.A.Raftery, *Biochemistry* **7**, 3269 (1968).

7 Dahlquist, F.W. and M.A.Raftery, *Biochemistry* **7**, 3277 (1968).

8 Dahlquist, F.W., and M.A.Raftery, *Biochemistry* **8**, 713 (1969).

9 Raftery, M.A., F.W.Dahlquist, S.M.Parsons, and R.G.Wolcott, *Proc. Natl. Acad. Sci. U.S.* **62**, 44 (1969).

10 Dahlquist, F.W., L.Jao, and M.A.Raftery, *Proc. Natl. Acad. Sci. U.S.* **56**, 26 (1966).

11 Rand-Meir, T., F.W.Dahlquist, and M.A.Raftery, *Biochemistry* **8**, 4206 (1969).

12 Dahlquist, F.W., T.Rand-Meir, and M.A.Raftery, *Biochemistry* **8**, 4214 (1969).

13 Dahlquist, F.W., T.Rand-Meir, and M.A.Raftery, *Proc. Natl. Acad. Sci. U.S.* **61**, 1194 (1968).

14 Osterberg, A.E., and E.C.Kendall, *J. Amer. Chem. Soc.* **42**, 2616 (1920).

15 Godchot, M., and M.Mousseron, *Bull. Soc. Chim.* **51**, 1279 (1932).

16 Yamana, S., *Tetrahedron* **21**, 709 (1965).

17 Umezawa, S., T.Tsuchiya, and K.Tatsuta, *Bull. Chem. Soc. Japan* **39**, 1235 (1966).

18 Wellman, K.M., W.Mungall, T.G.Mecca, and C.R.Hare, *J. Amer. Chem. Soc.* **89**, 3647 (1967).

Studies on O_2^- in biological systems

R. C. BRAY

Chester Beatty Research Institute, Institute of Cancer Research:
Royal Cancer Hospital, Fulham Road, London, S.W.3

O_2^-, THE SUPEROXIDE free radical anion, has long been regarded as a potential intermediate in biological oxidations. This paper summarizes recent EPR work in which O_2^- has been conclusively demonstrated in one biochemical system and good evidence obtained for it in two more. Bray and Knowles (1968) observed an unidentified EPR signal during steady-state oxidation of xanthine by oxygen catalysed by xanthine oxidase at pH 10 (Fig. 1). They used the rapid-freezing technique of Bray (1961, 1964). The signal gas has $g_\perp$ 2.00 and $g_\parallel$ 2.08. Knowles, Gibson, Pick and Bray (1969) reported more detailed studies on the same system. These g-values, although consistent with O_2^-, do not provide conclusive evidence for the identification of the signal-giving species, since O_2^- is known to show g-values highly dependent on the nature of the surrounding matrix. Knowles *et al.* (1969) showed that the intensity of the signal was more dependent on the concentration of the oxygen than on that of the enzyme as would be expected for a free, as opposed to an enzyme-bound, radical species. They further succeeded in obtaining an identical EPR signal from a purely inorganic system: the reaction between periodate ions and hydrogen peroxide (Fig. 2). Although data obtained with this system helped them very greatly in identification of the signal as O_2^- it is unfortunate that the mechanism of this model reaction is not well known. A much better understood system for generating O_2^- is the pulse radiolysis of oxygen-saturated aqueous solutions, which has been extensively studied in a number of laboratories by means of ultraviolet spectroscopy. Recently, Nilsson, Pick, Bray and Fielden (1969a) combined pulse radiolysis and rapid-

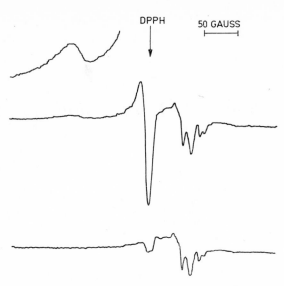

DPPH   50 GAUSS

Figure 1 EPR spectra at $-180\,^{\circ}C$ obtained on treating xanthine oxidase with oxygenated xanthine for 200 msec. at pH 10 and $20\,^{\circ}C$. *Centre:* sample prepared by the normal quenching method showing the O_2^- signal ($g_\perp$ 2.00, $g_{||}$ 2.08) together with molybdenum signals from the enzyme at higher fields; *top:* the same, at higher power and modulation, to show the $g_{||}$ component more clearly; *bottom:* the same sample after warming to about $-80\,^{\circ}C$ for 15 min; this treatment led to irreversible disappearance of the O_2^- signal but had little effect on the molybdenum signals. (Reproduced from Bray and Knowles, 1968)

freezing EPR and once again succeeded in obtaining an identical EPR signal (Fig. 2).

Despite this identity of the spectra in all the three systems which we have studied, it must be conceded that the spectrum is a relatively featureless one. Therefore, particularly in view of the confusing state of the literature on the stability of O_2^- (see below), we thought it desirable to carry out studies with oxygen-17, which were undertaken in collaboration with Dr. D. Samuel, Weitzmann Institute, Israel. Preliminary results are shown in Fig. 3 (Bray, Pick and Samuel, unpublished). Although all the hyperfine structure is not resolved, there is clear evidence for the presence of an 11-line spectrum with a splitting of 73 gauss, which could only arise from a di-nuclear oxygen radical. The spectrum is fully consistent with the much better resolved spectrum for O_2^- obtained by Tench and Holroyd (1968) using an entirely different system. They worked with O_2^- adsorbed on the surface of magnesium oxide and found A_{xx}: 77 gauss. The agreement between their value for the splitting and ours is quite surprisingly close considering how different the

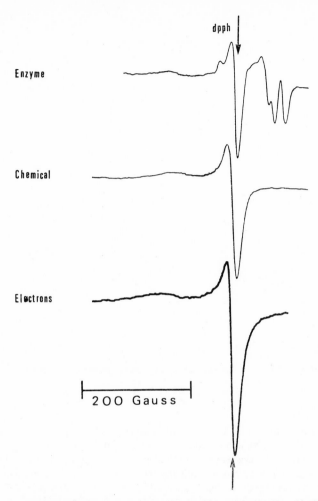

Figure 2 Comparison of EPR spectra of O_2^- obtained in various systems. In all cases rapid freezing was employed and the pH value was near to 10. Reaction times and the time from irradiation to freezing were all in the range 150–600 msec. Spectra were recorded at about $-170°$. *Top:* xanthine oxidase/xanthine/oxygen system (as in Fig. 1 but at a higher enzyme concentration); *centre:* reaction between $NaIO_4$ and H_2O_2; *bottom:* irradiation of oxygen-saturated $Ba(OH)_2$ with 4 MeV electrons. (Data from Knowles *et al.*, 1969, and from Nilsson *et al.*, 1969a)

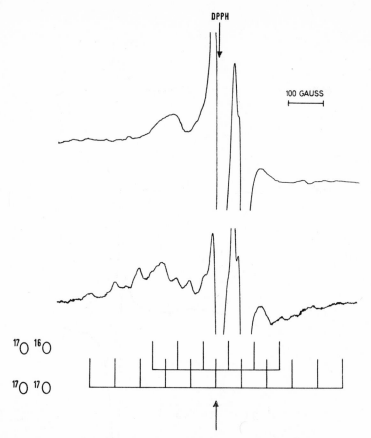

Figure 3 EPR spectra for O_2^- obtained with ordinary oxygen *(top)* and using oxygen enriched in ^{17}O *(bottom)*. Xanthine oxidase reduced with dithionite (Nilsson, Pick and Bray, 1969b) was treated with oxygen at pH 11 for 30 msec. and spectra were recorded at $-180°$. The enrichment of the oxygen was initially 90% but some slight contamination with ordinary oxygen may have occurred. The stick diagram corresponds to $A_\perp$: 73 gauss (Bray, Pick and Samuel, unpublished)

two systems are. The data thus seem to leave little room for doubt that the radical with $g_\perp$ 2.00, $g_{||}$ 2.08 produced by xanthine oxidase at pH 10 is indeed O_2^-.

The literature is confusing on the lifetime of O_2^- and is summarised in Table 1. Czapski and Dorfman (1964) measured a short lifetime for this radical in the neutral pH region and assumed that the decay constant they obtained applied in the alkaline region also. However, they also reported the presence of an unidentified long-lived species in the alkaline pH range, which

Table 1 Stability of O_2^- and related species

System	pH	Decay $t_{\frac{1}{2}}$ (for 10^{-5} M)	Order	References
Pulse radiolysis (U.V.)	5–7	0.006 sec.	2nd	Czapski and
	10	$\sim 0.2^a$	1st	Dorfman
	13	$\sim 5^a$	1st	1964
$XO_{red} + O_2$ (EPR)	10	$(0.01–0.2)^b$	—	Knowles
$IO_4^- + H_2O_2$ (EPR)	10	$(\gg 0.01)^c$	—	*et al.* 1969
Pulse radiolysis (EPR)	11.4	0.25	1st	Fielden *et al.*
	12.2	~ 0.7	—	(unpub.)

^{a} "Unidentified" species; spectrum same as O_2^-.

^{b} Lifetime not studied directly.

^{c} The yield (which was presumably governed by the lifetime) increased with pH in range 10–13.

showed first order decay kinetics and which (rather suspiciously) had the same U.V. spectrum as O_2^- itself. From the data with xanthine oxidase it is possible to put the halflife of the radical as observed by Knowles *et al.* (1969) in the range of about 10–200 msec. at pH 10. The lower limit is set by the limitations of the freezing method, since the time taken to freeze the samples and cool them to below $-80°$ is likely to be of the order of 10 msec. The upper limit of about 200 msec. is set by the rate at which the signal disappears when oxygen becomes exhausted from the system (Knowles, Pick and Bray, unpublished). These results seem incompatible with the short lifetime for O_2^- given by Czapski and Dorfman (1964). Knowles *et al.* (1969) also showed that the yield of radical in the periodate/hydrogen peroxide system increased with increasing pH. They assumed that in this system the yield was governed by radical lifetime. Their final conclusion was that the radical they detected by EPR must be identical with Czapski and Dorfman's long-lived species. Czapski and Dorfman had proposed various possible structures for this species but none of them seemed compatible with the EPR data. Knowles *et al.* (1969) therefore concluded that the long-lived species was O_2^- itself, and that this has a pH-dependent lifetime in the alkaline region. They proposed a mechanism to explain this. More recently, Fielden, Pick and Bray (unpublished) have carried out direct measurements on the lifetime of the radical using pulse radiolysis with rapid-freezing EPR and their results are fully consistent with the earlier conclusions.

Having established that xanthine oxidase does reduce oxygen to O_2^- and that at pH 10 this is stable enough to be trapped and detected, it was of interest to see if the method could be applied to other systems alos. The work of Komai, Massey and Palmer (1969) makes it clear that it is the flavin of the xanthine oxidase molecule which reacts with oxygen, not the iron, as had been thought earlier. Nevertheless it was of interest to study the oxidation at pH 10 of other reduced iron-sulphur proteins. Nilsson, Pick and Bray (1969) found that with spinach ferredoxin no signal was obtained. However, they did see a weak signal, apparently identical with O_2^-, during the oxidation of ferredoxin from Clostridium. These results are somewhat surprising since spinach ferredoxin is a 1-electron, and that from Clostridium a 2-electron acceptor. Nilsson et al. (1969b) also carried out some experiments on the peroxidase–oxidase system. In this, it was proposed some years ago (Yamazaki and Piette, 1963) that O_2^- is involved in a chain reaction and is formed by reaction of a dihydroxyfumarate radical with an oxygen molecule. The EPR work has now provided some direct evidence for the participation of O_2^- in the system.

The overall conclusion of Nilsson et al. (1969b) was that biochemical O_2^--formation does not depend on any specific enzyme structures and that it might arise in any system with a suitably low redox potential.

Notes added October 1969

(1)   Additional biochemical EPR work on O_2^- has now appeared in the literature. Beinert and Orme-Johnson (1969) have extended studies on iron-sulphur proteins while Ballou, Palmer and Massey (1969) have shown that reduced tetraacetyl riboflavin reacts with oxygen almost quantitatively to give O_2^-.

(2)   It is now recognised that the kinetic mechanism put forward by Knowles et al. (1969) to explain 1st order decay of O_2^- at high pH-values is not satisfactory.

Discussion

D. J. E. INGRAM   Have the g-values of the O_2^- resonances been checked at shorter wavelengths to make sure then are true g-values and not smeared hyperfine splittings?

R. C. BRAY   Measurements on O_2^- have been made only at 9 GHz.

References

Ballou, D., Palmer, G., and Massey, V. (1969) *Biochem. Biophys. Res. Comm.* **36**, 898.

Bray, R.C. (1961) *Biochem. J.* **81**, 189.

Bray, R.C., and Knowles, P.F. (1968) *Proc. Roy. Soc. A* **302**, 351.

Bray, R.C. (1964) in *Rapid Mixing and Sampling Techniques in Biochemistry.* Ed. Chance, Eisenhardt, Gibson and Lonberg-Holm, Academic Press, New York, p. 195.

Czapski, G., and Dorfman, L.M. (1964) *J. Phys. Chem.* **68**, 1169.

Knowles, P.F., Gibson, J.F., Pick, F.M., and Bray, R.C. (1969) *Biochem. J.* **111**, 53.

Komai, H., Massey, V., and Palmer, G. (1969) *J. Biol. Chem.* **244**, 1692.

Nilsson, R., Pick, F.M., Bray, R.C., and Fielden, M. (1969a) *Acta Chem. Scand.* **23**, 2554.

Nilsson, R., Pick, F.M., and Bray, R.C. (1969b) *Biochim. Biophys. Acta* **192**, 145.

Orme-Johnson, W.H., and Beinert, H. (1969) *Biochem. Biophys. Res. Comm.* **36**, 905.

Tench, A.J., and Holroyd, R. (1968) *Chem. Comm.*, p. 471.

Yamazaki, I., and Piette, L.N. (1963) *Biochim. Biophys. Acta* **77**, 43.

ESR probing of macromolecules. α-chymotrypsin. II[1]

D. J. KOSMAN[2] and L. H. PIETTE

Biophysics Laboratory, Department of Biochemistry and Biophysics
School of Medicine, University of Hawaii, Honolulu, Hawaii, 96822

Introduction

IN A PREVIOUS REPORT[3], we described the preparation of three new spin-labelled substrates, **1–3**, for α-chymotrypsin (α-ChT) and their use in probing the functional operation of this hydrolytic enzyme. α-ChT can be labelled at

$$\underset{\sim}{1} \quad R_1 NH\overset{O}{\overset{\|}{C}}CH_2CH_2\overset{O}{\overset{\|}{C}}OR_2$$

$$R_1 = \text{—} \langle \ \rangle \text{N-O}$$

$$\underset{\sim}{2} \quad R_1 O\overset{O}{\overset{\|}{C}}CH_2CH_2\overset{O}{\overset{\|}{C}}OR_2$$

$$\underset{\sim}{3} \quad R_1 O\overset{OH}{\overset{\|}{C}}C=\overset{HO}{\overset{\|}{C}}COR_2$$

$$R_2 = \text{—} \langle \bigcirc \rangle \text{—} NO_2$$

$$\underset{\sim}{4} \quad R_1 NH\overset{OH}{\overset{\|}{C}}C=\overset{HO}{\overset{\|}{C}}COR_2$$

the proteolytic Ser$_{195}$ via hydrolysis of a labelled substrate and isolation of the intermediate ES$_{SL}$ complex (acyl enzyme) by rapidly lowering the pH.

$$E + S_{SL} \overset{Km}{\rightleftharpoons} [ES_{SL}] \overset{k_2}{\longrightarrow} ES'_{SL} \overset{k_3}{\longrightarrow} E + P_{SL} + P$$

Correlation of changes in the extent of spin-label—enzyme interaction (as indicated by changes in rotational freedom of the label, i.e., the degree of immobilization) with alteration of the substrate structure yields unique information about the functional nature of the active region. Increasing immobilization of the nitroxide results in an increasingly assymmetric ESR spectrum. Furthermore, the deacylation rate of the acyl enzyme (k_3) can be directly measured by raising the pH to a catalytic value and observing the release of "free spin"[7], that is, the spin-labelled carboxylic acid which is now rapidly tumbling in solution unhindered by interaction with the enzyme. The effect on k_3 of structural modifications of both the substrate and enzyme also provides information about the catalytic function of α-ChT. By an analysis of the ESR spin-label spectra of the isolated acyl enzymes, the interaction of the spin labels with the protein was correlated with the structure of the label, the oxidation of methionine 180 and/or methionine 192[4], and the presence or absence of indole[5]. Dramatic changes in the ESR spectra indicating decreased immobilization of the substrate were observed for **2** after oxidation of the methionine 192 or upon addition of indole. Similar changes were reflected in **1** only after further oxidation of the methionine 180. **3** because of its rotational restrictions and its *cis* relationship to the hydrolytic serine and the (am) binding region yielded an ESR spectrum indicative of no secondary binding or restriction by the active site. Similar correlations were made with the rates of deacylation (k_3's). On the basis of the data, two pertinent conclusions were reached: 1) Met_{192} is involved in the definition of the aromatic ("ar") binding region[6] as is Met_{180} in the amide ("am") site, and 2) an enzymatic structural link exists between the "ar" and "am" binding regions which may be an integral part of the overall catalytic process.

We have now studied another substrate, **4**, which by virtue of the structural similarities and differences between it and the other substrates, provides important evidence concerning the binding specificity and, hence, location of each of the labels. The question is whether **4**, a maleic amide derivative, responds like the maleic ester spin label, **3**, or like the amide derivative **1**. That is, despite the rotational restrictions imposed on the molecule by its *cis* double bond, is **4** preferentially bound to a region other than that occupied by **3**, but identical to that which binds **1** namely the amide (am) region. Further, does **1** despite its flexibility, bind to the same region which **4**, because of its *cis* geometry, is almost forced to occupy. Confirmation of our original suggestion that **1** was bound to the "am" site and **2** the "ar" site is requisite to a strengthening of our operational hypothesis.

Experimental

Preparation of Acyl Enzymes

Alpha-chymotrypsin (0.5 ml, 2×10^{-4} M, 0.001 HCl) was added to phosphate buffer (0.5 ml, pH 9.7, 0.02 M). The pH of this solution was 7.8, $[\alpha\text{-ChT}] = 1.0 \times 10^{-4}$ M. To this was added substrate in acetonitrile (AcCN) (0.05 ml, 1.0×10^{-2} M). After *ca.* 2–3 mins glycine buffer (0.2 ml, 0.2 M, pH 2.5) was added. The pH of this solution was 3.0. The excess spin label (both substrate and hydrolysate) was removed by either immediate gel filtration (10 cm, Sephadex G-10 fine) or dialysis at pH 3.0, 0.2 M glycinate.

Deacylation

The ES' complex solution was dialyzed down to pH 3.0, 0.004 M. Addition of phosphate buffer (0.1 ml, pH 7.55, 0.2 M) to this solution (0.1 ml) brought up the pH to 7.2, 0.1 M in phosphate. Recording of the deacylation was begun at about $t = 20$ sec. At low pH, rapid changes of ionic strength had no effect on the spectra. The k_3's determined in this way were in agreement with the approximate k_3's determined spectrophotometrically (turnover rates).

Substrate

4 was prepared as described previously[2]; **4** has spectroscopic properties, mass spectrum, and elemental analysis consistent with its structure.

ESR Spectra

All ESR spectra were obtained with a Varian V-4502 system. Kinetics were performed in the cavity using a quartz aqueous mixing cell from Scanlon Glassblowing Co., Whittier, California.

Results

The spectrum of the acyl enzyme prepared from **4** is shown in Fig 1a. It is essentially identical to that derived from **1** and totally different from the spectrum recorded for **3** which indicated little protein—substrate interaction, i.e., a great deal of rotational freedom. As observed before for **1**, the addition

D. J. Kosman and L. H. Piette

of indole to the acyl enzyme from **4** induced an enzymatic structural change characterized by a definite tightening of the region surrounding the amide spin label (Fig. 1b). This effect is completely reversible. Upon removal of indole by dialysis, the original spectrum is obtained (Fig. 1c).

The acyl enzyme prepared from methionine-192-α-ChT sulfoxide yields a spectrum identical to that in Fig. 1a, but no indole effect is observed. This was again identical to the observation made with the acyl enzyme from **1** upon

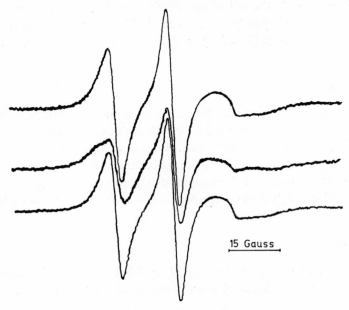

Figure 1 Spectra of acyl enzyme (a) and following addition of indole (b); (c) was recorded after 12 hours dialysis. (a) and (c) contain 4.5% v/v AcCN, (b) contains 1.0×10^{-2} indole and 4.5% v/v AcCN; pH 3.0, citrate, $I = 0.05$

Table I Deacylation rates of acyl enzymes, $k_3 \times 10^{2a}$

Substrate	1		2		3		4	
[Ind] \ Enzyme	NAT	Met-3-0	NAT	Met-3-0	NAT	Met-3-0	NAT	Met-3-0
0	2.20	2.36	0.32	1.37	0.04	0.13	1.98	2.02
0.01 M^b	0.68	2.32	1.53	1.43	0.12	0.14	0.98	2.01

[a] pH = 7.2, phosphate, $I = 0.1$, 4.5% v/v Ac CN.
[b] [Ind] = 100 [E], 4.5% v/v Ac CN.

oxidation of Met_{192}. Disulfoxide α-ChT (oxidized methionines 180 and 192) does, however, yield a modified acyl enzyme whose spectra (Fig. 2) is again identical to that obtained from **1** and is characterized by an increase in the rotational freedom of the nitroxide.

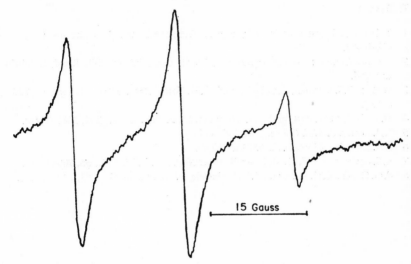

15 Gauss

Figure 2 Spectra of acyl enzyme from hydrolysis of **4** by α-ChT disulfoxide (methionines 180 and 192)

The rates of deacylation of two acyl enzymes (native and monosulfoxide) without and with indole (Ind) are presented in Table I with our previous data for 1–3[3]. The rate depressing effect of indole on the amide derivatives is contrasted with the accelerating effect on the ester derivatives.

Discussion

The salient feature of these new results is that the maleic amide responds as an amide (cf. **1**) rather than a maleic derivative (cf. **3**). **3** was originally chosen for study because Cohen[8] had suggested a *cis* double bond would orient the molecule out of the active site, or, more specifically, away from the "ar" region. His data suggested a *trans* relationship between the hydrolytic and aromatic sites. The corollary to this hypothesis is, however, a *cis* relationship between the hydrolytic and amide sites. The similarity in behavior between **1** and **4** strongly suggests this to be the case, and indicates that **1**, though highly flexible, does exhibit a specificity for binding at the "am" region. The encouragingly consistent picture drawn from the data indicates the validity of the spin-label approach to the study of the operation of α-Chymotrypsin and our study of this enzyme continues.

References

1 This research was supported by a grant from the National Cancer Institute, Grant CA-10977.
2 Present Address: MRC Laboratory of Molecular Biology, Hills Road, Cambridge, England.
3 Part I; D. J. Kosman, J. C. Hsia, and L. H. Piette, *Arch. Biochem. Biophys.*, **133**, 29 (1969).
4 H. Weiner, C. W. Batt, and D. J. Koshland, Jr., *J. Biol. Chem.*, **241**, 2687 (1966).
5 R. J. Foster, *J. Biol. Chem.*, **236**, 2461 (1961).
6 J. R. Knowles, *Biochem. J.*, **95**, 180 (1965).
7 L. S. Berliner and H. M. McConnell, *Proc. Natl. Acad. Sci. U.S.*, **55**, 708 (1966).
8 S. G. Cohen, L. H. Klee, and S. Y. Weinstein, *J. Am. Chem. Soc.*, **88**, 5302 (1966).

Electron nuclear double resonance of flavin and flavoprotein radicals

ANDERS EHRENBERG and L. E. GÖRAN ERIKSSON

Biofysiska Institutionen, Stockholms Universitet
Medicinska Nobelinstitutet, Karolinska Institutet, Stockholm, Sweden

JAMES S. HYDE

Analytical Instrument Division,
Varian Associates, Palo Alto, California

FOR SOME FLAVOPROTEINS it has been established that flavin radicals play a role as an intermediate in the enzymatic reaction (see e.g. Ref. 1). A photoreaction has been designed as a useful procedure for production of high yields of radicals in metal-free flavoproteins[2].

The non protein-bound flavin radicals were found to give ESR spectra with partly resolved hyperfine pattern[3,4], see Fig. 1. The isotropic coupling constants contributing to this pattern have been determined, by means of isotope and chemical substitution in the model compound lumiflavin, for the radical in the anionic[5,6], neutral[7] and cationic[8] form, as well as for the radical chelate[9].

In case of a flavoprotein radical in liquid solution no hyperfine structure is resolved[4], see Fig. 1. This demonstrates that the rotational relaxation of the protein is too slow to average out the anisotropic hyperfine interactions[4]. In the case of NADPH dehydrogenase with $M = 104,000$ τ_{rot} is of the order of 10^{-7} sec. In flavin radicals the largest anisotropic contribution should come from N(5) (see Fig. 2) which in the neutral and anionic radicals has an isotropic coupling of about 20 MHz. By analogy with the coupling of the trigo-

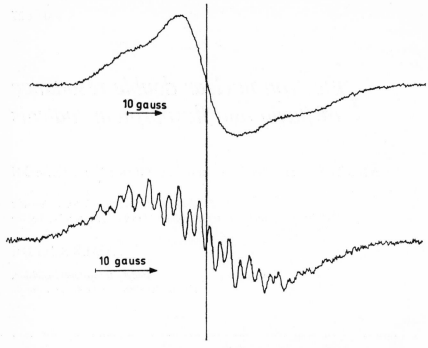

Figure 1 ESR spectra of radicals from NADPH dehydrogenase (top) and the free co-enzyme, flavin mononucleotide (FMN, bottom). pH 7 and 22 °C (Ref. 4)

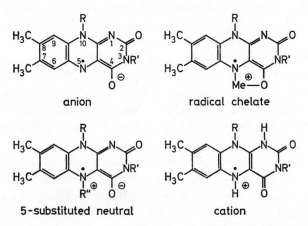

Figure 2 Flavin radical species studied by ESR and ENDOR in model systems

nal nitrogen of nitroxide radicals[10] the anisotropic contribution of N(5) in flavin should also be of the order of 20 MHz. The rotational speed of the radical (i.e. the protein molecule to which it is firmly attached) must be faster than this in order to average out the anisotropy. Since the protein is rotating slower than this in the liquid solution it is concluded that no major change will occur upon freezing of the protein solution, which is also found by experiment[4]. The ESR pattern of frozen solutions of free flavin radicals are similar to those of the flavoproteins radicals[4].

It was suggested that the hyperfine couplings of the protein bound flavin radicals could be studied by the ENDOR technique[4]. Alternatively it could conceivably be investigated by ESR on single crystals of the protein. The ENDOR approach should, however, be of a more general applicability. A suitable high-power ENDOR technique has been developed[11] and a review of this has been published[12]. This technique has now been applied to the proton ENDOR study of radicals of model flavin compounds and of flavoproteins[13-15].

The lumiflavin anionic radical and radical chelate give two ENDOR lines in liquid solution. These experiments were carried out in dimethylformamide (m.p. −61°C) at −70°C, i.e. in a very viscous solution or glass. With chemical and isotope substitution the two lines could be assigned to the methyl groups in positions 8 and 10, see Fig. 3[14]. The methyl groups are

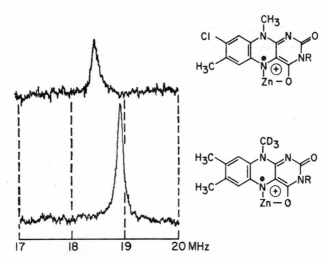

Figure 3 Parts of ENDOR recordings of flavin radical chelates. Dimethylformamide at −74°C. The upper signal is originating from $CH_3(10)$ and the lower one from $CH_3(8)$. The free proton frequency is 13.7 MHz (Ref. 14)

freely rotating around the C–C axes and have low enough anisotropy in their hyperfine tensors to give sharp ENDOR signals under the experimental conditions. Ring protons have much more anisotropic hyperfine interaction and are broadened beyond detectability because of the slow mobility of the radicals. This difference in behaviour of methyl and ring protons has been investigated in detail for the tris(p-tolyl)methyl radical[16]. It would conceivably be possible to obtain a narrow ENDOR signal from the ring proton in position 6 of the flavin radical at higher temperatures, but the decrease of the overall ENDOR sensitivity and of the stability of the radicals did not allow such experiments.

The model radicals tris(p-tolyl)methyl was also investigated in frozen glass and polycrystalline medium[16]. In the former case the methyl groups gave bell shaped ENDOR signals, whereas in the latter case a line shape typical for

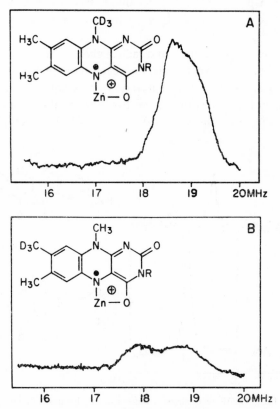

Figure 4   Parts of ENDOR spectra of isotopically substituted lumiflavin radical chelates. Dimethylformamide at about −160 °C. The free proton frequency is 13.7 MHz (Ref. 14)

axially symmetrical couplings of a randomly oriented sample (powder ENDOR) was obtained. The flavin radicals in frozen solution also gave methyl signals of the powder ENDOR type, for which assignments could be made by isotope substitution, as shown in Fig. 4. Centred at the free proton frequency a so called matrix ENDOR signal was obtained, which will be discussed below. On the wings of this matrix ENDOR signal weakly coupling protons in positions 7 and 9 appear for which tentative assignments could be made[14].

In riboflavin the 10-methyl group of lumiflavin is substituted, so that its rotation is hindered. Hence the ENDOR signal is broadened and only the 8-methyl signal is detected. In succinate dehydrogenase (SD) the prosthetic group FAD is covalently linked to the protein. From the enzyme an SD-riboflavin peptide can be prepared and has been investigated by the ENDOR technique[17]. In this case also the 8-methyl signal is away, proving that the point of covalent attachment of the peptide must be just on the 8-methyl group[17].

High yields of radicals were produced in the two flavoproteins NADPH dehydrogenase and *Azotobacter* flavoprotein by photoreaction and dithionite reduction, respectively[14,15]. The type of ENDOR spectrum obtained illustrated for the former protein in Fig. 5. The appearence of the 8-methyl signal demonstrates that the binding of the flavin to the protein, which in these cases is not covalent, does not hinder the free rotation of the methyl group[13,14]. For NADPH dehydrogenase the 8-methyl coupling is determined to be 11.2 MHz and for the *Azotobacter* flavoprotein 8.4 MHz[15]. For lumiflavin, by ESR and ENDOR, the corresponding coupling has been measured to be close to 11 MHz for the anionic radical[6,14] and 7 MHz for the neutral radical[7,15]. This shows that the radical in the NADPH dehydrogenase is of the anionic type and in the *Azotobacter* flavoprotein most closely corresponds to the neutral radical[15]. This is in agreement with their light absorption characteristics.

The matrix ENDOR signal arises through electron-nuclear dipolar interaction. A model has been put forward[16] showing that nearby protons within a radius of 6 Å about the radical would contribute, the more remote ones to the center, and the closer ones to the wings of the signal. When H_2O is replaced by D_2O the matrix ENDOR signal of NADPH dehydrogenase is decreased by about 40%, see Fig. 5. The matrix signals were normalized with respect to the size of the methyl signals. This result shows that the immediate surrounding of the flavin in this enzyme is partly hydrophilic with exchangeable protons and partly hydrophobic[14]. In case of the *Azotobacter* flavo-

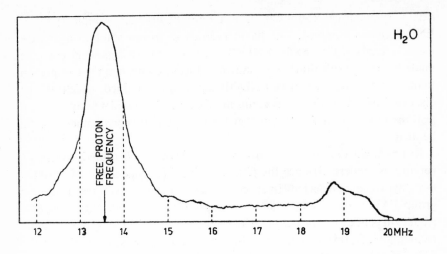

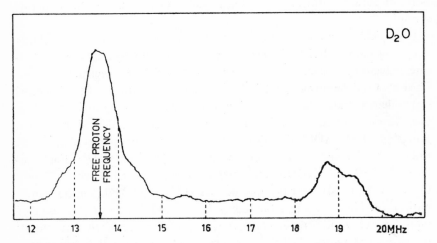

Figure 5 ENDOR recordings at −120°C from flavin radicals conjugated to NADPH dehydrogenase (Ref. 15)

proteins it was found that D_2O abolished the matrix signal nearly completely, suggesting that in this case the flavin group is in very little contact with hydrophobic parts of the protein[15].

Acknowledgements: The work was supported by grants from U.S. Public Health Service (AM-05895), the Swedish Statens Medicinska Forskningsråd, Statens Naturvetenskapliga Forskningsråd and Wallenbergstiftelsen.

Discussion

(Contribution to the discussion to the paper by Prof. Ehrenberg)

D.J.E. INGRAM What microwave power level was required to saturate your E.S.R. signals for ENDOR work?

A. EHRENBERG   For the simple flavin radicals the microwave power applied was about 30 mW. For the ENDOR spectra of the flavoproteins we used about 5 mW.

This is the optimum power for the matrix ENDOR, whereas the optimum for the methyl ENDOR is somewhat higher.

References

1   Slater, E. C. S. (Ed.), *Flavins and Flavoproteins*, Elsevier, Amsterdam, 1966.
2   Massey, V., and Palmer, G., *Biochemistry* **5** (1966) 3181.
3   Ehrenberg, A., *Acta Chem. Scand.* **14** (1960) 766.
4   Ehrenberg, A., *Arkiv Kemi* **19** (1962) 97.
5   Eriksson, L. E. G., and Ehrenberg, A., *Acta Chem. Scand.* **18** (1964) 1437.
6   Ehrenberg, A., Müller, F., and Hemmerich, P., *European J. Biochem.* **2** (1967) 286.
7   Müller, F., Hemmerich, P., Ehrenberg, A., Palmer, G., and Massey, V., *European J. Biochem.* **14** (1970) 185.
8   Ehrenberg, A., Hemmerich, P., and Müller, F., in preparation.
9   Müller, F., Eriksson, L. E. G., and Ehrenberg, A., *European J. Biochem.*, **17** (1970) 539.
10   Griffith, O. H., Cornell, D. W., and McConnell, H. M., *J. Chem. Phys.* **43** (1965) 2909.
11   Hyde, J. S., *J. Chem. Phys.* **43** (1965) 1806.
12   Hyde, J. S., in A. Ehrenberg, B. G. Malmström, and T. Vänngård, *Magnetic Resonance in Biological Systems*, Pergamon Press, Oxford, 1967, p. 63.
13   Ehrenberg, A., Eriksson, L. E. G., and Hyde, J. S., *Biochim. Biophys. Acta*, **167** (1968) 482.
14   Eriksson, L. E. G., Hyde, J. S., and Ehrenberg, A., *Biochim. Biophys. Acta*, **192** (1969) 211.
15   Ehrenberg, A., Hyde, J. S., Eriksson, L. E. G., and Walker, W., in preparation.
16   Hyde, J. S., Rist, G. H., and Eriksson, L. E. G., *J. Phys. Chem.* **72** (1968) 4269.
17   Walker, W. H., Salach, J., Gutman, M., Singer, T. P., Hyde, J. S., and Ehrenberg, A., *FEBS Letters*, **5** (1969) 237.

Magnetic resonance studies of model molecules of vitamin B_{12}

H. A. O. HILL, K. G. MORALLEE

Inorganic Chemistry Laboratory, Oxford

G. COSTA G. PELLIZER

Istituto di Chimica, Università di Trieste

A. LOEWENSTEIN

Physical Chemistry Dept., Technion, Haifa

THE MOST STRIKING FEATURE of the vitamin B_{12} coenzyme is the presence of the cobalt–carbon bond. This group may be involved in the biological reactions or might influence other parts of the molecule making them suited for the biological function.

The Oxford group studied thoroughly the changes in physical properties of cobalamins and cobinamides (Fig. 1) with varying axial ligands and a good amount of information is now available on the reciprocal influence of the ligands in these compounds (1).

Cobalt complexes of the following tetradentate ligands

bis (acetylacetone)ethylenedi-iminate   (BAE) (Fig. 2a)
bis (salicylaldehyde)ethylene-iminate   (Salen) (Fig. 2b)
bis (dimethylglyoximate) (DH_2) (Fig. 3a)
bis (diacetylmonoxime-imino)propane 1,3   (DOH) (Fig. 3b)

retain some important features of vitamin B_{12} and have been proposed as models for it (2), (3).

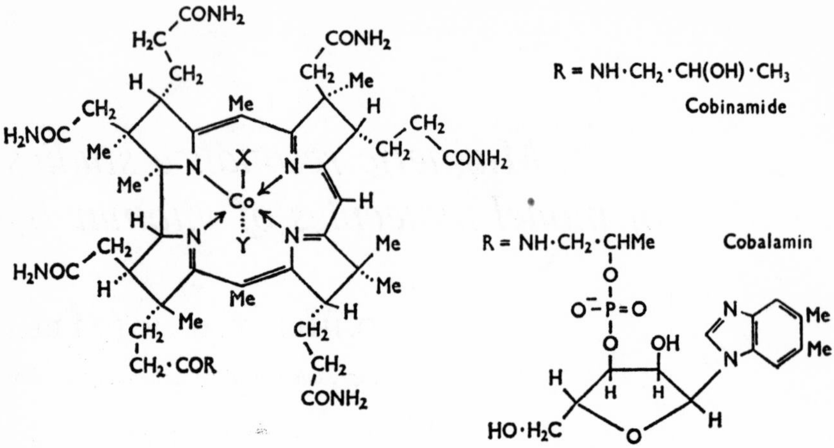

Figure 1

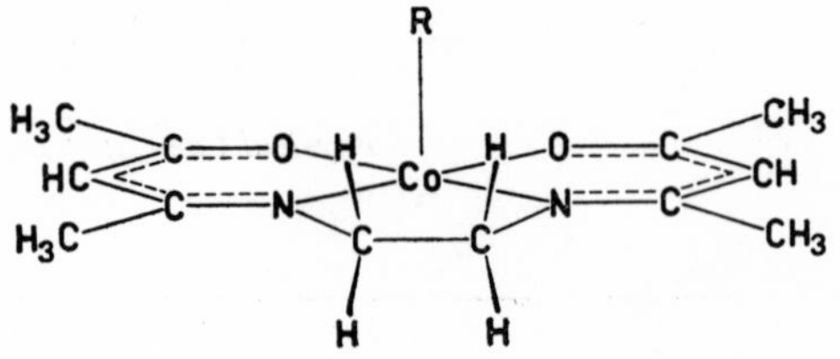

(A)

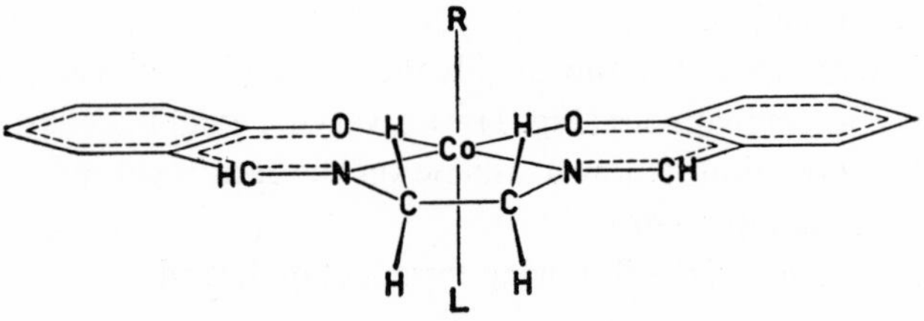

(B)

Figure 2

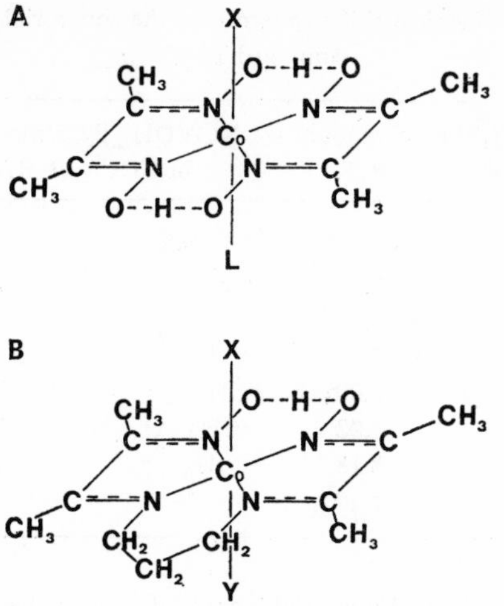

A

B

Figure 3

Some aspects of ligand–ligand interactions could be studied more conveniently in these compounds and we report here some of the results obtained by n.m.r. and e.p.r. spectroscopies.

Axial ligand–equatorial ligand interactions

a) Transmission of the Electronic Effects

The proton resonances of corrin ring C(10) hydrogen in cobalamins (Table 1) (4), of bisdimethylglyoximate methyls in $XCo(DH)_2Py$ and $XCo(DH)_2PPh_3$ complexes (Table 2) (5), of BAE methene and methyls in five coordinate RCoBAE complexes (Table 3) (6) depend on the axial ligands.

Table 1 Corrin C(10) proton chemical shifts (p.p.m.) in cobalamins. Solvent D_2O. Standard DSS.

Compound	τ value
Ethylcobalamin	4.12
Methylcobalamin	4.12
Vinylcobalamin	4.05
Hydroxocobalamin	3.92
Aquocobalamin	3.72

Table 2   Proton chemical shifts (p.p.m.) for the equatorial ligand methyls.
Standard TMS

XCo(DH)$_2$PPh$_3$ compounds in CH$_2$Cl$_2$ at R.T.		XCo(DH)$_2$Py compounds in CDCl$_3$ at R.T.	
X	τ value	X	τ value
NO$_2$	7.71	Br	7.61
CN	7.79	I	7.63
Cl	7.99	CH$_2$CF$_3$	7.82
Br	8.00	CH$_3$	7.87
I	8.02	(CH$_2$)$_2$OC$_2$H$_5$	7.88
CH$_3$	8.18	C$_2$H$_5$	7.88
C$_2$H$_5$	8.19		

Table 3   Equatorial ligand protons τ values
in XCoBAE compounds. Solvent CDCl$_3$

X	methene proton	methyl protons
C$_2$H$_5$	4.92	7.93; 7.99
CH$_3$	4.88	7.91; 7.97
C$_2$H$_5$	4.76	7.89; 7.92
C$_6$H$_5$	4.58	7.84; 7.86
pOCH$_3$C$_6$H$_4$	4.59	7.84; 7.86
pCH$_3$C$_6$H$_4$	4.59	7.84; 7.87
pIC$_6$H$_4$	4.58	7.84; 7.85
pBrC$_6$H$_4$	4.57	7.84; 7.86
pCNC$_6$H$_4$	4.52	7.82
pNO$_2$C$_6$H$_4$	4.49	7.82

   In all these series the resonances move to lower field with increasing electron-withdrawing power of the axial ligand. If the shielding constants of the
equatorial ligand protons are due mainly to variations in the diamagnetic
term, the observed results may reflect changes in the ground state electron
density of the equatorial ligand due to transmission of the electronic effect of
the axial ligands *via* the cobalt.

   Other properties of the axial ligand may, however, influence the chemical
shifts. Though in a series (7) of CoBAE compounds with substituted phenyls
as the fifth ligands, the BAE methene and methyl proton resonances move
(Table 3) to low field with increasing electron withdrawing power of the

substituent, as measured by its Hammett σ function, electronic transmission may not be the only factor responsible of the observed effects. In particular the order of the chemical shifts is the same as that of the dipole moments of the substituted benzenes. However, we estimated roughly the ring current and the electric dipole effects (7) and came to the conclusion that the transmission of the electronic effect is the dominant factor in the BAE *methene* proton chemical shift variation.

b) Ring Current Effects

The axial ligand ring current is one of the reasons why the methene and methyls protons resonate in PhCoBAE at significantly lower field than in vinyl CoBAE (Table 3). But the most remarkable effect is shown on ethylene bridge protons (Fig. 4). By adding pyridine to methyl, ethyl or vinyl CoBAE

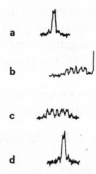

a

b

c

d

Figure 4   Ethylenic bridge proton n.m.r. spectra: a CH_3CoBAE in $CDCl_3$, b CH_3CoBAE in $CDCl_3$ + pyridine, c PhCoBAE in $CDCl_3$, d PhCoBAE in $CDCl_3$ + pyridine

the spectral pattern of these protons, which constitute an AA'BB' system, changes, as if the difference in their shielding constants increased, and becomes like that of pentacoordinate phenyl CoBAE compounds. We attribute this to pyridine or phenyl ring current having different effects on protons which are above and below the equatorial ligand plane (Fig. 2). On adding pyridine to the phenyl and para substituted phenyl CoBAE compounds the spectral pattern of this AA'BB' system collapses to a singlet, the difference in shielding constants caused by one axial ligand being now compensated by the other.

Axial ligand–axial ligand interactions

a) Proton chemical shifts of axial ligands

The chemical shifts of the Lewis base protons in cobalamins (4) and $XCo(DH)_2Py$ (5) depend on the other axial ligand. The C(7) benzimidazole and α, β, γ pyridine protons resonate at lower field, i.e. nearer and nearer to the free base value, with decreasing electron withdrawing power of the other axial ligand. This suggests that factors such as anisotropic shielding by the cobalt and from the equatorial ligand π system are important in determining the protons chemical shifts of the Lewis base.

The observed variations may reflect a reduction in anisotropic shielding associated with an increase in the Co-N distance.

b) Equilibrium constants

In compounds of vitamin B_{12} group the axial ligand–axial ligand interactions were studied mainly by infrared spectroscopy and equilibrium constants (1). As the axial ligand places more negative charge on the equatorial plane (i.e. in the order H_2O, CN^- $\cdots$ $CH_3CH_2^-$) so the other axial ligand–metal interaction becomes weaker.

In the XCoBAE compounds (6) (7) we could utilise the ^1H n.m.r. spectra to calculate the constants (Table 4) for the equilibrium

$$XCoBAE + L \rightleftharpoons XCo(BAE)L$$

Table 4 Formation constants for the equilibrium $XCoBAE + L \rightleftharpoons XCo(BAE)$ L in $CDCl_3$ at R.T.

X =	C_2H_5	CH_3	C_2H_3	C_6H_5	$pCH_3OC_6H_4$	$pCH_3C_6H_4$	pIC_6H_4	$pBrC_6H_4$	$pCNC_6H_4$	$pNO_2C_6H_4$
(L = py) K (mole·l^{-1}) =	1	9	12	5.3 ± 0.7	6.1 ± 0.8	5.0 ± 0.6	19 ± 2.5	20 ± 2.5	54 ± 7	85 ± 13
(L = piperidine) K =				5.4 ± 0.7			18 ± 2.5	20 ± 3	56 ± 11	83 ± 20

The effect of organic ligand has not the same trend as its electron withdrawing power when X = C_2H_5, CH_3, C_2H_3 and C_6H_5 and L = pyridine. When X = p-substituted phenyl and L = pyridine and piperidine the effects other than the electronic transmission are weaker and the trend of equilibrium constants is the same as the Hammett σ-functions of X, which we interpret as if increasing electron density on cobalt makes it a weaker Lewis acid, therefore weakening the interaction with the base.

The possibility of pentacoordination in XCoBAE and XCo Salen compounds led to the discovery of an equilibrium between six and five coordinate forms even for cobinamides and cobalamins (8). The amount of five coordinate form increases with decreasing electron-withdrawing power of the remaining axial ligand, still in excellent agreement with other results shown so far.

Physical properties of the cobalt atom

To the study the influence of the ligands on the cobalt atom we used the ^{59}Co n.m.r. spectra and the e.p.r. spectra of Co(II) derivatives.

a) ^{59}Co N.M.R. Spectra

The ^{59}Co resonances depend strongly on the atoms directly bonded to cobalt (Table 5).

Table 5 ^{59}Co chemical shifts (p.p.m. $\times 10^{-3}$) relative to saturated aqueous solution of K$_3$Co(CN)$_6$

CH$_3$CoBAE^a	-7.3	ICo(DH)$_2$py^a	-6.0
CH$_3$CoBAE py^b	-7.2	CH$_3$Co(DH)$_2$PPh$_3$^{a}	-2.8
C$_2$H$_5$CoBAE^a	-7.5	CH$_3$Co(DH)$_2$ py^a	-3.6
C$_2$H$_5$CoBAE py^b	-7.3	C$_2$H$_5$Co(DH)$_2$PPh$_3$^{a}	-2.9
C$_2$H$_3$CoBAE^a	-6.8	CF$_3$CH$_2$Co(DH)$_2$ py^a	-4.0
C$_2$H$_3$CoBAE py^b	-6.95	[CH$_3$Co(DOH)Mid]ClO$_4$^{d}	-4.1
PhCoBAE^a	-7.2	[CH$_3$Co(DOH)Im]BPh$_4$^{d}	-4.2
PhCoBAE py^b	-7.05	[CH$_3$Co(DOH)H$_2$O]BPh$_4$^{d}	-4.6
pBrC$_6$H$_4$CoBAE^a	-7.2	PhCo(DOH)I^a	-4.25
pBrC$_6$H$_4$CoBAE py^b	-7.1	vinyl cobalamin^e	-4.1
pNO$_2$C$_6$H$_4$CoBAE^a	-7.3		
pNO$_2$C$_6$H$_4$CoBAE py^b	-6.95		
CH$_3$Co(BTFAE)^a	-7.3		
CH$_3$Co(BTFAE) py^b	-7.3		
CH$_3$Co salen H$_2$O^{c6}	-7.2		

Solvents: ^{a} dichloromethane, ^{b} pyridine, ^{c} *N,N'*-dimethylformamide, ^{d} acetone, ^{e} water
Abbreviations: BTFAE = bis(trifluoroacetylacetone)ethylendiiminato, Ph = phenyl,
            Im = imidazole, Mid = *N* methylimidazole, py = pyridine,
            PPh$_3$ = triphenylphosphine.

The results show a clear distinction between two kinds of equatorial ligands: those bonded to the cobalt by four nitrogens and those bonded by two nitrogens and two oxygens, the resonances in the former complexes being at significantly higher fields than in the latter.

The axial ligands also affect the chemical shifts. In $CH_3Co(DH)_2Py$ the resonance is at 2.4×10^3 p.p.m. higher field than in $ICo(DH)_2Py$, whilst the difference between resonances in $CH_3Co(DH)_2PPh_3$ and $CH_3Co(DH)_2Py$ is 0.8×10^3 p.p.m. and that between resonances of $[CH_3Co(DOH) Im] BPh_4$ and $(CH_3Co(DOH)H_2O] BPh_4$ is 0.4×10^3 p.p.m. Moreover in the BAE complexes there are very little differences between the five and the six co-ordinate species, thus reflecting a weak bond between cobalt and pyridine consistent with the low formation constants of the six coordinate compounds.

When the cobalt nearest neighbours are the same, as in XCoBAE complexes with different organic X ligands, the differences in resonances are very small.

The ^{59}Co chemical shifts suggest that the complexes with the same nearest neighbour atoms are the most relevant models; in particular the DOH complexes, which have the same charge as cobalamins, appear the most promising.

Table 6   Data from e.p.r. spectra of Co(II) compounds

Planar ligand	Nitrogenous axial ligand(s)	$g_{||}$	$g_\perp$	$A_{||Co} \times 10^4$ cm^{-1}	$A_{||N} \times 10^4$ cm^{-1}
BTFAE[10]	Py	2.016	2.30	84	13.5
			2.45		
BAE[10]	Py	2.018	2.39	92	14.7
Salen[10]	Py$_2$	2.033	2.34	76	11.8
(DH)$_2$[10]	Py$_2$	2.016	2.27	80	15.3
Corrin[9]	Py	2.004	2.32	103	17.5
Corrin[9]	Bz, Py	2.011	2.28	102	13.4

Bz: 5,6 dimethylbenzimidazole

The parameters (table 6) derived from the e.p.r. spectra are consistent with a description of the complexes as low-spin d^7 complexes in which the single unpaired electron is in the $d_z{}^2$ orbital. Superhyperfine structure is observed on the $A_{||Co}$ lines; triplets when one nitrogenous ligand is coordinated, quintets when two. A detailed analysis is in progress.

References

1   R.A.Firth, H.A.O.Hill, J.M.Pratt, R.G.Thorp and R.J.P.Williams, *J. Chem. Soc. (A)* 2428 (1968); H.A.O.Hill, J.M.Pratt and R.J.P.Williams, *Chemistry in Britain,* **5**, 156 (1969).

2   G.Costa, G.Mestroni, E. de Savorgnani, *Inorganica Chimica Acta* **3**, 323 (1969) and references there reported.

3   G.N.Shrauzer, *Accounts of Chemical Research* **1**, 97 (1968).

4   H.A.O.Hill, B.E.Mann, J.M.Pratt and R.J.P.Williams, *J. Chem. Soc. (A)* 564 (1968).

5   H.A.O.Hill and K.G.Morallee, *J. Chem. Soc. (A)* 554 (1969).

6   H.A.O.Hill, K.G.Morallee, G.Pellizer, G.Mestroni and G.Costa, *J. Organometal. Chem.* **11**, 167 (1968).

7   H.A.O.Hill, K.G.Morallee and G.Pellizer, *J. Chem. Soc. (A)* 2096 (1969).

8   R.A.Firth, H.A.O.Hill, B.E.Mann, J.M.Pratt, R.G.Thorp and R.J.P.Williams, *J. Chem. Soc. (A)* 2419 (1968).

9   S.A.Cockle, H.A.O.Hill, J.M.Pratt and R.J.P.Williams, *Biochim. Biophys. Acta,* 1969, **157**, 686 (1969).

10 S.A.Cockle, H.A.O.Hill and Z.G.Morallee, to be published.

The helix-coil transition
in polypeptides studied by NMR

F. CONTI

Istituto di Chimica-Fisica, Università di Roma, Italy

POLYPEPTIDES and peptide-like compounds have been studied by means of NMR, and a number of papers on this subject have recently appeared in the literature[1-19]. The importance of these studies is enhanced by the fact that polypeptides can be considered as model-compounds for proteins. Attempts have been made to identify the resonance peaks of specific groups of amino-acid residues located in any protein-like compound by comparison with data referring to the free aminoacids. Other authors tried to relate the chemical-shift of some functional groups such as the αCH and NH protons of the peptide bond with a specific conformation of the polypeptide chain.

Preliminary data seemed at first to be most encouraging, so that the use of NMR in problems concerning protein-like compounds has progressively increased. Recently, however, the increased number of researchers involved in this problem, as well as technological improvement of the NMR instrumentation, has shown that such preliminary results, when more critically considered, do not justify completely the first enthusiasm concerning the power of this technique applied without particular care to this particular class of compounds. The purpose of the present paper is to discuss, by examining both data already published in the literature and those obtained in our laboratory, the validity and the limitations of the results obtainable with this method with respect to conformational studies in peptide-like compounds in solution.

By minimization of conformational potential energy as a function of dihedral rotation angles around non-rigid bonds of the skeleton, it has been shown[20-22] that in α-helix structure, the right-handed conformation, in the case of L aminoacids, is preferentially stabilized, when compared with a left-handed α-helix, through intramolecular interactions of the Van der Waals type. More recently the idea has been suggested[23] that only five energetically allowed conformational states exist for an aminoacid residue in a poly-peptide chain. These five conformations, through which the concept of a "stereochemical code" has been introduced, have been labelled: *Ra, b, c, d, La* (Fig. 1). An identical repetition of *Ra* states for consecutive aminoacid

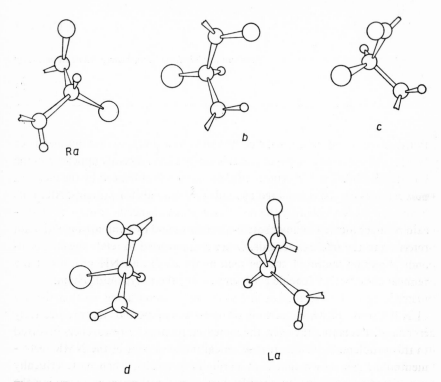

Figure 1 Energetically allowed conformations states for an amino acid residue in a poly-peptide chain

residues generates a right-handed alpha-helix (R_α), while an identical repetition of *La* states generates a left-handed alpha-helix (L_α). A like repetition, in the above specified sense, of *b*, *c* or *d* codons, generates three other helical structure; respectively the three helices β, γ and δ. These helices, correspond-ing to minima of the overall conformational potential energy for the whole structure, are obtained in a homopolypeptide by imposing the equivalence condition between consecutive aminoacid residues[20-21]. A close inspection of the local geometry of a polypeptide chain (including the β–C atom), in correspondence with the different energetically allowed conformations, shows marked differences in the environment of some proton of the skeleton, such as the αCH and the NH protons (Fig. 2). It is particularly striking that

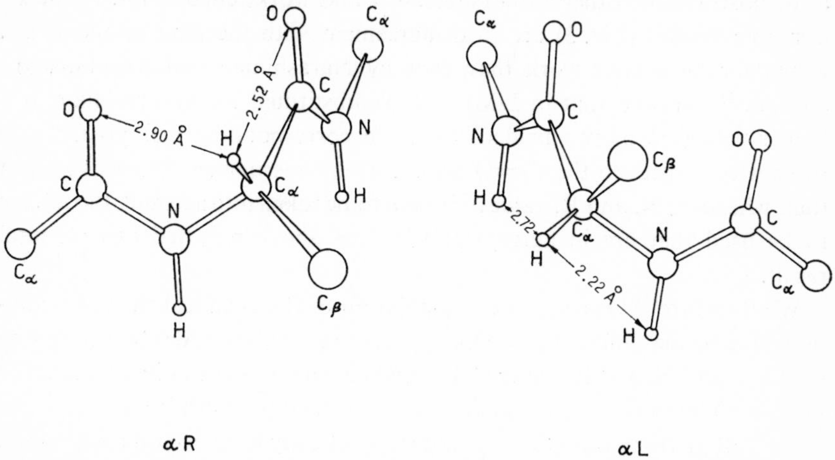

α R α L

Figure 2 Ra and La conformations of L-amino acid residues helical polypeptide chains

nearest-neighbours C═O are at different distances from the αCH and N—H protons, and have different orientations with respect to the αC—H and N—H bonds. This fact, as in the case of amides[24], should give rise to a different chemical-shift for αCH and N—H protons in the different helical conformations[4].

The first and, until now, most cited paper is that by Narasimhan and Rogers[24], who exmined the magnetical non-equivalence of the methyl groups in various differently N-substituted amides. This non-equivalence was attributed by the AA to a magnetic anisotropy of the amidic C═O group, and by means of a semi-empirical calculation they determined two magnetic susceptibility sets which justified the experimental values. On the other hand these AA left the sign of the chemical shift for the two methyl groups *cis* or *trans* with respect to the C═O bond undecided. Sternlicht and Wilson[18], later on, adopting the susceptibility values that involve a downfield (more deshielded) resonance for the group in the *cis* position with respect to the C═O than its corresponding trans group, extended this calculation to homopolypeptides and obtained a theoretical downfield shift (0.4 ppm) for an αCH proton in an Rα structure relative to the same proton in an Lα structure. This result has been considered by the AA in agreement with preliminary experimental data given by Ferretti as a private communication and referring to two typical α-helical polypeptides poly-benzyl-L-glutamate (PBG) (Rα) and poly-benzyl-L-aspartate (PBA) (Lα). However, the experimental results referring to these two polypeptides (in TFA/CDCl₃ mixture) as demostrated by other authors (ref. 6, 7 and works cited in ref. 1) and by our own studies (Fig. 3) are in disagreement with the data of these AA. Moreover, in a later work than that by Narashiman and Rogers, other authors[25] showed that in DMF the N-substituted methyls resonate at a higher field (are more shielded) than the corresponding methyls that are trans with respect to the C=O group. As a consequence, it seems evident that, whenever N. and R. values[24] have to be taken into account, those have to be used that are in agreement with the above reported experimental result.

We therefore[26] carried out a calculation of the chemical shift contributions due to the different C=O groups along a polypeptide chain adopting the "correct" N. and R. data[24]. In these calculations the point dipole along the C=O bond has been assumed, in agreement with other Authors[24] to be at 1.21 Å from the C atom, that is, practically localized on the O atom. Results thus obtained have been tabulated (Table 1).

Table 1   Chemical shifts contributions relative to an α CH protons
of an amino acid residue

	Ra	b	c	d	La
Ra	0.60	0.40	0.43	0.33	0.09
b	0.64	0.47	0.48	0.42	0.14
c	0.61	0.39	0.43	0.33	0.09
d	0.62	0.46	0.51	0.40	0.17
La	0.58	0.39	0.43	0.33	0.09

From the calculation it appears that the largest contributions to the αCH shift arise from the nearest neighbour C=O dipoles. All other further removed dipoles, in whatever conformation one may arrange them, turn out to give much smaller contributions, that are, in general, quite negligible (0.01 ppm) considering the approximate character of the whole calculation.

However, the results thus obtained, though justifying the different shielding in the case of the two helical structures, respectively Rα and Lα, are in

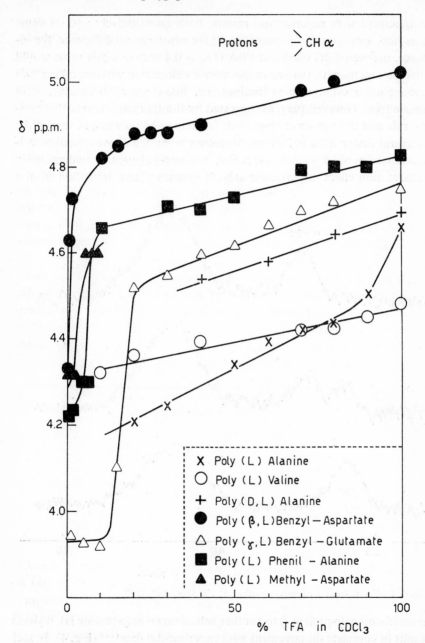

Figure 3 Chemical shifts of the α-CH protons of different polypeptides in CDCl₃/TFA solvent mixtures. The δ values are in ppm from the TMS used as an internal reference at 60 MHz and 27°C

disagreement with experimental results, both quantitatively and, in some cases, also qualitatively. In fact (Fig. 3) the experimental difference, for instance, between PBG (Rα) and PBA (Lα) is 0.4 ppm and this value would coincide exactly with the above mentioned calculation without taking into account other eventual contributions, i.e. side-chain contributions. Side-chain effects, however, play, as suggested by Bradbury and others[6], a decisive role and their order of magnitude is likely to be the same as that of the structural factor itself (0.2 ppm). Moreover when the above described calculation is applied to *poly*-L-proline, a homo-polypeptide undergoing a mutarotation effect in particular solvent systems with a transition from a

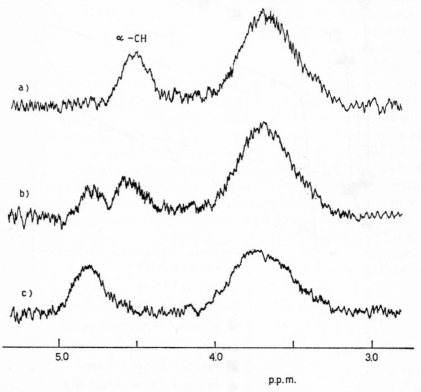

Figure 4 NMR spectra of poly-L-proline at 100 MH$_2$ and 27 °C

helical form (*poly*-proline I) to another helical form (*poly*-proline II), it yields results in complete disagreement with experimental data[11] (Fig. 4). In fact the form I (with peptide bonds in *cis* conformation) results more deshielded than the form II (with peptide bonds in *trans* conformation).

These results are rather difficult to explain on the basis of the data available at present. This behaviour is likely to be due, at least partially, to a different solvent effect in each of the two different *poly*-proline structures. In this respect it is perhaps worth noting that, while helical *poly*-proline I consist of a relatively compact helical structure, *poly*-proline II is less folded, and therefore more accessible to the solvent molecules. Furthermore, in Gramicidine S[27,28] the proline residue gives a resonance peak with a chemical shift quite different from those corresponding to both forms I and II.

It seems, therefore, evident that the chemical shift is affected by the contributions due to the local conformation of the aminoacid residue, the solvent adopted, as well as to the nature of the side chain.

Let us consider the contribution due to the chemical nature of the aminoacid residues. NMR data referring to various aminoacids, *di*- and *tri*-peptides have been reported in the literature.

The preliminary results obtained in the case of glycine dipeptides with various aminoacids in aqueous solvents[29] should imply that the chemical shift of the αCH proton of an aminoacid residue in a peptide chain is very little dependent on the nature of the nearest neighbour aminoacid residues. As a consequence we can suppose that different homopolypeptides in the same conformation, when dissolved in the same solvent, present different chemical shift values for the αCH protons depending on their various side chains. In this connection (Table 2) it is of importance to compare the chemical shifts of the αCH proton of different aminoacids and of the corresponding *homo*polypeptides in TFA (in this latter solvent a random coil conformation is assumed). It must be emphasized that these values, both in the case of aminoacids, and in the corresponding homopolypeptides, are affected by the TFA solvent effect.

Table 2

Compounds	α-CH Aminoacids in TFA	α-Ch Polypeptides in TFA	$\Delta\alpha$-CH Aminoacids in TFA	$\Delta\alpha$-CH Polypeptides in TFA
Glycine	4.20 ppm	4.40 ppm	0.00 ppm	0.00 ppm
Alanine	4.46 ppm	4.68 ppm	0.26 ppm	0.28 ppm
Valine	4.31 ppm	4.48 ppm	0.11 ppm	0.08 ppm
Leucine	4.36 ppm	4.60 ppm	0.16 ppm	0.20 ppm
Phenylalanine	4.60 ppm	4.82 ppm	0.40 ppm	0.42 ppm
Histidine	4.91 ppm	5.08 ppm	0.71 ppm	0.68 ppm

A determination of the αCH proton chemical shift contribution due to the chemical nature of the compound seems possible.

Leaving aside an absolute variation, due to peptide bond formation and to the presence of ionized groups in the aminoacids, the relative differences for the αCH proton chemical shifts coincide taking glycine and *poly*-glycine as references respectively, in the case of aminoacids and polypeptides. These relative differences may therefore be ascribed mainly to the influence of the different side chains, and the relative values so found can therefore be considered to characterize the influence of the different side chains on the αCH proton chemical shift although the limited number of experiments calls for a trend rather than a proof.

As to the solvent contributions the effects connected with the dissolution of a polypeptide in a particular solvent are many and they arise from the kind of solvent adopted, the chemical nature of the particular aminoacid residue of the polypeptide, and eventually, from the specific conformation of the latter. A direct comparison of the various data is in most cases extremely difficult owing to the different solubility of the various polypeptides in the different solvents, and the different stability of their conformations in the same solvent.

The solvent systems adopted (with respect to their influence on the conformation of a dissolved peptide) can be divided into two general classes: "helix supporting" and "coil supporting". This division, however, also depends on the nature of the polypeptide itself, so that the same solvent system may behave as helix supporting for a given polypeptide and as coil supporting for another.

The more commonly employed solvents to study helix-coil transition are:
a) D_2O (pH-induced transition)
b) Mixtures of helix-supporting solvents (e.g. $CDCl_3$ etc.) with coil-supporting solvents (e.g. dichloroacetic acid DCA, trifluoroacetic acid TFA etc.) (composition-induced transition).

In both cases, an up-field shift of the αCH signals is obtained under the effect of the helix-coil transition. The most striking difference that can be observed, is that the entity of the shift is different when transitions occur in aqueous solvents than when they occur in organic solvents[2]. In fact, while in the case of aqueous solvents the variation amounts to 0.1 ppm, in the case of $CDCl_3$/TFA or $CDCl_3$/DCA mixtures it amounts to 0.4 ppm.

A number of considerations have to be made in this respect. First, all polypeptides that have been studied in D_2O at various pH values[2] are compounds possessing an ionizable group in their side chain (—COOH or —NH$_2$

i.e. *poly*-L-lysine and *poly*-L-glutamic acid). Therefore a chemical shift contribution could be obtained due to ionization of the COOH and NH_2 groups. However, the magnitude of such a contribution ought to be smaller as the ionizable group is relatively far away from the αCH proton.

Second, in the case of solvents such as TFA and DCA two hypotheses have been suggested. According to one of these hypotheses, in analogy with simple model systems such as amides, TFA and DCA or like solvents could lead to a protonation of peptide bonds[30]. This hypothesis turns out to be in disagreement with both CD[31] and NMR studies[1,5]. A second, more satisfying explanation implies peptide-solvent hydrogen bond formation. In any case, both protonation and hydrogen bond formation involve a αCH chemical shift variation to more paramagnetic frequency resonance values.

Furthermore, in terms of previously introduced stereochemical codons[23] it has to be considered that, while in a helix an identical codon is being repeated along the chain, in a random coil all codons may be present at the same time, yielding a statistical distribution in which the different codons will have a relative population density determined by their relative energy, as well as by the solvent itself and by the temperature. It can be deduced that the chemical shift variation of a random coil will be, a sum of two distinct contributions, one arising from the progressive solvation, the other arising from a change in the relative population of the different conformational states (*Ra, b, c, d, La*), the latter being also a function of the chemical nature of the side chains.

This accounts at least partially, for the existing αCH proton chemical shift difference for helix-coil transition when in aqueous or organic solvents. The magnitude of this effect in the case of polypeptides can be shown by αCH proton chemical shift variation in *poly*-L-valine and in *poly*-D-L-alanine (0.3 ppm) which hold a random-coil conformation through the whole TFA percentage interval considered (10–100%) (Fig. 3).

As far as only the structural factor contribution is involved, the *La* codon should result in least-shielded conformation with respect to the αCH proton relative to all other allowed conformations. For an *La* conformation therefore, one should expect the helix-coil transition to cause a diamagnetic shift or at least no paramagnetic shift. This disagrees with the experimental results (Fig. 3).

In general the αCH proton chemical shift for the polypeptides we have studied shows two different types of behaviour as far as helix-coil transitions are concerned. A class of compounds (e.g. PBG, PBA, *poly*-L-phenylalanine (PFA) etc.) show a chemical-shift variation, in a very narrow range of the solvent mixture composition (Fig. 3). On the other hand, other compounds (such as *poly*-L-alanine, PLA) show a smooth change of these parameters in a much wider range of solvent mixture variation. This has been related to change in helix content, such a change being higly cooperative in the first case and gradual and continuous in the other.

Besides, this behaviour could be related to the different solvation accessibilities of the different helices and coils depending on the different chemical nature of their aminoacid residues. Solvation is in general related to the accessibility of the solvent to the solute molecule. It is evident that, for different polypeptides, a different accessibility to the solvent of the side chain is obtained due for instance to steric hindrance as well the type of structure; i.e. an ordered one (helix) or a disordered one (random coil).

Data relating to *poly*-L-alanine (Fig. 3) can be connected with this hypothesis. In *poly*-L-alanine, unlike other polypeptides, the αCH proton chemical shift undergoes a variation also when the helical conformation is conserved (as shown by Moffit's b_0 parameter)[5]. The possibility of detecting the helix

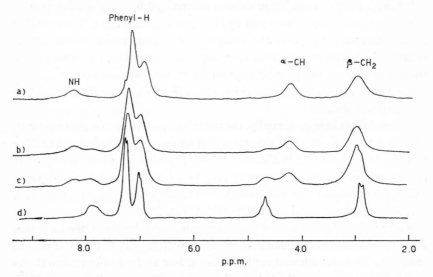

Figure 5 NMR spectra of poly-L-phenylalanine in different mixtures of $CDCl_3$-TFA: a) 2% of TFA; b) 4%; c) 6%; d) 10%

and the coil resonances in the interval in which the transition takes place seems to depend on this accessibility to the solvent. In fact while in some cases (e.g. PFA, Fig. 5) two distinct peaks can be detected whose intensity varies as a function of the helix percentage (this being, in its turn, a function of the solvent mixture composition) in other cases only one peak is observed, varying in position.

The difference between these two cases can be related to the average life-times of the various compounds. In particular in the case of $\tau > 1/2\pi\Delta\nu$ where $\Delta\nu$ stands for the chemical shift difference of αCH protons in the helix and in the coil form respectively, two distinct peaks will be detected. When $\tau < 1/2\pi\Delta\nu$ a single peak will be obtained representing the weighted average of the two conformations.

This effect shows itself in a particularly striking way in *poly*-N-methyl-L-alanine[32] (Fig. 6).

In $CDCl_3$-TFA traces (yielding the helix conformation only) (Fig. 6a), a singlet is obtained for the N–CH_3 group ($\delta = 3.0$ ppm) and a doublet for CH_3 ($\delta = 1.5$ ppm; $J = 7$ Hz). The αCH proton gives a broad quadruplet ($\delta = 5.1$ ppm). By increasing the TFA percentage (coil supporting solvent) (Fig. 6b, c) the three peaks of the above specified resonances split, each given a multiplicity of signals.

This experiment shows, moreover, that, according to what has been specified above, in the coil too, in relative population densities of the different codons are a function of the solvent mixture composition. In fact a change can be observed of the relative intensities in the multiplets of the various resonance peaks.

From all the above considerations it follows that first of all it is generally impossible to compare the chemical shift value for a given aminoacid residue in a homopolypeptide chain with the value referring to the same residue in a heteropolypeptide chain without taking into account all the possible contributions. In this respect the case of Gramicidine S can be examined. In it a comparison of αCH proton chemical shifts of the different residues with the corresponding chemical shift of the same residues in homopolypeptide chains leads to completely erroneous conclusions[33,27,28], if these results ar not made compatible with shift and coupling constant data relative to different groups of the molecule.

Second, the αCH proton chemical shift turns out to be dependent on several other parameters, besides the structural factor, such as the very nature of the aminoacid and the solvent system adopted. Only through a deeper knowledge of these contributions a separation between this structural effect, strictly connected with particular geometrical conformations of an aminoacid residue, and the others will be obtained. It has to be underlined that the effect of the solvents we considered, the effect of the side chains and the structural factor are all of the same order of magnitude.

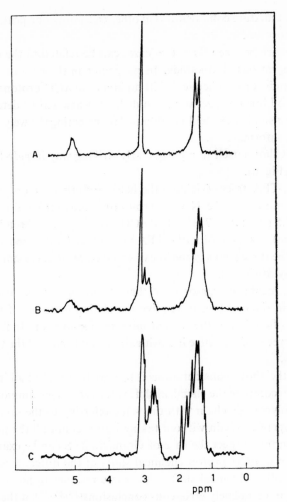

Figure 6 NMR spectra of poly-N-methyl-L-alanine in a) CDCl$_3$ with TFA traces; b) CDCl$_3$ with 10% TFA; c) CDCl$_3$ with 40% TFA

Furthermore, data referring to the conformations of the proline are diffi-cult to interpretate. Calculation of the structural term contribution to the chemical shift yields a result not in agreement with the experimental data.

It has to be underlined that the effect of the solvents, the effect of the side chains and the structural factor we considered, are all of the same order of magnitude.

Finally we have attempted to interpret the NMR spectra of *poly*-N-Me-L-

Alanine[32] and polysarcosine[1] aided by parallel calculations on the conformational energies. Taking into account *cis* and *trans* imide bond conformations we have studied the sterically allowed conformations on the basis of evaluation of conformational energy.

Three minima are present for a TT, TC, CT, CC consecutive imidic groups conformation[32], two of them connected by a low potential energy barrier (~2 kcal/mole) which may be characterized by the same ψ angle of about 260° and two different ϕ angles (respectively ~50° and ~240°). A comparison of the energy maps with the NMR results shows a correspondence between the multiplicity of the possible local conformers an that of the resonance peaks of CH_3, NCH_3, CH protons in mixed solvents.

The possible existence of 2×4 local conformers should produce in the NMR spectra eight peaks for each proton. Two different classes of conformers relative to the *cis* or *trans* conformation of the peptide bond, each split in four different conformers by rotation about the ϕ and ψ angles, should be recognized in the case of NCH_3 protons as well in the other groups, this

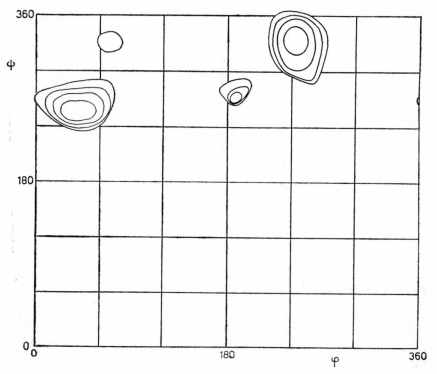

Figure 7   Conformationa ιenergy diagrams of the poly-N-methyl-L-alanine

later complicated by spin–spin coupling. This agree with experimental results. Finally the inspection of the spectra suggests that the more possible helical structure of the polymer should be characterized by a position for N–CH$_3$ and αCH groups in a deshielded zone and for C–CH$_3$ groups in a shielded zone compared to the other sterically allowed conformations.

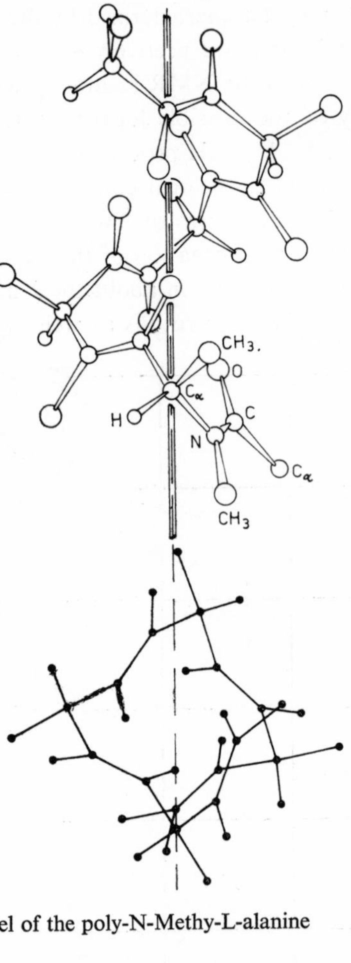

Figure 8   Proposed model of the poly-N-Methy-L-alanine

The inspection of the conformational energy diagrams of the polymer (Fig. 7) suggests a structure characterized by: $\phi = -16°$; $\psi = -90°$ with the imidic bond in trans conformation, corresponding to a monomer repeat equal to 1.67° and a number of monomers for turn equal to 3.82 (Fig. 8).

The above body of results, when also fully analyzed by a check of their consistency, show that a proton chemical shift analysis of polypeptides gives significant information about their conformation and therefore the claim of futility made recently for such an approach[34] is not justifiable.

Acknowledgment

The author would like to thank Prof. A.M.Liquori for his helpful discussions.

Bibliography

1   F.A.Bovey, Lecture presented at the *International Symposium on Marcomolecular Chemistry*, in Brussels, 1967 and references therein.
2   E.M.Bradbury, C.Crane-Robinson, H.Goldman and H.W.E.Rattle, *Biopolymers* **6**, 851 (1968).
3   D.I.Malborough, K.G.Orrel and H.N.Rydon, *Chem. Comm.* No. 21, 518 (1965).
4   J.L.Makley, D.H.Meadows and O.Jardetzky, *J. Mol. Biol.* **27**, 25 (1967).
5   W.E.Stewart, L.Mandelkre and R.E.Glick, *Biochemistry* **6**, 143 (1967).
6   E.M.Bradbury, C.Crane-Robinson and H.W.Rattle, *Nature* **216**, 862 (1967).
7   E.M.Bradbury, C.Crane-Robinson, H.Goldman and H.W.E.Rattle, *Nature* **217**, 812 (1968).
8   E.M.Bradbury and H.W.E.Rattle, *Polymer* **IX**, 201 (1968).
9   E. M. Bradkury, and B. G. Carpenter, C. Crane–Robinson and H. W. Rattlee *Nature* **220**, 69 (1968).
10 F.Conti and A.M.Liquori, *J. Mol. Biol.* **33**, 853 (1968).
11 F.Conti, M.Piattelli and P.Viglino, *Biopolymers* **7**, 411 (1969).
12 D.H.Meadows, J.L.Markley, J.S.Cohen and O.Jardetzky, *Proc. Nat. Acad. Sci. U.S.A.* **58**, 1307 (1967).
13 J.H.Bradbury and H.A.Sheraga, *J. Am. Chem. Soc.* **88**, 4240 (1966).
14 J.H.Bradbury and P.Wilairat, *Biochem. Biophys. Res. Commun.* **29**, 34 (1967).
15 J.S.Cohen and O.Jardetzky, *Proc. Nat. Acad. Sci. U.S.A.* **60**, 92 (1968).
16 A.Kowalsky, *Biochemistry* **4**, 2382 (1965).
17 K.Wüthrich, R.G.Shulman and J.Peisach, *Proc. Nat. Acad. Sci. U.S.A.* **60**, 373 (1968).
18 H.Sternlicht and D.Wilson, *Biochemistry* **6**, 2881 (1967).
19 C.C.McDonald and W.D.Phillips, *J. Am. Chem. Soc.* **89**, 6332 (1967).
20 A.M.Liquori, *Chimica delle Macromolecule*, Ed. C.N.R. 209 (1963).
21 P. de Santis, E.Giglio, A.M.Liquori and A.Ripamonti, *Nature* **206**, 406 (1965).
22 A.M.Liquori, *J. Polymer Sci.* C **12**, 209 (1966).
23 A.M.Liquori, P. de Santis, A.L.Kovacs and L.Mazzarella, *Nature* **211**, 10309 (1966).
24 P.T.Narashiman and H.T.Rogers, *J. Chem. Phys.* **63**, 1385 (1959).
25 L.A.La Planche and H.T.Rogers, *J. Am. Chem. Soc.* **85**, 3728 (1963) and references therein.
26 F.Conti and M.Piattelli, in preparation.

27   F. Conti, *Nature* **221**, 777 (1969).
28   A. Stern, A. A. Gibbons and L. C. Craig, *Proc. Nat. Acad. Sci. U.S.A.* **61**, 735 (1968).
29   A. Nakamura and O. Jardetzky, *Proc. Nat. Ac. Sci. U.S.A.* **58**, 2212 (1967) and references therein.
30   S. Hanlon and I. M. Klotz, *Biochemistry* **4**, 37 (1965).
31   F. Quadrifoglio and D. Urry, *J. Phys. Chem.* **71**, 2364 (1967).
32   F. Conti and P. de Santis, in press in *Biopolymers*.
33   A. M. Liquori and F. Conti *Nature*, **213**, 235 (1968).
34   See Editorial footnote of ref. 27.

Studies on the tertiary and quaternary structure of cyclic antibiotic polypeptides

A. STERN, W. A. GIBBONS and L. C. CRAIG

Rockefeller University, New York **10021**, *U.S.A.*

SINCE THE DISCOVERY of the gramicidins and tyrocidines by Dubos and Hotchkiss (1) over thirty years ago these polypeptides have been used as model substances on which to test various theories and techniques for biochemistry. Gramicidin and tyrocidine were at first thought to be pure crystalline substances but with the development of countercurrent distribution we showed that they were mixtures of several closely related peptides (2) and thus demonstrated that fractional crystallization and the plane rule alone could not be relied upon to establish purity.

Their true molecular weight could not be reliably determined by physical methods because of strong self association but by developing a method called "partial substitution" which made use of CCD (3, 4), Battersby and Craig unequivocally revealed their true minimum molecular size.

Drs. R. L. M. Synge and A. J. P. Martin used them in developing their famous separation methods, partition chromatography and paper chromatography. Consden and collaborators by partial hydrolysis and paper chromatography showed that gramicidin-S, a peptide isolated by Gauze and Brazhnikova (5), contained a pentapeptide sequence Val Orn Leu Phe Pro (6). This was shown unequivocally to be only half the cyclic molecule by our partial substitution method (3). The decapeptide gramicidin-S became a model for ir studies by Abbott and Ambrose (7) and for x-ray crystallographic studies by Hodgkin and Oughton (8). Its structure and conformation were studied by Schwyzer and Sieber (9) by synthesis. More recently we have used

Table 1

1940	Gramicidin and Tyrocidine	Dubos and Hotchkiss
1944	Gramicidin-S	Gauze and Brazhnikova
1947	Sequence Val · Orn · Leu · Phe · Pro	Consden, et al.
	Shown in Gramicidin-S	
1948	Gramicidin-S shown to diffuse as a decapeptide	Pedersen and Synge
1948	These peptides shown to be mixtures by C.C.D.	Craig, et al.
1951	Molecular Weights of Gramicidin S-A and	Battersby and Craig
	Tyrocidine A established unequivocally by	
	C.C.D. and shown to be decapeptides	
1954	Sequence of Tyrocidine A	Paladini and Craig
1957	Gramicidin S-A synthesized	Schwyzer and Sieber
	A conformation suggested	
1957	X-ray crystallographic studies	Hodgkin, et al.
	A conformation suggested	

it as well characterized model to develop our method of this film dialysis (10). A summary of the development up to 1957 is given in Table 1.

A number of theories concerning these peptides and particularly gramicidin-S have been proposed and discarded. These will not be discussed in this talk. They together with the brief history just given, however, do show the need for considerable caution in the interpretations derived from new techniques until they are well confirmed by experience with known models. A striking example comes from the possible misinterpretation of ORD and CD spectra (11) which came to be quite generally accepted as revealing the degree of helicity in macromolecules. Our finding that gramicidin S-A and other cyclic peptides gave this typical spectra came as an unwelcome surprise and partly because of this it became extremely important to determine the exact conformation of gramicidin S-A. A more critical assessment became possible because of the high resolution of the 220 MHz spectrometer. The interpretations of the NMR data, however, were supported by much other data bearing on purity, stability, etc., which were derived from the ultracentrifuge, CCD, thin film dialysis, tritium exchange and ir spectroscopy. The final conformation derived is shown in Fig. 1 (12). It is essentially the antiparallel pleated sheet conformation proposed earlier by Hodgkin and Oughton (8) and by Schwyzer and Sieber (9). The NMR results have been confirmed by Ohnishi and Urry (13).

The type of data obtained by NMR at 100 MHz are shown in Fig. 2. Here there are resonances obviously assignable to three regions, the amide protons, the C_α protons and the side chain protons. Data obtained at 220 MHz are

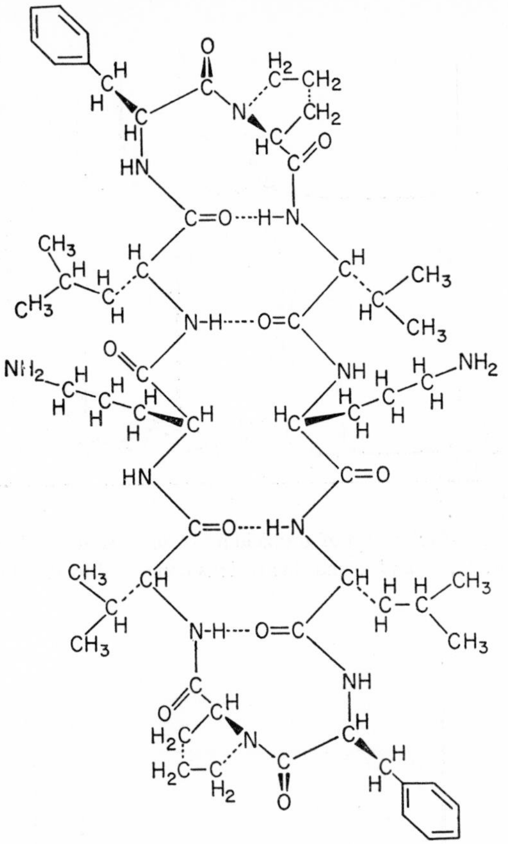

Figure 1 Schematic drawing of model with ω angles all 0 except Leu $\cong$ 10°. Φ angles of Val, Orn, and Leu = 30°, Phe = 150°. ψ angles of Val, Orn and Leu = 0, Phe = 150°, and Pro = 130°. Dashed lines O --- H indicate hydrogen bonds

shown in Fig. 3. It was possible to assign the amide resonances to individual amino acid residues by comparative data from single amino acids, different field strengths, decoupling, solvent perturbation, temperature studies and rates of proton-deuterium exchange. Time does not allow tracing the exact path taken in arriving at the final assignment.

Integration indicated four amide and 5 C_α protons. But since there are known to be 5 pairs of different amino acid residues the C_2 axis of symmetry previously postulated is confirmed because of the magnetic equivalence of the protons in the pairs.

The coupling constants J_{NC} of the three most upfield amide protons are

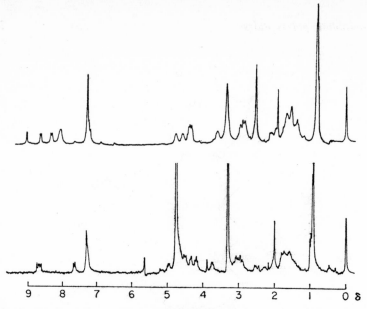

Figure 2 (Lower) 100-Mc spectra of gramicidin S-A in CD₃OD, and (upper) in DMSO-d₆. Resonances: ●, □, × and + indicate regions connected by decoupling

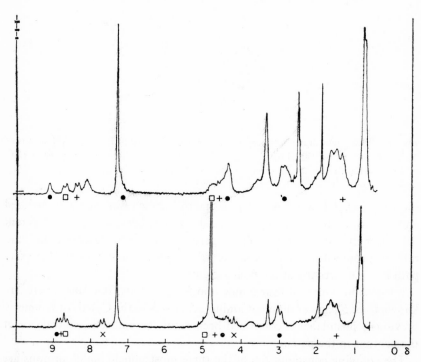

Figure 3 (Lower) 220-Mc spectra of gramicidin S-A in CD₃OD, and (upper) in DMSO p-₆. An additional a peak ppears at 8.98

similar and of the order of 8.5–9 cps whereas the most downfield one, clearly assignable to Phe, is much smaller. On the basis of data in the literature it can be deduced that the ϕ dihedral angles of the two Phe residues approach that expected from a cis configuration while the ϕ angles of the Val, Orn, Leu residues correspond to the trans configuration.

At this point it was possible to construct a model with Corey–Pauling–Koltun models which was consistent with all the data including those derived from the rates of proton–deuterium exchange. The presence of the four hydrogen bonds indicated in Fig. 1, which involve the valine and leucine residues as shown by time and temperature studies, have been confirmed by proton–titrium exchange in dialysis experiments carried out by Laiken *et al.* (14). All the known data are entirely consistent with this conformation. Earlier theoretical calculations of Scheraga and coworkers based on energy minimization have now been recalculated and also found to be consistent with this formula.

Many problems in biochemistry deal with molecular interactions or association of some sort. The tyrocidines (15) are excellent models for this type

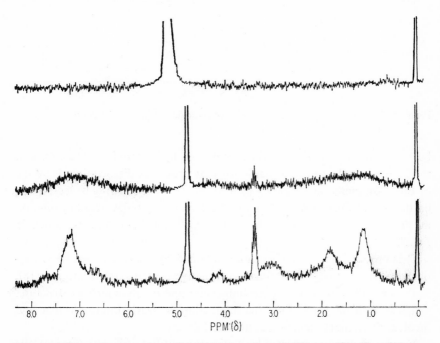

Figure 4 Effect of methanol on 60-MHz nmr pattern of 8% tyrocidine B at 25°. Top = 100% D_2O, middle = 80% D_2O–20% CD_3OD, bottom = 50% D_2O–50% CD_3OD

of study since their association properties have been well characterized by CCD, the ultracentrifuge, ord, cd and thin film dialysis. It was, therefore, of considerable interest to find that the association had a marked effect on NMR resonance as shown in Fig. 4 (16). The association had previously line broadening observed in water is represented by methanol. Independent proof of the dissociation is obtained by thin film dialysis, Fig. 5. Here the rate

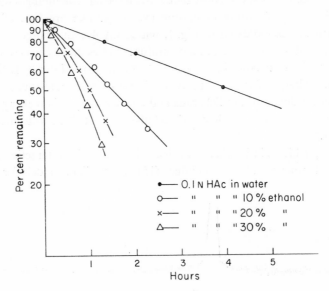

Figure 5 Thin-film dialysis escape patterns of tyrocidine B in various solutions

been shown by us (17) to be largely due to hydrophobic interaction and therefore alcohol or other organic solvents promotes dissociation. Here the of diffusion increases strikingly as alcohol is added showing a decrease in particle size. In aqueous solution the line broadening is decreased by dilution as shown in Fig. 6 and also by increasing the temperature as shown in Fig. 7.

These results are similar to those obtained by numerous workers, particularly Macdonald and Phillips when native proteins are denatured (18). In this case the interaction is intramolecular and involves tertiary structure. Whereas with the tyrocidines the interaction is clearly intermolecular and involves quaternary structure almost exclusively.

This research was supported by National Institutes of Health Grant 02493 and United States Public Health Service Grant AM-02449.

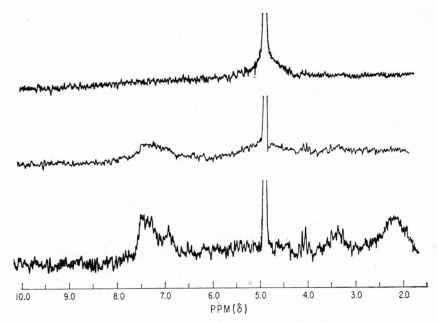

Figure 6 Effect of concentration on 100-MHz nmr spectrum of tyrocidine B in D_2O at 60°. Top pattern = 6%, middle pattern = 2%, bottom pattern = 1%

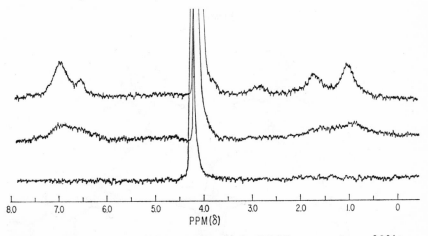

Figure 7 Effect of changing the temperature on the 60-MHz nmr spectrum of 8% tyrocidine B in D_2O. Top pattern = 100°, middle pattern = 80°, lower pattern = 45°

Discussion

P. DE SANTIS The conformation of Gramicidin S presented by Dr. Craig, derived on the basis of NMR data, looks very similar to that found in our laboratory by means of conformational calculations.

We started from the consideration that the sterically allowed structures of a polypeptide are not very different from those which may be derived from the stable local conformations of amino acid residues. Then we tried to derive the most stable structure of a polypeptide by the best sequence of conformers corresponding to the local energy minima (Liquori stereochemical codons) taking into account geometrical constraints (ring closure, sulphur

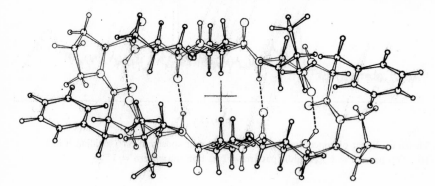

Figure 1

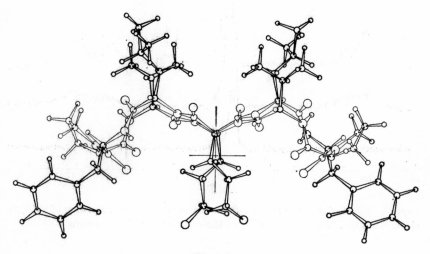

Figure 2

bridges, etc.) and experimental evidences which reveal some features of the structure. The choice may be improved by minimization of conformational energy of the molecule as a whole with the physico-chemical informations and geometrical conditions as constraints using different mathematical techniques.

In the case of Gramicidin SA we selected a number of codon sequences with an end to end distance less than 10 Å imposing conformational equivalence of the two chemical equivalent of the molecule. This condition ensures the presence of a dyad axis if the closure is reached.

During the refining of these structures, on the basis of the optimization of Van der Waals energy, hydrogen bond energy and ring closure, using simultaneous steepest descent processes which revealed very operative in this connection, the NMR experiments of Schwyzer were reported which prove that in Gramicidin SA the Phe and Orn side chains should be closed.

This condition oriented the search on the model schematically reported in Figs. 1 and 2. As may be seen the conformation looks very similar to that presented by Dr. Craig.

Bibliography

1 R.D.Hotchkiss and R.J.Dubos, *J. Biol. Chem.*, **132**, 791 (1940).
2 L.C.Craig, J.D.Gregory and G.T.Barry, *Cold Spring Harbor Symposia Quantitative Biol.*, **14**, 24 (1949).
3 A.R.Battersby and L.C.Craig, *J. Am. Chem. Soc.* **73**, 1887 (1951).
4 A.R.Battersby and L.C.Craig, *J. Am. Chem. Soc.*, **74**, 4023 (1952).
5 G.F.Gauze and M.G.Brazhnikova, *Lancet*, **247**, 715 (1944).
6 R.Consden, A.H.Gordon, A.J.P.Martin and R.L.M.Synge, *Biochem. J.*, **41**, 596 (1947).
7 N.B.Abbott and E.J.Ambrose, *Proc. Roy. Soc.*, (London), A**219**, 17 (1953).
8 D.C.Hodgkin and B.M.Oughton, *Biochem. J.*, **65**, 752 (1957).
9 R.Schwyzer and P.Sieber, *Helv. Chim. Acta*, **40**, 624 (1957).
10 L.C.Craig, *Science*, **144**, 1093 (1964).
11 L.C.Craig, *Proc. Acad. Sci.*, **61**, 152 (1968).
12 A.Stern, W.A.Gibbons and L.C.Craig, *Proc. Nat. Acad. Sci.*, **61**, 734 (1968).
13 M.Ohnishi and D.W.Urry, *Biochem. Biophys. Res. Comm.*, **36**, 194 (1969).
14 S.L.Laiken, M.P.Printz and L.C.Craig, *Biochemistry*, **8**, 519 (1969).
15 M.A.Ruttenberg, T.P.King and L.C.Craig, *J. Am. Chem. Soc.*, **87**, 4196 (1965).
16 A.Stern, W.A.Gibbons and L.C.Craig, *J. Am. Chem. Soc.*, **91**, 2794 (1969).
17 M.A.Ruttenberg, T.P.King and L.C.Craig, *Biochemistry*, **5**, 2857 (1966).
18 C.C.Macdonald and W.F.Phillips, *J. Am. Chem. Soc.*, **89**, 6332 (1967).

High resolution nuclear magnetic resonance studies of alamethicin and valinomycin

E. G. FINER, H. HAUSER and D. CHAPMAN

*Biophysics Department, Unilever Research Laboratory Colworth/Welwyn,
The Frythe, Welwyn, Herts, England*

ALAMETHICIN AND VALINOMYCIN are both cyclic antibiotics which have the property of transporting ions across natural mitochondrial and artificial membranes[1]. We have studied these molecules by high resolution NMR, our studies having two aspects; an investigation of the properties of the antibiotics in solution, and a study of their interaction with phospholipids. The latter aspect is important because both artificial and natural membranes contain a large proportion of phospholipid.

Valinomycin is a cyclic depsipeptide containing 12 residues, 6 α-amino acids and 6 α-hydroxy acids. These are arranged in three groups of four residues, such that the molecule has a threefold axis of symmetry, the sequence being (D-val → L-lac → L-val → D-hyisoval)$_3$. Valinomycin induces ion transport across both mitochondrial membranes and black lipid films, the latter being single artificial bimolecular layers of lipid (normally phospholipid) separating two aqueous compartments[2]. This antibiotic can discriminate between different alkali metal cations, e.g. transport of K^+ is much more efficient than of Na^{+}[1]. Thus an artificial membrane containing valinomycin shows K^+/Na^+ selective permeability, and in this respect resembles nerve cell membranes. The mechanism by which valinomycin induces ion transport is not yet fully understood, but the antibiotic is known to form complexes with the ions it is able to transport; these complexes are soluble in hydrophobic media, like the interior of a lipid membrane[1].

We have studied the NMR spectra of some of these alkali metal complexes dissolved in $CDCl_3$, together with their infra red spectra, in order to obtain information (using chemical shift and coupling constant data) about the structure of the complexes in solution. Our results are in agreement with those recently published by other workers[3], and so will not be described in detail here. Essentially, the valinomycin ring forms a crown with alternate vertical peptide bonds and horizontal ester bonds; the ion is gripped in the centre by six ester carbonyls pointing inwards (sixfold ion-dipole coordination). Hydrogen bonds between peptide NH's and carbonyls link adjacent vertical regions of the backbone, and the sidechains of the residue give the complex a hydrophobic exterior.

Alamethicin is another cyclic antibiotic, containing 18 or 19 peptide residues. The amino acid composition[4] is

$(GluN)_2$ $(Glu)_1$ $(Pro)_2$ $(Gly)_1$ $(Ala)_2$ $(Val)_2$ $(Leu)_1$ $(2\text{-methyl Ala})_{7\,or\,8}$.

The molecule has no overall symmetry[4(c)] and its spectrum in solution is rather complex[5] and yields little or no information about the conformation of the molecule. Like valinomycin, alamethicin complexes metal ions, but in this case with little discrimination[1]. It is interesting, however, in that it induces into black lipid films some of the electrical properties of nerve membranes. Mueller and Rudin[6] have shown, for example, that action potentials can be produced in some circumstances. Some of the interesting electrical properties shown by alamethicin are due to its aggregation properties, which we have studied by NMR[5] and other techniques[7].

The interaction of alamethicin, valinomycin and related molecules with phospholipids[8] has proved to be of great interest. We shall restrict our discussion here to our studies of the interaction of alamethicin with aqueous dispersions of phosphatidyl choline (PC) and phosphatidyl serine (PS).

CH_2OCOR_1 R_1, R_2 are long saturated
| or unsaturated alkyl chains
$CHOCOR_2$
|
$CH_2O\overline{P}O_2OR_3$

Phosphatidyl serine (PS): $R_3 = CH_2CHCOO^-Na^+$

                                     $^+NH_3$

Phosphatidyl choline
(PC, lecithin): $R_3 = CH_2CH_2\overset{+}{N}(CH_3)_3$

In organic solvent, the principal features of the NMR spectra of these molecules are a large peak from the alkyl chain CH_2 protons, a smaller peak from the methyls terminating the chains, and a sharp peak from the $\overset{+}{N}Me_3$ group in PC. The situation in water is a little more complicated because of the aggregation properties of these molecules[9]—they aggregate to form concentric spheres of bilayers, the bilayers being separated by water. The alignment of the phospholipid molecules in the bilayers is such that the polar head groups lie at the bilayer–water interface, while the long hydrocarbon

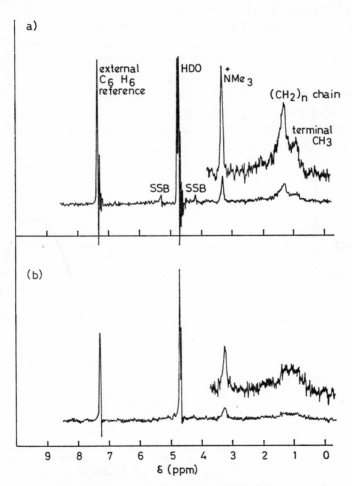

Figure 1 (a) 60 MHz NMR spectrum of a sonicated 1 % D_2O dispersion of egg yolk lecithin. (b) As (a), but with the addition of alamethicin in a molar ratio lecithin:alamethicin 100:1

chains lie in the bilayer interior. There is enough internal motion of the molecules within the bilayers to give an NMR spectrum only about 500 Hz broad[10] (much less than the width of the spectrum of a solid). Sonication of the particles reduces their size sufficiently to enable them to tumble (by Brownian motion) fast enough to give the extra line narrowing which will give a high resolution spectrum[10] (Fig. 1).

If a small quantity of alamethicin is incorporated in the phospholipid dispersion, the high resolution spectrum of the phospholipid is considerably reduced in intensity (i.e. lines are broadened out to such an extent that they are lost in the baseline). This is illustrated for PC in Fig. 1; the effect on a PS spectrum is even more dramatic, the high resolution signal being almost completely removed by alamethicin in a molar ratio of several hundred PS to one alamethicin. This is shown in Fig. 2, where we plot the integral of the high resolution PS hydrocarbon chain signal against molar ratio of PS: alamethicin. The increase in signal at small concentrations of alamethicin is probably due to a detergent effect of the polypeptide increasing the efficiency of the sonication process.

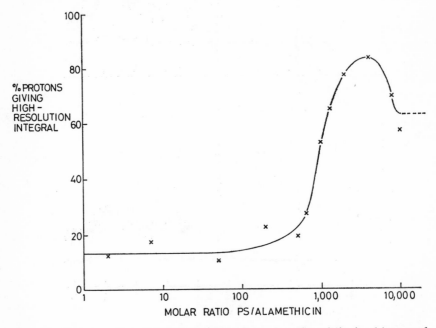

Figure 2 Variation of integral of high resolution PS hydrocarbon chain signal (expressed as a percentage of the total number of chain protons present) with molar ratio [PS]/[alamethicin] for sonicated 1 % D_2O dispersions of ox-brain PS

The origin of the loss of signal lies in a reduction in the mobility of the phospholipid molecules. Column chromatography and ultracentrifuge experiments[5] have shown that this is not due to an increase in particle size, and it is therefore the *internal* motion of the molecules which is being reduced. X-ray scattering experiments[7] and electron microscopy indicate that alamethicin causes a loss of the multilamellar structure which is characteristic of aqueous dispersions. We therefore conclude that alamethicin is breaking down the phospholipid bilayer structure to form a tightly-bound phospholipid-polypeptide aggregate which does not give a high resolution spectrum. We were able to calculate[5] from the shape of the curve shown in Fig. 2, on the basis of a simple equilibrium between PS molecules in the bilayer state and in the lipid-polypeptide aggregate, that one molecule of alamethicin affects about 600 molecules of PS; it seems clear that some sort of cooperative effect must be present.

We believe these findings to be important from several points of view. Not only are they relevant to ion transport models and phenomena in natural membranes, but they are also significant in the general field of lipid-peptide and lipid-protein interactions, which must be understood if we are to unravel the complexities involved in the structure and function of membranes. In addition, the effect of a small amount of antibiotic on the arrangement of a large number of phospholipid molecules may well be relevant to the antibiotic action of alamethicin and related compounds.

References

1 B. C. Pressman, *Federation Proc.* **27**, 1283 (1968).
2 P. Mueller and D. O. Rudin, *Biochim. Biophys. Res. Commun.* **26**, 398 (1967).
3 V. T. Ivanov, I. A. Laine, N. D. Abdulaev, L. B. Senyavina, E. M. Popov, Yu. A. Ovchinnikov and M. M. Shemyakin, *Biochem. Biophys. Res. Commun.* **34**, 803 (1969).
4 (a) C. E. Meyer and F. Reusser, *Experientia* **23**, 85 (1967).
   (b) F. Reusser, *J. Biol. Chem.* **242**, 243 (1967).
   (c) B. S. Hartley and J. Payne, personal communication (to be published).
5 H. Hauser, E. G. Finer and D. Chapman, to be published.
6 P. Mueller and D. O. Rudin, *Nature* **217**, 713 (1968).
7 D. Chapman, R. J. Cherry, E. G. Finer, H. Hauser, M. C. Phillips, G. G. Shipley and A. I. McMullen, *Nature*, 1969 (in the press).
8 E. G. Finer, H. Hauser and D. Chapman, *Chem. Phys. Lipids*, 1969 (in the press).
9 A. D. Bangham, M. M. Standish and J. C. Watkins, *J. Mol. Biol.* **13**, 238 (1965).
10 S. A. Penkett, A. G. Flook and D. Chapman, *Chem. Phys. Lipids* **2**, 273 (1968).

The structure of collagen

G. E. CHAPMAN, I. D. CAMPBELL
and K. A. McLAUCHLAN

Physical Chemistry Laboratory, South Parks Road, Oxford

THE KNOWN STRUCTURES of proteins and enzymes have been deduced from X-ray diffraction studies on single crystals or highly oriented fibres. Such studies yield coordinates of the heavy atoms but it is impossible to locate the hydrogen atoms. From the positions of the heavy atoms and previous knowledge of bond lengths it is possible to construct a model of the protein although a unique solution is not always obtained. An example of this imprecision is found in the collagen molecule in which X-ray studies have shown that the basic structure is a triple helix, each strand of which contains glycine as every third amino acid residue, but cannot yield the secondary structure directly. Of the proposals put forward as to the nature of this structure the most favoured have been the Collagen II structure[1], the Ramachandran two-bonded structure[2] and a recently suggested form which we shall refer to as Collagen IIa[3]. It is useful, therefore, to obtain a method for the determination of the structure of proteins which yields direct measurement of the orientations of chemical bonds relative to some axis in the molecule. Furthermore, a method able to detect and characterise hydrogen bonds would be advantageous. Both of these requirements are met by a method[4] involving the observation of the quadrupole splittings in the N.M.R. spectra of deuterium nuclei substituted by exchange into peptide linkages in the backbone of the protein. It is suitable for application to any fibrous protein and indeed any other biological material which can be obtained in an oriented form, e.g. DNA.

The principle of the method is that all deuterium nuclei in equivalent en-

343

vironments in the structure yield identical N.M.R. spectra. For each type of deuterium environment the N.M.R. spectrum consists of a simple doublet whose splitting is $3/2h \cdot e^2 q_{zz} Q$ if the applied magnetic field is sufficiently high, where Q is the nuclear quadrupole moment and q_{zz} is the component of the electric field gradient of the N–D bond in the magnetic field direction.

$$q_{zz} = q_{11} [\cos^2\alpha_1 - {}^1/_2 (1 - \eta) \cos^2\alpha_2 - \tfrac{1}{2} (1 + \eta) \cos^2\alpha_3]$$

where q_{11} is the field gradient in a molecule-fixed 1-direction, α_n is the angle between the nth principle axis of this tensor and the z-axis and η is a parameter which represents the asymmetry of the tensor in the 1-direction. From this equation it is evident that the observed quadrupole splittings are sensitive both to the nature of the bond and to the angles it makes with the magnetic field direction.

There are many proteins which may be substituted by deuterium simply by exchange. However, if the fibre axis is made coincident with the magnetic field direction, all equivalent deuterons attached to the backbone of the protein yield identical doublets whereas side-chain ones may be expected to have little regularity in orientation and should contribute only a broad background spectrum. In order to translate the observed splittings into angular information it is necessary to know the molecular parameters in the equation above. We assume that those determined for the trans deuteriums in d_4-urea are applicable: $q_{11} = 210.8$ kHz and $\eta = 0.139$,[5] where the 1-axis is parallel to the C–O bond and the 3-axis is perpendicular to the plane of the peptide group. It is convenient to express the angles α_n in terms of the spherical polar angles θ and ϕ defined by the relations $\cos \alpha_1 = \cos \theta$, $\cos \alpha_2 = \sin \theta \cos \phi$, $\cos \alpha_3 = \sin \theta \sin \phi$.

To test the assumption as to the magnitudes of the molecular parameters a sample of a simple α-helical molecule was studied. This was 300,000 m.w. poly-γ-benzyl-L-glutamate which had been deuterated by treatment with anhydrous deutero-trifluoracetic acid and dissolved in dichloromethane. In the presence of an applied field this phase is known to orient so that the helices are parallel to the field[6]. For the α-helix, working from know co-ordinates[7], $\theta = 14.1°$ and $\phi = 0°$. From these figures a splitting of 289.4 kHz was predicted and one of 288.3 kHz observed in a spectrometer operating at about 50 kilogauss with a superconducting magnet.

The case of collagen is rather more complicated. Each of the three most likely models of its structure contain three different orientations of N–D bonds with respect to the fibre axis and so should yield three doublets. There are also three distinct hydrogen-bonded states of the peptide hydrogens: the short hydrogen-bond, as in the α-helix (state A), a long hydrogen-bond (state B) and no formal hydrogen bond (state C). Types B and C almost certainly involve an increase in the quadrupole coupling constant from the urea value, as is evident in D_2O as the degree of hydrogen-bonding is decreased[8]. We estimate the maximum possible limits of this increase to be 10% for state B and 50% for C. Another apparent complication is that it is impossible to orient the fibres accurately along the magnetic field direction and this implies that a distribution of q_{zz} values is sampled and leads to a line broadening proportional to

$$\left[\left(\frac{dq_{zz}}{d\theta} \right)_\varphi^2 + \left(\frac{dq_{zz}}{d\varphi} \right)_\theta^2 \right]^{\frac{1}{2}}$$

This effect actually yields extra information therefore, as the relative linewidth is a calculable parameter.

It is now possible to construct a table of predicted splittings and linewidths for each of the three models from their coordinates:

Model	θ	ϕ	State	Splitting (kHz)	Relative broadening
Collagen II	71	36	C	102–153	0.90–1.35
	81	30	C	136–204	0.47–0.71
	83	53	A	157	0.39
Collagen IIa	70	42	C	101–152	0.97–1.45
	70	27	C	91–137	0.96–1.44
	85	38	B	147–162	0.29–0.32
Ramachandran	83	53	B	157–173	0.39–0.43
	86	39	B	151–166	0.25–0.27
	73	38	C	107–161	0.84–1.26

The sample studied was a cylindrical piece of well-oriented bovine achilles tendon that had been immersed in acidified D_2O for three months, removed and partially dried. Its spectrum was recorded as the sum of 300 accumulated spectra and consisted of three doublets of separations 157 kHz, 127 kHz and 117 kHz. The two smaller doublets consisted of apparently equally broad lines whilst the outer doublet lines were much sharper. These observations appear to favour the Collagen IIa model and it is concluded that the major part of the molecule has this structure.

G.E.C. thanks the Medical Research Council for a maintenance award.

References

1   A. Rich and F. H. C. Crick, *J. Mol. Biol.* **3**, 483 (1961).

2   G. N. Ramachandran, *Treatise on collagen* **1**, 103 (Academic Press, 1967).

3   W. Traub, A. Yonath and D. M. Segal, *Nature* **221**, 914 (1969).

4   G. E. Chapman, I. D. Campbell and K. A. McLaughlan, *Nature*, to be published.

5   T. Chiba, *Bull. Chem. Soc. Japan* **38**, 259 (1965).

6   S. Sobajima, *J. Phys. Soc. Japan* **23**, 1070 (1967).

7   R. D. B. Fraser, T. P. MacRae and A. Miller, *Acta. Cryst.* **17**, 813 (1964).

8   P. Waldstein, S. W. Rabideau and J. A. Jackson, *J. Chem. Phys.* **41**, 3401 (1964).

A case of high proton mobility at low temperatures in human dental hard tissues

E. FORSLIND

The Royal Institute of Technology, Stockholm 70

N. MYRBERG

Umeå University, Umeå

B. ROOS

The University of Stockholm

U. WAHLGREN

The University of Stockholm

FROM A SERIES of investigations based on proton magnetic resonances and dielectric relaxation in water adsorbed on hydrophilic clays (D. M. Anderson, A. Jacobsson and E. Forslind 1966–1967; A. Jacobsson and E. Forslind, 1968–1969) it was inferred that the spectral band narrowing at low temperatures observed in human dental enamel and dentine (Myrberg, 1968) might be due to exchange narrowing effects and a theoretical investigation was indicated to explore the possibility.

The present paper summarizes some pertinent observations on human dental hard tissues made by Myrberg (loc. cit.) at the NMR Research Group of the Institute of Technology in Stockholm. It further reports some preliminary results from a theoretical study of the hydroxyl interactions in the apatite lattice, carried out by B. Roos and U. Wahlgren.

Human dental enamel from 14 clinically sound permanent teeth collected in low fluoride areas was dissected free by removing the dentine with the aid of diamonds and burs, using running water as a cooling medium. Purity of enamel was checked by the use of ultraviolet light.

The NMR wide line bandwidth, defined as the square root of the second moment, M_2, of the proton absorption band of dental enamel has been measured as a function of the temperature with the results shown in Fig. 1. The measurements were performed in a Varian V-4200 B spectrometer at fixed frequency 14.8 MHz. The second and fourth moments were evaluated from formal band shape analyses based on the expansion of the absorption band derivative in a series of the linear harmonic oscillator wave functions, the ground state of which represents a gaussian distribution that is adapted to the normalized experimental non-gaussian band shape by the addition of oscillator overtones.

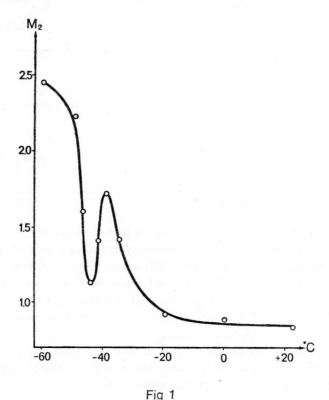

Fig 1

Figure 1

Figure 1 reveals two remarkable features of the proton bands. The first is the persistance of low bandwidths down to temperatures below −40°C and the second is the appearance of a pronounced minimum at about −45°C. The small bandwidth indicates weak coupling and high mobility or extreme isolation of the protons, or both, even at low temperatures. This observation seemed at first to be at variance with a suggestion by Little and Casciani (5) that the water in dental hard tissue is essentially coupled to the apatite lattice. Under normal coupling conditions the water would then be expected to prolduce a very broad proton resonance band. As it turns out, the hydroxy coupling is, however, far from normal.

In a recent discussion of the properties of various apatites, Young and Elliott (6) have made a detailed study of the OH–Ca coordination group in the apatite lattice based on the X-ray structure redetermination of the hydrody-apatite made by Kay, Young and Posner (7).

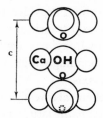

Figure 2

They find a minimum distance between the oxygens along the *c*-axis of about 2.75 Å while the maximum distance is about 4.13 Å (Fig. 2). There is also an intermediate oxygen separation possible of about 3.44 Å, corresponding to the O–O distance of a long and rather weal hydrogen bond. These oxygen distances obviously vary with the thermal lattice vibrations which may induce the hydroxyls to move through the plane of the three C,a ions to take up more or less short-lived equilibrium positions on either side of the Ca plane, eventually giving rise to a certain amount of hydrogen bonding between the hydroxyls. In order, however, to arrive at an estimate of the energy and stability of the hydrogen bonds corresponding to the minimum and intermediate oxygen separations shown in Fig. 2, it is necessary to consider the Ca–OH interaction in some detail.

To a first approximation the situation may be summarized as follows.

The positions of the Ca-ions is expected to produce a polarization of the OH^- oxygens very close to an sp^2 type of hybridization. The three lone pair electron hybrids, each capable of donating one electron to the adjacent calciums, could contribute to a d^2p trigonal pyramid calcium hybrid. This is the most probable configuration suitable to link pairs of oxygens of the neighbouring PO_4 tetrahedra to the hydroxyl as may be inferred from the PO_4 oxygen valence angles. The corresponding Ca–O distance is computed from X-ray data (6) to 2.386 Å, the sum of the ionic radii being, according to Pauling, 2.39 Å.

The covalent contribution to the configuration just described may be expected to resonate with the remaining Ca ions of the lattice. The strong trigonal polarization of the OH^--group will, however, persist and prevent a strong polarization in the c-axis direction, reducing the strength of the intermolecular hydrogen bonds along the group axis and excluding the intermediate and long spacing bonds as energetically unfavourable. Since the c-axis direction is the only possible direction of the hydrogen bonding between hydroxyls it follows that the lifetimes of the direct hydroxyl interactions will be very short at body temperature. Provided the thermal lattice excitation can, as expected, induce hydroxyl transitions through the plane of the Ca atoms, the frequency of the hydroxyl passage through the plane, the inversion frequency, will be correspondingly high, facilitating an eventual transport of hydroxyls along the c-axis.

The ab-initio quantum mechanical investigation of the hydroxyl ion transport mechanism in the hydroxyapatite lattice, now going on at the Institute of Theoretical Physics, University of Stockholm, makes use of the LCAO-MO-SCF method, with Gaussian functions as basis functions. The computation is done with the molecular program IBMOL on an IBM 360/75 computor.

The investigation is carried out in two steps: first the equilibrium position of one OH^- in the field of the rest of the lattice is calculated, then the interaction between three OH^- in the field of the remaining lattice. In the case of the three OH^- calculation, two of the ions will be held in fixed positions, and the third, middle one will be moved through a Ca plane.

To find the correct lattice potential a program was written, which makes use of the Ewald method for finding the Madelung energy and the potential in any point of the lattice with any number of ions subtracted.

The lattice was supposed to be purely ionic, with Ca^{2+}, $(PO_4)^{3-}$ and OH^- ions. The $(PO_4)^{3-}$ ion was divided into phosphorus and oxygen, with charge $+0.5$ on the phosphorus and -0.875 on the oxygen. The charge distribution was taken from an earlier calculation on $(PO_4)^{3-}$. The potential was, however, not very sensitive to variations of the internal charge distribution of P and O in the $(PO_4)^{3-}$ group.

In the first step of the calculation, which is now ready, the potential of the lattice with one vacant OH^--position was calculated, and an SCF-calculation of one OH^- in three different positions along the OH^--axis was performed. The total energy obtained for these three points was fitted to a parabola to estimate the minimum total energy and thus the equilibrium position of the OH^-.

The potential was then calculated along the OH^- axis and along four axes parallel to it, covering three unit cells, with the vacant OH^--position in the middle unit cell. The density of the points calculated was 20 points per axis per unit cell. The parallel axes were placed in a plane bisecting one of the angles of a Ca-triangle; axis 1, three a.u. from the OH^--axis, axis 2, one a.u. from the OH^--axis, axes 3 and 4, one and three a.u., respectively, on the other side of the OH^--axis. Due to the C_3-symmetry of the lattice this implies that the potential is known on thirteen axes, twelve off and one coinciding with the OH^--axis.

Forty-four point charges were then distributed in $1\frac{1}{2}$-unit cells around a vacancy (cell 000 and half of cell 001 if the z-axis is taken parallel to the OH^--axis) in order to simulate the crystal field. The positions of the point charge were chosen in an arbitrary way, most of them, however, coinciding with ion positions but some of them randomly distributed to compensate for the other ions and the remainder of the lattice. The values of the point charges were then calculated by least square error techniques using the potential in 155 points. This gave, along the OH^- axis and the axes 2 and 3, deviations from the lattice potential which were within 0.005 a.u. Since the deviations were all rather equal and of the same sign, the effective difference, but for a constant factor, was about 0.002 a.u. Along the axes 1 and 4, the deviation was larger, ~ 0.02 a.u. The SCF-calculation gave an equilibrium distance of 0.613 a.u. from the Ca plane, which should be compared with the experimental value, 0.640 a.u. (neutron diffraction, ref. 7). The results of these calculations are summarized in Table 1.

In the second step of the calculation, which will be reported in detail later, some preliminary results are now ready.

In the calculations of the lattice potential and in the potential fitting, the same methods as in the first step were used, and the accuracy of the fitting procedure attained the same order of magnitude.

In the SCF-calculations the equilibrium distance obtained between the OH^- group and the Ca-plane was too short or about 0.4 a.u. below the plane. This deviation from experimental results was supposed to be due to the Van der Waal's interaction between the Ca^{2+}-ions and the OH^--ion. A

Table 1

Basis set: $(O/7.3)\,(H/4.1)$ contracted to
$$\langle O/4.2\rangle\ \langle H/2.1\rangle$$
OH^- total energy, $E_0 = -75.237544$

Points of SCF-calculation:

$R\,(Ca\!-\!O) = 0.30\ \ 0.60\ \ 0.90$ a.u.

$(R\,(Ca\!-\!OX) = R\,(Ca\!-\!O) + 1.808779$ a.u.$)$

Result

$R\,(Ca\!-\!O)$	E
0.30	-87.731283
0.60	-87.734335
0.90	-87.731763

Parabola fitting gives $R\,(Ca\!-\!O)_{eq} = 0.613$ a.u.
expr. $R\,(Ca\!-\!O) = 0.640$ a.u. (neutr. diff.)

Charges on O and H according to Mullikens population analysis (gross atomic population) at $R\,(Ca\!-\!O) = 0.60$:

$$Q(O)\!: \ -1.17 \text{ a.u.} \qquad Q(H) = +0.17 \text{ a.u.}$$
$$1 \text{ a.u.} = 0.52917\,\text{Å}$$

rough estimate of this interaction was therefore added to the potential curve, giving rise to a new equilibrium distance of 0.63 a.u.

The present results of the theoretical investigation support the conclusions drawn in the introductory part of this paper regarding the weakness of the interhydroxyl hydrogen bonding. It appears that, under the assumptions made of purely ionic interactions, the polarization of the hydroxyl oxygen in the direction of the c axis decreases in the field of the surrounding Ca ions. This is a prerequisite for the predominance of electronic charge s-character at the proton sites needed to enhance the efficiency of the indirect tensorial spin coupling mechanism. The detailed investigation of these polarization effects is in progress. The calculations will moreover be extended to cover the substitution of fluorine for oxygen.

Acknowledgments

The authors are indebted to Professor Inga Fischer-Hjalmars for encouraging interest.

References

1 Forslind, E., *Acta Polytechnica 3*, No. 5, 115 (1952).
2 Forslind, E., *Proc. 2nd Intern. Congr. on Rheology*, Oxford 1953, p. 50.
3 Forslind, E., *Swedish Com. and Concr. Res. Inst. Proc.* **16**, 1952.
4 Anderson, D.M., and Low, P.F., *Soil Scu. Soc. of Am. Proc.* **22**, 99, 1958.
5 Little, M.F., and Casciani, F.S., *Archs oral Biol.* **11**, 565, 1966.
6 Young, R.A., and Elliott, J.C., *Archs oral Biol.* **11**, 699, 1966.
7 Kay, M.I., Young, R.A., and Posner, A.S., *Nature* (London) **204**, 1050, 1964.
8 Anderson, D.M., Jacobsson, A., and Forslind, E., Progress reports for the years 1966–1967 to be published.
9 Jacobsson, A., and Forslind, E., Progress reports for the years 1968–1969 to be published.
10 Myrberg, N., *Proton Magnetic Resonance in Human Dental Enamel and Dentine.* Translations of the Royal Schools of Dentistry, Stockholm and Umeå, No. 14, 1968.

^{13}C *NMR spectroscopy of biopolymers**

PAUL C. LAUTERBUR[†], E. JAMES RUNDE,
and BENETT L. BLITZER

Department of Chemistry, State University of New York at Stony Brook
Stony Brook, New York 11790, U.S.A.

RECENTLY IT HAS BECOME apparent that new experimental techniques and improved apparatus will permit the application of ^{13}C nuclear magnetic resonance spectroscopy to very complex chemical systems, including many of biological interest.[1] Some general considerations relevant to such applications will be outlined below, some predictions will be made, and a first attempt to observe the natural-abundance ^{13}C spectrum of a protein will be described.

The field of ^{13}C NMR spectroscopy has developed slowly over the part thirteen years. Because the magnetic moment of ^{13}C is only about one-quarter as large as that of ^{1}H, and because its natural abundance is only 1.1 per cent, ^{13}C NMR spectra are thousands of times weaker than ^{1}H NMR spectra, and genuine applications to structural chemical problems have been rare. In several laboratories, however, continuing experimental and theoretical studies have gradually revealed a number of relationships between ^{13}C chemical shifts (and coupling constants) and molecular structures and properties. The development of double resonance ("spin decoupling") techniques[1(c),2] increased the usefulness of ^{13}C NMR spectroscopy many-fold by eliminating the confusion of overlapping multiplets and by making pos-

---

* Preliminary results of this work were presented at the Third International Conference on Magnetic Resonance in Biological Systems, Warrenton, Va., October 16, 1968.

† Alfred P. Sloan Fellow 1965–1969.

sible signal enhancements by the nuclear Overhauser effect. Further large increases in sensitivity resulted from the development of multiple-scan techniques employing signal averaging devices,[3] and from the introduction of large magnets which permitted the use of large samples at high magnetic fields. The imminent development of ^{13}C Fourier Transform spectroscopy[4] promises to increase sensitivities by another one to two orders of magnitude, making ^{13}C NMR spectra about as readily obtainable as were ^{1}H spectra a few years ago.

In the light of these developments, the ^{13}C NMR spectroscopy of biopolymers appeared to hold considerable potential interest. Despite great promise, and some impressive achievements,[5] the ^{1}H NMR spectroscopy of such compounds is faced with rather frustrating difficulties. Even in the smaller proteins, only a few ^{1}H resonances can usually be resolved, and the rest overlap to give uninterpretable broad peaks. Selective replacement of most of the ^{1}H in the molecule with ^{2}H, when possible, permits more detailed spectral analyses,[6] but the broadening of the resonances in proteins of higher molecular weight or in more rigid structures will still limit the usefulness of the spectra. ^{13}C NMR spectroscopy holds the promise of at least partial relief from these difficulties. The ^{13}C chemical shift range of about 200 ppm is about 20 times larger than that for ^{1}H, and, although detailed predictions are difficult, the ^{13}C resonances may often be narrower than those of ^{1}H in the same molecule because of the smaller dipole–dipole broadening experienced by ^{13}C. Furthermore, the ^{13}C resonances can be reduced to single lines by noise decoupling of the ^{1}H, but ^{1}H resonances are usually multiplets. It seems reasonable, therefore, to hope that ^{13}C NMR spectra of proteins and other biopolymers will often exhibit greater effective resolution than do ^{1}H NMR spectra of the same substances. There is also the attractive possibility of preparing selectivity ^{13}C labeled molecules by chemical and biosynthetic means, permitting the detailed assignment of complex spectra, and simplifying studies of structural perturbations, such as interactions with substrates and inhibitors. Isotopic tracer experiments with labile or complex substances will also be possible, since neither degradation nor isolation of the molecule may be necessary before determination of the locations of the labels.

To illustrate the opportunities that may be offered by this technique, and to guide our experimentations and efforts at assignment of spectra, we have simulated the spectra of a large number of compounds, including all of the common amino acids and many proteins. The ^{13}C chemical shifts for the amino acids were obtained from the literature,[4,7,8] from experiments in our own laboratory, and by application of additivity relationships.[7a,8] The carbon–hydrogen one-bond spin–spin coupling constants were usually estimated with the aid of published data and appropriate rules.[9] The calculations were carried out with a computer program that generated line spectra from tables of shifts, couplings, and amino acid compositions, and then performed convolutions with a line shape function (usually Lorentzian) to obtain simulated spectra with any desired line width.[10] Figure 1 shows the simulated spectra obtained for three proteins, assuming Lorentzian line shapes with full widths of 24 Hz (about 1 ppm). The abscissa is calibrated in ppm to high field of the carbon disulfide resonance. No structural effects are included in the simulations; it has been assumed that all amino acid residues of a given kind have identical spectra, so that even the effects of primary structure are not included. If the protein sequence has little effect on the ^{13}C chemical shifts, the spectra may resemble rather closely the actual spectra of fully denatured proteins, and should provide a useful first approximation to the general appearance of the spectra of the native forms.

The general impression produced by the simulations, even at the relatively low resolution shown in these examples, is moderately encouraging. The carbonyl carbons form a well-separated group centered around 18 ppm, the guanidinium carbon of arginyl residues gives a peak at about 37 ppm, various aromatic carbons have resonances in the region between 35 and 80 ppm, and the α carbons give peaks in the lower part of the high-field aliphatic carbon region. Within each group there is a considerable amount of structure, differing from one protein to another.

The effect of decoupling, to eliminate all splittings of the ^{13}C resonances by ^{1}H, is simulated in Fig. 2 for an arginine-rich protamine. The spectrum in the 120–190 ppm region, containing α-carbons and aliphatic side-chain carbons, is considerably simplified by decoupling. Enhancement of the peaks by the nuclear Overhauser effect will accompany decoupling, but will not otherwise affect the appearance of the spectra unless different peaks are enhanced to different extents. This may occur, but it is not yet possible to estimate the degree to which it will be important in decoupled polymer spectra.

At the time this work was begun, the observation of resonances of single carbons in protein solutions, at typical maximum concentrations of about 0.01 molar, could be shown to be impossible in experiments of reasonable duration with the instruments available to us. This circumstance made it necessary to begin by developing techniques for obtaining greater sensitivity by sacrificing some resolution. In hen's egg white lysozyme, there are 155 carbonyl groups, counting both those in the main polypeptide chain and those

RIBONUCLEASE A
AMINO ACID SET   2   LORENTZIAN   HALFWIDTH=0012   FREQUENCY =   25,143 MHZ
                                                  UNDECOUPLED

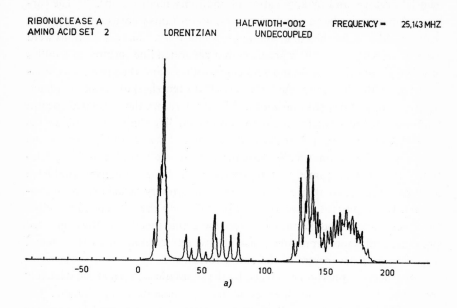

a)

BOVINE SERUM ALBUMIN
AMINO ACID SET 2   LORENTZIAN   HALFWIDTH= 0012   FREQUENCY =   25.143 MHZ
                                                  UNDECOUPLED

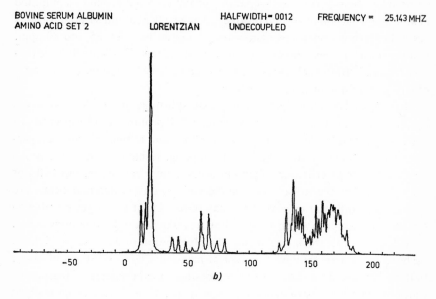

b)

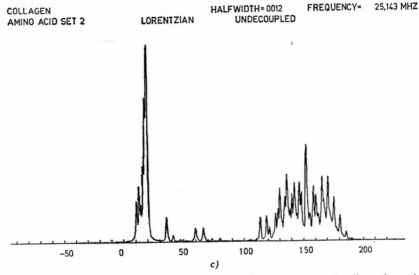

COLLAGEN   HALFWIDTH= 0012   FREQUENCY=   25.143 MHZ
AMINO ACID SET 2   LORENTZIAN   UNDECOUPLED

c)

Figure 1   Simulated 25.1 MHz NMR spectra of (a) bovine pancreatic ribonuclease A,
(b) bovine serum albumin, and (c) rabbit skin collagen

in the side chains of glutamyl, glutaryl, asparagyl and aspartyl residues. The
total carbonyl group concentration in a 0.01 molar solution of the protein is
therefore 1.55 molar, and the combined resonances of all such groups, combined as a single peak, should be readily observable.[11] Similar considerations
apply to other groups of resonances in the spectra. This general approach
to the observation of complex spectra will retain its usefulness even as much
more sensitive spectrometers become available. There will always be solutions of interest that are too dilute to permit the detection of single lines, and
those for which broad lines preclude analyses based upon the identification
of completely resolved lines.

The three spectra in Fig. 3 show the effect, on simulated lysozyme spectra,
of three different filtering functions. The broadest, for which the full line
widths are 200 Hz (about 8 ppm), gives a spectrum in which five partially-
resolved bands can still be identified. The first, at 10–20 ppm, contains the
carbonyl resonances; the second, between 30 and 45 ppm, contains those
from the arginyl guanidinium carbons, the tryosyl hydroxyl-bearing carbons,
and the substituted ring carbons in phenylalanyl residues; the third, between
45 and 90 ppm, contains the rest of the aromatic carbons, and the fourth and
fifth, between 120 and 190 ppm, consist of α-carbon and aliphatic side chain
resonances.

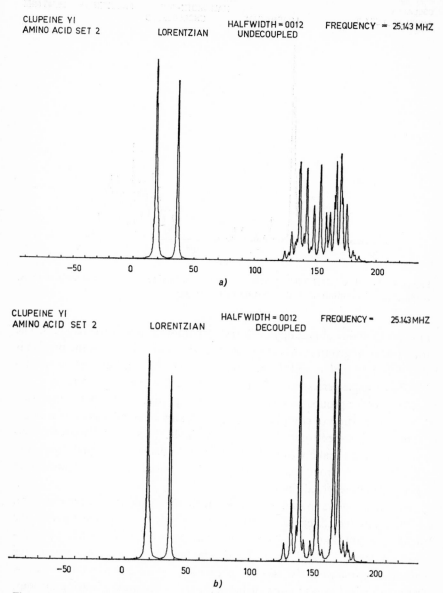

Figure 2 Simulated 25.1 MHz ^{13}C NMR spectra of clupeine YI (Pacific Herring): (a) undecoupled, (b) decoupled

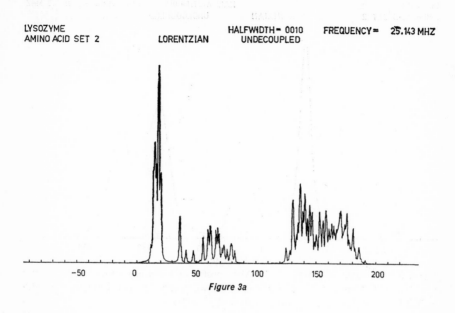

LYSOZYME
AMINO ACID SET 2   LORENTZIAN   HALFWIDTH = 0010   FREQUENCY =   25.143 MHZ
                                            UNDECOUPLED

Figure 3a

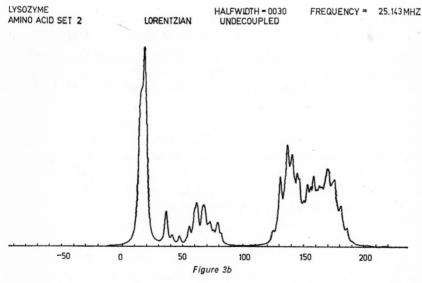

LYSOZYME
AMINO ACID SET 2   LORENTZIAN   HALFWIDTH = 0030   FREQUENCY =   25.143 MHZ
                                            UNDECOUPLED

Figure 3b

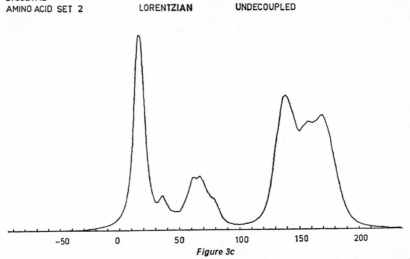

Figure 3   Simulated 25.1 MHz ^{13}C NMR spectra of hen's egg white lysozyme after application of different filtering functions: (a) full width at half height 20 Hz, (b) full width at half height 60 Hz, (c) full width at half height 200 Hz

In order to obtain a spectrum to compare with such simulated spectra, we used a Varian HR-100 spectrometer, equipped with a 25.1 MHz RF Unit and a C-1024 signal averager. In the initial series of experiments, the C-1024 ramp output was used to sweep the magnetic field with the probe sweep coils, and the C-1024 was triggered by a peak of the quartet from a 55% ^{13}C enriched sample of methyl iodide in a 1 mm capillary. The combination of a sharply curved baseline and interference by the low-field tail of the methyl iodide reference signal made it impossible to obtain reliable signals in the high-field portion of the spectrum. Recognizable peaks, however, could be seen in the low-field portion, as shown in Fig. 4. The peaks there, in the range 0 to 60 ppm, are superimposed on a sloping baseline, but the features in the simulation are clearly observable. This spectrum was the result of 200 50 second scans on a 10% solution in a 5 mm sample tube, digitally filtered with a 100 Hz half-width filter function.

An improved system for obtaining spectra averaged over a great many scans, free from baseline curvature, has been developed.[12] This computer-controlled spectrometer system is now in use, and much-improved spectra are expected in the near future. It seems clear that improvements in NMR spectrometers and in signal-processing techniques will soon make the ^{13}C NMR spectroscopy of biopolymers a practical reality.

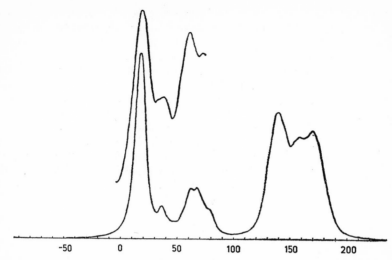

Figure 4 Upper trace: Actual partial 25.1 MHz ¹³C NMR spectrum of hen's egg white lysozyme, 5 mm O.D. sample tube, 10% solution in H_2O, pH 4, 25°, 200 50 sec. scans, digitally filtered to 200 Hz full width at half height. Lower trace: Simulated spectrum from Fig. 3(c) on the same scale

Acknowledgements:

This work was partially supported by a Grant-in-Aid from the Graduate School of the State University of New York at Stony Brook, by the National Science Foundation, and by the Petroleum Research Fund of the American Chemical Society. Data acquisition and processing were greatly aided by support for the Department of Chemistry IBM 1800 Computer Center by the New York State Science and Technology Foundation and by Air Force Cambridge Research Laboratory, Contract F-19-(628)-69-C-0082.

References

1 For reviews and references, see (a) P.C.Lauterbur, Chapt. 7 in *Determination of Organic Structures by Physical Methods*, ed. F.C.Nachod and W.D.Phillips, Academic Press, N.Y., 1962; (b) J.B.Stothers, *Quart. Rev.*, **19**, 144 (1965); (c) F.J.Weigert, M.Jautelet, and J.D.Roberts, *Proc. Nat. Acad. Sci. (U.S.)*, **60**, 1152 (1968).
2 For a review of the principles and early work see J.D.Baldeschwieler and E.W.Randall, *Chem. Rev.*, **63**, 81 (1963).
3 See R.R.Ernst, *Advances Mag. Res.*, **2**, 1 (1966), for a review and detailed discussion.
4 W.Horsley, H.Sternlicht, and J.S.Cohen, *Biochem. Biophys. Res. Comm.*, **37**, 47 (1969).

5 A good selection of examples may be found among the other papers in this volume. See also E. M. Bradbury and C. Crane-Robinson, *Nature*, **220**, 1079 (1968) for a recent review.

6 J. L. Markley, I. Putter, and O. Jardetzky, *Science*, **161**, 1249 (1968).

7 (a) W. Horsley and H. Sternlicht, *J. Am. Chem. Soc.*, **90**, 3738 (1968); (b) H. Sternlicht, private communication.

8 (a) Ref. 1(a), pp. 492f.; (b) D. M. Grant and E. G. Paul, *J. Am. Chem. Soc.*, **86**, 2984 (1964).

9 A. W. Douglas, *J. Chem. Phys.*, **40**, 2413 (1964) and references therein.

10 E. J. Runde, J. L. Ackerman and P. C. Lauterbur, to be published.

11 This was originally demonstrated during preliminary experiments by one of us (P.C.L.) at the Spectrospin laboratories of Bruker Physik in Zurich, Switzerland during November, 1967, without proton decoupling.

12 E. J. Runde and P. C. Lauterbur, to be published.

Mg²⁵ *nuclear resonance as a probe of* Mg⁺⁺ *complex formation**

J. A. MAGNUSSON† and A. A. BOTHNER-BY

Department of Chemistry, Carnegie Mellon University,
4400 Fifth Avenue, Pittsburgh, Pennsylvania

Abstract

Observations of the Mg²⁵ nuclear magnetic resonance of solutions of magnesium chloride in water are reported. Addition of certain substances known to complex with Mg⁺⁺ ion cause an observable, pH dependent broadening of the resonance signal. With certain assumptions these observations may be interpreted in terms of complex formation constants. The formation constants for complexes with citric acid, adenosine monophosphate, dimethyl pyrophosphate, adenosine triphosphate, and epinephrine are estimated. The values are in rough agreement with expectation or with measurements based on alternative methods.

MANY ENZYMATIC REACTIONS display a requirement for magnesium ion.[1] The exact role of the magnesium is not known, although it seems probable that it is complexed in some way with the enzyme or substrate or both. Complexes of magnesium with a variety of organic ligands have been studied extensively by indirect means.[2] Kinetics of complex formation and dissociation have

* From the Mellon Institute, Carnegie-Mellon University, Pittsburgh, Pennsylvania 15213. This work was supported in part by NSF Grant GP 6122 and used facilities provided by National Institutes of Health Grant FR-00292.

† Present address: Department of Chemistry, Washington State University, Pullman, Washington 99163.

likewise been observed, and these processes are relatively slow, compared, for example to those of calcium ion.[3] The possibility exists of observing complexing of magnesium ion directly by nuclear magnetic resonance line-width measurements on Mg^{25}. In this paper we report on preliminary investigations of the utility of this method.

Uncomplexed metal or halide ions in solution are, on the average, symmetrically hydrated, and the electric field at the nucleus departs little from spherical symmetry. When an agent which complexes with the ion is introduced, the symmetry of the electric field may be reduced. If the nucleus possesses an electric quadrupole moment, the interaction of the moment with the electric field gradient provides a mechanism for relaxation of the nucleus. In the limit of short rotational correlation times, the relaxation time for Zeeman states is given by

$$T_2^{-1} = C(eqQ)^2 \, \tau_r$$

where T_2 is the transverse relaxation time, eqQ is the quadrupole coupling constant in the complex, and τ_r is the rotational correlation time. In addition complexing may cause a change in the magnetic shielding of the nucleus causing a shift in the resonance line position. Broadening by complexing agents was observed in Na^{23} resonance by Wertz and Jardetzky[4,5,6] and these studies have recently been refined and extended by Cope[7] and by James and Noggle.[8]

If a small amount of complexing agent is added to an excess of the ions to be observed, there will, in general, be an exchange of ions between the complexed and uncomplexed states. The exchange may be fast or slow, and the complexed state may or may not be significantly different in resonance frequency from the uncomplexed. In general, if exchange is not too slow, a broadening will be observed in the magnetic resonance signal of the ion. Expressions for the degree of broadening in the four limiting cases have been derived by Swift and Connick[9], and are:

Case A: Chemical shift in complex large compared to line-width, slow exchange

$$\Delta v \approx p_c/\tau_c$$

Case B: Chemical shift in complex large compared to line-width, fast exchange

$$\Delta v \approx p_c\tau_c \, (v_c - v_u)^2$$

Case C: Line width much greater than chemical shift, slow exchange

$$\Delta v \approx p_c/\tau_c$$

Case D: Line width much greater than chemical shift, fast exchange

$$\Delta\nu \approx p_c/T_{2c}$$

In these expressions $\Delta\nu$ is the increase in line-width observed when complexing agent is added, p_c is the fraction of ion present in the complexed state, τ_c is the average residence time of ion in the complexed state, ν_c and ν_u are the resonance frequencies of the ion in the two states, and T_{2c} is the transverse relaxation time of the complexed ion. Stengle and Baldeschwieler[10] have made ingenious application of this phenomenon to the study of biopolymers. They observed that the resonance of Cl$^-$ ions was broadened in the presence of macromolecules to which mercuric ion was bound. Case D (relaxation time in the complex very short, fast exchange) was demonstrated to hold in this case. The observed broadening could be interpreted in terms of number of binding sites and rotational correlation time for the macromolecule. The technique has been extended to several biologically relevant cases[11]. Bryant[12] has recently reported exploratory studies on the use of Ca^{43} magnetic resonance to investigate complexing of Ca^{++} with macromolecules, and has concluded that conditions are favorable for such studies.

Natural magnesium contains 10% of ^{25}Mg, which has a nuclear spin of $\frac{5}{2}$ and thus possesses a quadrupole moment. In this paper we report exploratory measurements of the ^{25}Mg magnetic resonance of magnesium ions in aqueous solution and of the effect of added complexing agents on the line widths

Experimental procedures

The spectrometer used for all magnesium-25 resonance studies has been described previously.[13] The instrument is a Varian DP-60 spectrometer modified for radiofrequency swept experiments. The magnetic field strength was locked to a water sample resonance at 60.000000 MHz and the magnesium-25 resonances were observed at 3.671929 MHz. The radiofrequency was supplied by a General Radio GR-1164 frequency synthesizer. A Princeton Applied Research lock-in amplifier was used with a 1000 Hz audio modulation of the field to stabilize the base line. The frequency synthesizer and the 60 MHz source signal were phase-locked to prevent relative drift of observing and locking frequencies.

2 M magnesium chloride solution was used as the standard for determining line broadening interactions. 5 ml samples were prepared by dissolving the compound of interest in the stock solution. Spectra were recorded in the absorption mode. The line width was measured at half-height, and includes

some contribution from field inhomogeneity (estimated at <2 Hz). At least three spectra were taken of all samples, and more than three were taken when the noise level caused wide variations in the measured line widths. All spectra were obtained at radiofrequency power levels below saturation. Single measurements of line width had a precision of approximately $\pm 5\%$.

The pH measurements were obtained with a Leeds & Northrop Model 7664-Al pH meter equipped with a Sargent S-30070-10 miniature electrode. Measurements were taken directly on the 2 M magnesium chloride solutions and were to ± 0.05 pH units. pH changes in the sample were produced by additions of hydrochloric acid or sodium hydroxide. Sodium hydroxide addition often produced precipitates of magnesium hydroxide, and sample tubes were shaken for several minutes to dissolve the precipitate. The highest workable pH was approximately 8; at higher pH magnesium hydroxide precipitated irreversibly.

Results

With the magnetic field locked on water proton resonances at 60.000000 MHz resonances of magnesium-25 cation were observed at 3.671929 MHz. Preliminary measurements revealed a slight concentration dependence of linewidth. The variation in the 1.0 to 4.0 M concentration range is linear within experimental error, the observed line-width increasing from 5 to 9 Hz. This width (and all subsequently discussed, include some contributions from magnetic field inhomogeneity).

The pH of a 3.0 M magnesium chloride solution was varied from 0.2 to 8.0 by addition of hydrochloric acid or sodium hydroxide. No change in linewidth was detected.

No chemical shift change was detected in either of the above cases or in any of the experiments discussed below. The precision of this measurement was within 3 Hz; smaller shifts would not have been detected. The precision of chemical shift determination was checked by observing the potassium-39 resonance of aqueous potassium chloride. The chemical shift changes with concentration reported by Deverell and Richards[14] were confirmed.

Exploratory experiments were carried out to survey the effect produced on the magnesium-25 resonance by potential ligands. The amount of ligand added was 0.1 to 1.0 g. No line-width change was observed with the following compounds: tetrahydrofuran, 1.2-dimethoxyethane, N,N-diethylaniline, pyridine, ethylene glycol, pyridoxal hydrochloride, phenylalanine, dioxane, ammonium bromide, or sodium dodecyl sulfate. Many other compounds affected the magnesium resonance slightly or not at all. When the compounds added were quite insoluble, a saturated solution was used. At pH less than 7.0, ethylenediaminetetraacetic acid did not change the line width of the magnesium resonance. However, at pH greater than 7.0, broadening of the resonance line did occur indicating complexing of magnesium cation with the tri- or tetra-anion.

Among the compounds initially surveyed, two caused marked line broadening: citric acid and glucose-6-phosphate. The phosphate was chosen because of the possible requirement for phosphate-magnesium chelation in many biological reactions. Subsequently, many phosphate esters were found to broaden the magnesium line. Other chemical species found to influence the magnesium resonance line width were fluoride ion, epinephrine, and sulfhydryl carboxylic acids. The nature of these interactions has not yet been established. Most of the compounds investigated were weak acids. The line width of the magnesium resonance in the presence of these weak acids was a function of the pH. The resonance was therefore observed as a function of pH, with range 1.0 to 8.0.

Carboxylic acids

The behavior of the Mg^{++} resonance in the presence of carboxylic acids as a function of pH may be illustrated using citric acid as an example. A plot of Mg^{++} resonance line width versus pH for a solution 2 M in MgCl$_2$ and 0.14 M in citrate is given in Fig. 1. The appearance of the curve is reminiscent of that of a titration curve. A crude analysis may be made as follows:

Inflection in the curve may be seen at pH 2.0, 2.75 and 3.5. These may be assumed to represent complete formation of the complex between magnesium and citrate mono-, di-, and tri-anion, respectively. If the magnesium forms no complex with unionized citric acid, the complex with the mono-anion is half formed at pH 1.6, judging from the increase in line width. This means that the citrate is present to the extent of ~0.07 M as citric acid and ~0.07 M as complex. From the known[15] first ionization constant of citric acid, the concentration of mono-anion at pH 1.6 is calculated to be 0.0024. Uncomplexed magnesium is present to the extent of $2.00 - 0.07 = 1.93$ M. The formation constant for the complex

$$K_1 = \frac{[Mg^{++}CitH_2^-]}{[Mg^{++}]\,[CitH_2^-]}$$

is thus estimated at 16. Similar treatments yield association constants of

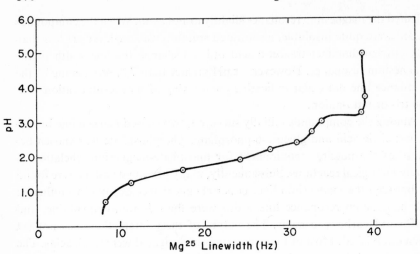

Figure 1   Titration of 2 M MgCl$_2$ containing 0.14 M citrate

4.4×10^3 and 6.6×10^5 for complex formation with the dianion and tri-anion respectively. These values are compatible with the values of 14, 1.1×10^3, and 7.9×10^4 reported for the corresponding complexes with calcium[2], and 25, 1.6×10^2, and 3.3×10^4 reported[17] for the beryllium complexes. This derivation assumes that the observed broadening will be proportional to the amount of complex present, *i.e.* that the broadening will be given by an equation of the form

$$(\nu) = \Sigma\, p_i V_i \qquad\qquad (1)$$

where p_i is the fraction in the ith (free or complexed) form and V_i is an "effective line width". Examination of the expressions given by Swift and Connick reveals that the law will be obeyed exactly if broadening arises from case D (fast exchange, quadrupolar relaxation broadening). In the other cases it will be obeyed only if τ_c, the mean residence time of a Mg nucleus in the complex, is constant under the conditions of the experiment.

Measurements of the kinetics of complex forming reactions by Eigen and others[3,18] suggest that this will in fact be the case. Dissociation of Mg^{++} complexes appear to be essentially a first-order process. In interpreting the results of experiments like the above, we will assume this to be true, and treat observed broadening in terms of fractional complexed species, and effective line broadenings. From the line widths at the points of inflection, it may be deduced that the effective line widths for the complex of Mg^{++} with mono-, di-, and tri-anion are 270, 360, and 450 Hz respectively.

Table 1 displays effective line widths for a number of carboxylic acids, along with the complex formation constants, where known. In all cases the acid should be nearly completely complexed. Temperature dependence studies would be required to determine whether the low effective broadening in, for example, glycine complex is due to slow exchange, accidental low quadrupole coupling constant in the complex, accidental small chemical shift, or some combination of these.

Table 1

Carboxylic acids	V_{eff} (Hz)	log K_1	Ref.
Citric acid (trianion)	490	–	
Malic acid	240	1.55	a, g
Salicylic acid	160	4.70	b
cis-Aconitic acid	150		
trans-Aconitic acid	80	(1.50)	c
Aspartic acid	110	2.43	d
Threonine	80	–	
α-Ketoglutaric acid	90	–	
Succinic acid	50	1.2	e, g
Glycine	25	(1.4)	f
Glycylglycine	20	1.06	g

a   R.K.Cannan and A.Kibrick, *J. Am. Chem. Soc.*, **69**, 2314 (1938).
b   (In 75% dioxane) L.G. van Uitert and W.C.Fernelius, *J. Am. Chem. Soc.* **76**, 375 (1959).
c   Value for Ca^{++} complex. J.Schubert and A.Lindenbaum, *J. Am. Chem. Soc.* **74**, 3529 (1952).
d   R.F.Lamb and A.E.Martell, *J. Phys. Chem.* **57**, 690 (1953).
e   R.K.Cannan and A.Kibrick, *J. Am. Chem. Soc.* **60**, 2314 (1938).
f   (Average value for Ca^{++}) C.W.Davies, *J. Chem. Soc.* 277 (1938); C.W.Davies and C.N.Waind, *J. Chem. Soc.* 301 (1950); C.A.Colman-Porter and C.B.Mock, *J. Chem. Soc.* 4363 (1952), of reference g.
g   C.B.Mock, *Trans. Fara. Soc.* **47**, 292, 297 (1951).

Phosphate and polyphosphate ester

Figures 2 and 3 display the line width dependence on pH for adenosine-monophosphate (AMP) and adenosine triphosphate (ATP) respectively. In the case of AMP there appears to be a single complex forming step. The approximate treatment used in the case of citrate may be expressed concisely as

$$pK_f \approx pH - pK_a - pMg^{++} \qquad (2)$$

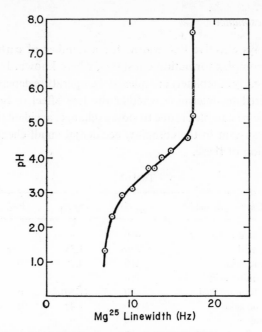

Figure 2 Titration of 2 M MgCl$_2$ containing 0.0367 M AMP

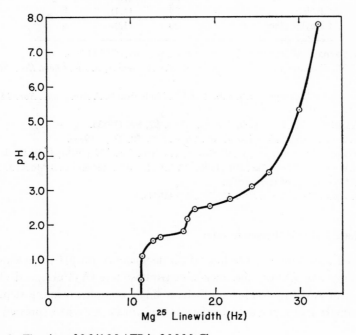

Figure 3 Titration of 0.041 M ATP in 2 M MgCl$_2$

This equation is approximately valid when half of the potential ligand is in the form of Mg complex. Substituting the value of pK_a given by Martell and Schwarzenbach,[19] and estimating the point of half conversion to be at pH 3.75, we obtain $K_f = 130$. Martell and Schwarzenbach obtained the value of 50 for this K_f. 2-Glycerophosphate gave a titration curve very similar to that of AMP, but with pH 3.5 at the half-conversion point. The pK_a given by Clarke *et al.*[20], was 6.6 which leads to a value of K_f of 630. Clarke *et al.*[20] reported a value of 400, in reasonable agreement.

The behavior of ATP is anomalous, in that the line-width continues to increase above pH 4.0, for reasons which are not clear. Taking the values pK_a and pK_f measured by Martell and Schwarzenbach or by Phillips *et al.*[21] one predicts that the ligand will be half converted at pH $\approx$ 2.2, which however appears to be at a point of inflection, where the final complex is only beginning to form. Decreasing the given log K_f by one-half log unit to 3.5 would seem reasonable, and displace the half conversion point to pH 2.75, on a horizontal portion of the curve. The anomalous appearance of the curve, however, suggests that the criterium of constant effective line width may not be met, and caution is indicated. In addition the influence of ionic strength on the magnitude of the equilibrium constants should be very large.[21]

Effective broadenings for a variety of other phosphate esters are listed in Table 2. In the case of the barium and calcium salts competition between the metals for the complex forms is possible.

Table 2

Phosphates	V_{eff} (Hz)
Pyrophosphate	350
o-phosphoserine	810
o-phosphothreonine	840
Calcium phosphocholine bromide	460
Barium ethylphosphate	280
Calcium glycerophosphate	530
Glucose-1-phosphate, dipotassium	640
Glucose 6-phosphate, barium	550
Diethyl phosphate	150
Dimethyl pyrophosphate	320
2-Pyridylmethyl phosphate	360
Adenosine 5'-phosphate	570
Adenosine triphosphate	1300

Epinephrine

The dependence of Mg^{25} line width on pH in the presence of epinephrine is shown in Fig. 4. Complexing is detected with half-conversion at pH ~ 6.3. Since the pK_a of adrenaline hydrochloride can hardly be higher than about 5.0 or 5.5, this suggests that the ionization leading to the effective ligand is not the loss of a proton from the ammonium group, but rather the ionization of a phenolic hydroxyl. Assuming pK_a ~ 10 for this ionization, a formation constant of 2.5×10^3 for the complex is deduced. The effective line width of the complex is 850 Hz.

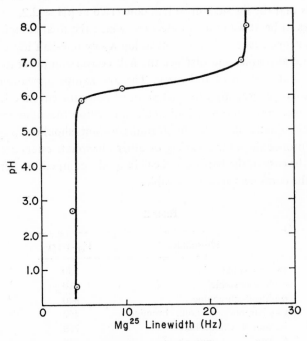

Figure 4   Titration curve for epinephrine in 2 M $MgCl_2$

The association of epinephrine and AMP has been demonstrated by Weiner and Jardetzky[22], using an NMR technique. Since Mg^{++} interacts with AMP, this appeared to be an appropriate system to study for evidence of formation of ternary complexes. At pH 5.0 a mixture of all three gave a Mg line width characteristic of a solution of AMP alone. At pH 7.8 the line width was characteristic of a solution of epinephrine alone. This suggests the formation of ternary complexes exclusively, with an effective line-width equal to that of the complex with epinephrine. More detailed investigation is needed, however, before any firm conclusion can be drawn.

Fluoride ion

Addition of potassium fluoride to magnesium chloride solutions caused line broadening which was not linear in concentration with the fluoride ion. A solution containing magnesium chloride, potassium fluoride, and glucose-1-phosphate yielded line widths in excess of the sum of the broadenings produced by potassium fluoride or glucose-1-phosphate separately.

Conclusions

The usefulness of Mg^{25} resonance line-width measurements to detect complexing of Mg^{++} with organic ligands depends on the system. Although oxalate forms strong[2] complexes with Mg^{++}, no broadening is observed, presumably because of slow exchange. If the dissociation rate of complex is constant under the conditions of the experiment, and broadening is observed, it is possible to deduce values for complex formation constants.

It should be possible to extend, refine, and elaborate these methods in various ways. For example the determination of T_1 and T_2 separately could help cast light on the kinetic processes involved. Competition in complex formation between different metals may be investigated. Enriched Mg^{25} and time-averaging could increase sensitivity to allow observations at biologically endurable concentrations of Mg^{++}. The method appears worthy of further investigation.

Acknowledgements:

We are grateful to Dr. B. L. Shapiro for discussions of this topic and to Dr. R. G. Bryant for kindly allowing us to read his manuscript before publication.

References

1 *Enzymes* by M. C. Dixon and E. C. Webb, Academic Press, New York, 1964, pp. 421–6.
2 *Stability Constants*, J. Bjerrum, G. Schwarzenbach, and L. G. Sillen, Special Publication No. 6, the Chemical Society, London (1957).
3 H. Diebler, M. Eigen and G. G. Hammes, *Z. Naturf. B*, **15**, 554 (1960); D. N. Hague and M. Eigen, *Trans. Faraday Society*, **62**, 1236 (1966); M. Eigen and G. Maass, *Z. Physik. Chem.*, **49**, 163 (1966).

4   J.E.Wertz and O.Jardetzky, *J. Chem. Phys.* **25**, 357 (1956).
5   O.Jardetzky and J.E.Wertz, *Arch. Biochem. Biophys.* **65**, 569 (1956).
6   O.Jardetzky and J.E.Wertz, *J. Am. Chem. Soc.* **82**, 318 (1960).
7   F.W.Cope, *J. Gen. Physiol.* **50**, 1353 (1967).
8   T.L.James and J.H.Noggle, *J. Am. Chem. Soc.* **91**, 3424 (1969); *idem, Proc. Nat. Acad. Sci.* **62**, 644 (1969).
9   T.J.Swift and R.Connick, *J. Chem. Phys.* **37**, 307 (1962).
10 T.R.Stengle and J.D.Baldeschwieler, *Proc. Natl. Acad. Sci.* **55**, 1020 (1966).
11 R.G.Bryant, *J. Am. Chem. Soc.* **91**, 976 (1969); A.G.Marshall, *Biochemistry* **1**, 2450 (1968).
12 R.G.Bryant, *J. Am. Chem. Soc.* **91**, 1870 (1969).
13 D.G.Davis and D.E.Wisnosky, paper presented at the 9th Experimental NMR Conference, 1968, Pittsburgh, Pennsylvania.
14 C.Deverell and R.E.Richards, *Mol. Phys.* **9**, 551 (1966).
15 C.W.Davies and B.E.Hoyle, *J. Chem. Soc.* 4134 (1953).
16 R.G.Bates and G.D.Pinching, *J. Am. Chem. Soc.* **71**, 1274 (1949).
17 R.Danley, J.Feldman and W.F.Neumann, AEC Report, contract W-7401-eng-49.
18 Cf. reviews by N.Sutin, *Ann. Rev. Phys. Chem.* **17**, 119 (1966) and by E.M.Eyring and B.C.Bennion, *Ann. Rev. Phys. Chem.* **19**, 129 (1968).
19 A.E.Martell and G.Schwarzenbach, *Helv. Chim. Acta* **39**, 653 (1956).
20 H.B.Clarke, D.C.Cusworth and J.P.Datta, *Biochem. J.* **58**, 146 (1954).
21 R.C.Phillips, G.Philip and R.J.Rutman, *J. Am. Chem. Soc.* **88**, 2631 (1966).
22 N.Weiner and O.Jardetzky, *Arch. exp. Path. u. Pharm.* **248**, 308 (1964).

HFS-zero-field
magnetic resonance spectroscopy

M. BIRKLE* and G. SCHOFFA

Universität Karlsruhe

THE HYPERFINE STRUCTURE zero field electron magnetic resonance, further on called HSF-zero field resonance, is a special method within the usual electron paramagnetic resonance. In the case of zero-field resonance we have no external magnetic field and therefore we can investigate only substances having a splitting of the spin degenerated electronic levels due to an interaction with the surroundings. In HFS-zero-field resonance this splitting is due to the hyperfine coupling between the unpaired electron spin and nuclear spins. Considering free radicals in organic compounds with $S = \frac{1}{2}$ there can be no spin–spin splitting and in all cases investigated here the orbital angular momentum is quenched by the molecular field. Under these two restrictions all the systems in HFS-zero-field resonance can be described by a spin Hamiltonian of the form

$$\mathscr{H} = \sum_i ST_iI_i$$

S and I_i are the electronic and nuclear spin operators, respectively. T_i is the hyperfine coupling tensor of the nucleus i. Since in zero-field resonance we have no external magnetic field, the Hamiltonian contains no terms involving a preferred axis of quantization, and the energy levels of the systems are independent of its orientation with respect to the laboratory frame. Therefore all the three principal values of the HFS coupling tensors can be deter-

---

* Present adress: Siemens AG, Karlsruhe.

mined by a single spectrum of the polycrystalline substance so that we are able to investigate hyperfine coupling tensors and radiation damages in organic compounds without tedious growing of single crystals. On the other hand there are two essential disadvantages: first, the knowledge of the electronic g-tensor is lost, and second, we have considerable losses in sensitivity. All systems treated here are radicals of the form R_1—$\dot{C}H$—R_2 with different groups R_1 and R_2, they are produced by 1 MeV electron irradiation of the polycristalline substances. The HFS-zero-field spectrum of these radicals is expected to lie in a frequency range below 250 Mc/s and our spectrometer operates in the range from 10 Mc/s to 200 Mc/s. In spite of the fact that one can use larger sample volumes due to smaller dielectric losses one has to expect considerable losses in sensitivity due to the lower frequencies in comparison with usual X-band spectrometers.

The easiest example of HFS-zero field spectra we find at radicals with noninteracting groups R for example in the malonic acid radical or in the barbituric acid radical. In this case of radicals with one proton interaction the solution of the eigenvalue problem leads for anisotropic coupling tensor to four energy levels where any transition is allowed. So we get a six line spectrum in the range between 10 Mc/sec and 80 Mc/sec. Figure 1 shows as an example for one proton interaction the spectrum of the barbiturate acid radical.

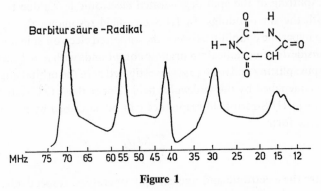

Figure 1

The two lines at lowest frequencies are situated close together and in general they are only poorly resolved. The frequencies of three lines are sufficient to calculate directly the three principal values of the coupling tensor. If the levels are nondegenerated, the radical has no first order magnetic moment and no first order interaction with local fields so that we get linewidths in the order of 1 Mc/sec to 3 Mc/sec corresponding to the calculation of second order perturbation of Cole, Kushida and Heller. This estimation is valid only for nondegenerated states.

In most radicals the unpaired electron interacts with more than one nucleus and often the coupling tensors are diagonal in different systems of coordinates. If we have n protons and m nuclei with Spin equal 1 interacting with the unpaired electron there are $2 \cdot 2^n \cdot 3^m$ zero field levels. Although the number of possible transitions is very large, there are such considerable differences in the intensities between these transitions, that the number of significant lines is relatively small and often spectra are rather well resolved even in regions with a large density of lines. The method of interpreting spectra with a great number of lines or with unresolved bands of lines is the simulation of the spectra with their relative intensities starting with estimated or approximated values of the coupling tensors and then the fitting of the calculated spectra to the experimental ones, varying the estimated tensor values. This procedure is only possible using digital computers. A program for spectra simulation for several nuclei with spin $\frac{1}{2}$ has been published by Lefebvre and Maruani and in cooperation with us Lefebvre and Oliver have developped a method for fitting, including also nuclei with spin 1. These calculations have shown that in first approximation the line frequencies are determined by the values of the coupling tensors whereas the line intensities are very sensitive to the positions of the diagonal systems of the tensors.

If we have only two protons interacting with the unpaired electron the solution of the eigenvalue problem leads to four doubly degenerated levels. We find a six line spectrum too as in the case of one-proton-interaction. As there are only transitions between degenerated levels the line widths are different from spectrum to spectrum. In line width calculation first order perturbation with local fields must be considered too.

The spectra of radicals with three interacting protons show great differences depending on the magnitude of the coupling tensors. An example for a strong coupled proton in α-position and two weak coupled protons will be shown on Fig. 2.

The spectrum of diglycolic acid consists of six broad lines. The coupling tensor determines the position of the broadened lines and in first approximation this tensor can be evaluated in the same manner as in the case of one proton interaction. The couplings of the two protons in γ-position act as a second order perturbation on the interaction between the α-proton and the electron and produce the broadening and a small shift of some lines. The simulation has started with values for the α-coupling tensor evaluated from the experimental spectrum. The coupling of the two protons, presumed equal and isotropic, has been varied between 7.3 and 12 Mc/s. The magnitude of this coupling determines the shape of the broadened lines. Therefore it can be evaluated by fitting the line shapes and then the exact values of the α-coupling tensor can be determined by a similar fitting of the line frequencies.

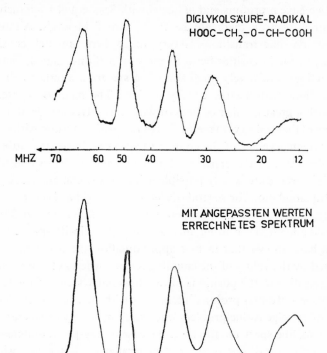

DIGLYKOLSÄURE-RADIKAL
HOOC–CH$_2$–O–CH–COOH

MHZ   70   60   50   40   30   20   12

MIT ANGEPASSTEN WERTEN
ERRECHNETES SPEKTRUM

MHZ   70   60   50   40   30   20   12

Figure 2

The spectrum calculated with the correct values is shown on Fig. 2 and coincides well with the experimental one. If the couplings of the γ-protons are greater than 14 Mc/s some of the broadened bands of lines will be split into resolved lines and if there are coupling tensors of protons in β-position of the same order of magnitude as the α-coupling tensor we find a wide spread spectrum with a great number of lines. As an example of this case Fig. 3 shows the zero field spectrum of succinic acid.

Figure 3 shows the spectrum calculated with high field EPR values and the experimental one. The line positions coincide very well, indicating that the high field values are rather correct. Some differences we find in the intensities

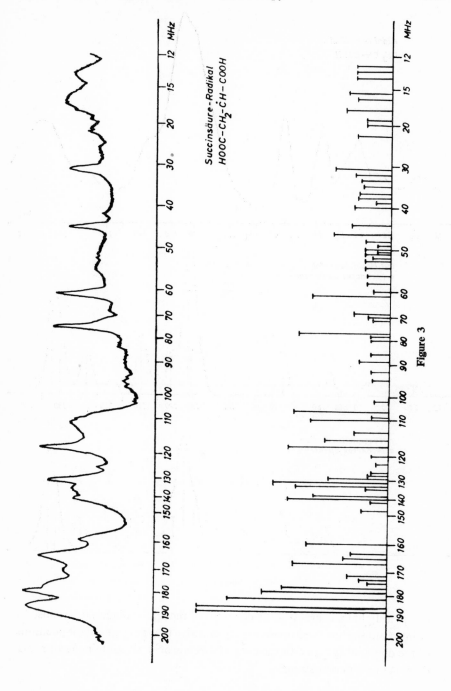

Succinsäure-Radikal
HOOC—CH₂—ĊH—COOH

Figure 3

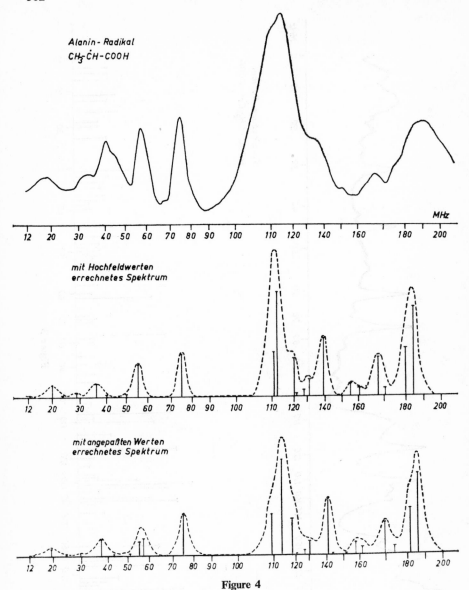

Figure 4

due to a slight inaccuracy of the positions of the tensor diagonal systems. As pointed out the line intensities are very sensitive to differences of the positions of the diagonal systems. In the case of succinic acid all coupling tensors are diagonal in different systems.

Broadened bands of lines we get also from radicals having strong coupled groups which are only slightly anisotropic as in the example of the alanin-radical in Fig. 4. In the calculation it is presumed, that the couplings of the methylprotons are equal and isotropic. The isotropic part of the tensor derived from the fitted spectrum is equal to the value derived from flow-system experiments.

Until now we have considered radicals having only proton interaction. In biological research, compounds having amino-groups are of great interest, and so we have examined a series of amino-acids and related compounds. From high field EPR experiments it is known that irradiated amino-acids often show radicals of the following from:

$$—NH—\dot{C}H—$$

As an example for the zero field spectra of such radical species you see in the upper part of Fig. 5 the zero field bands of the acetylglycine radical and below a fitted spectrum.

Miyagawa, Kurita and Gordy have investigated this radical in high field. In their paper the nitrogen is supposed to broaden the high field lines. There is also a zero-field study by Mangiaracina whose interpretation disagrees with the high field one. He postulated two radicals with slightly different α-proton couplings, while the nitrogen coupling is assumed to be negligible. The zero-field spectrum presented here is somewhat different from that of Mangiaracina and it can be fitted on the bases of a proton and a nitrogen coupling. From high field studies of similar compounds it is known that the nitrogen coupling is one order of magnitude smaller than the α-proton coupling, so we have a situation similar to that of diglycolic acid. The α-coupling can be determined in first approximation from the three highest bands of the spectrum whereas the nitrogen acts as a perturbation producing the splitting of each line into 9 unresolved lines and causing a slight shift of the band center. If we use the program of Lefebvre and Oliver a rough know-ledge of the nitrogen tensor is sufficient for getting exact values of the α-coupling tensor. The Table I shows the results of this procedure from 4 amino acids. The nitrogen couplings are in all four cases approximately equal to the high field values of glycylglycine.

The reliability of this procedure can be illustrated by the following fact: if we replace the nitrogen tensor by an isotropic tensor having the same trace this leads to a change of the α-coupling tensor in the order of $\frac{1}{2}$ Mc/s.

A more complicated example is the creatinine radical. Ueda has shown that there are two nitrogen nuclei and one proton coupled with the electron spin

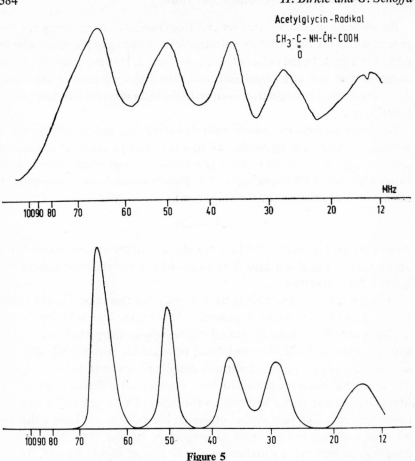

Figure 5

density in the radical. Only the nitrogen splitting at a particular orientation could be determined. The spectrum simulated with the aid of the high field coupling values does not agree with the experimental zero field spectrum. The spectrum of irradiated creatinine is shown in Fig. 6 together with a simulated one, calculated on the following bases: we introduce two nitrogen tensors, that of nitrogen 1 being taken as equal to the nitrogen tensor used for the amino acids and that of nitrogen 2 being twice smaller. This last assumption is suggested by a Hückel calculation performed by Ueda.

With these assumptions about the nitrogen couplings one can determine the α-proton coupling tensor, aiming at best fitting. The calculated spectrum gives good coincidence with the experimental one. It even shows the hump which is characteristic of the band at highest frequency.

Table I Principal values[a] (in Mc/s) of the α-proton coupling tensor

	From high-field spectra			From zero-field spectra before correction[b]			From zero-field spectra after correction[c]		
	T_{xx}	T_{yy}	T_{zz}	T_{xx}	T_{yy}	T_{zz}	T_{xx}	T_{yy}	T_z
Glycylglycine	−46.7	−82.5	−24.1	−51.7	−79.6	−22.7	−48.8	−76.7	−24.2
Carbamylglycine	−61.5	−73.0	−28.0	−48.0	−78.0	−20.0	−45.4	−74.1	−22.6
Glycocyamine	−	−	−	−48.0	−80.4	−21.6	−45.1	−77.6	−23.1
Acetylglycine	−47.3	−75.1	−27.8	−51.1	−78.0	−21.7	−47.0	−76.3	−23.0

[a] Signs chosen from the theoretical arguments.   [b] ±1 Mc/sec.   [c] ±1 Mc/s.

Zero-field resonance studies can present an alternative route towards the parameters of the Spin-Hamiltonian. However, except in the case of radicals with exceptionally simple structure the zero-field spectrum consists of a very large number of lines, with intensities and positions which cannot be expressed in a simple analytical form. If the Spin-Hamiltonian is known from

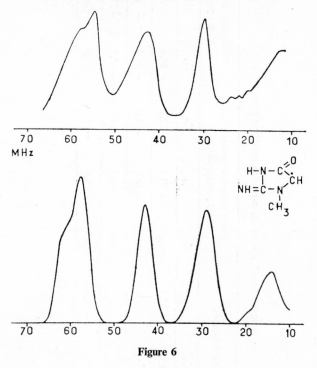

Figure 6

high field experiments, or if we can get a rough knowledge about the structure of the radical from zero-field spectrum it is possible to simulate such a spectrum. A fitting procedure using a computer-program then leads to the correct values. Even in cases of a great number of different couplings giving a wide spread spectrum with a great line density the agreement of the spectrum calculated by high-field values with the observed spectrum is a very good check for the correctness of the high-field values, because the zero-field spectrum is about one order of magnitude more sensitive to deviations of the hyperfine-Spin-Hamiltonian than high-field spectra. If we have a rough knowledge about the radical and if we are interested in accurate values of the coupling tensors then zero-field resonance is an adequate method, without growing and alignment of single crystals.

Publications about HFZ-zero-field resonance

1 McConnell, H.M., Thompson, D.D., and Fessenden, R.W., *Proc. Nat. Acad. Sci.* **45**, p. 1600–1601, 1959.

2 Cole, T., Kushida, T., and Heller, H.C., *J. Chem. Phys.* **38**, p. 2915–2923, 1963.

2 Heller, H.C., *J. Chem. Phys.* **42**, p. 2611–2612, 1965.

4 Lefebvre, R., and Maruani, J., *J. Chem. Phys.* **42**, p. 1496–1502, 1965.

5 Mangiaracina, R.S., *Rad. Res.* **26**, p. 343–352, 1965.

6 Birkle, M., and Schoffa, G., *J. Chem. Phys.* **49**, p. 3191–3193, 1968.

7 Birkle, M., and Schoffa, G., *Zeitschrift für Naturforschung* **23a**, S. 918–926, 1968.

8 Birkle, M., and Schoffa, G., *Zeitschrift für Naturforschung* **24a**, S. 1093–1103, 1968.

9 Geiger, A., Birkle, M., Oliver, G., and Lefebvre, R., *Mol. Phys.* **17**, No. 1, p. 101–103, 1969.

10 Birkle, M., Schoffa, G., Oliver, G., and Lefebvre, R., will be published in *J. Magn. Resonance*, 1969.

Simplified x-band EPR spectroscopy at liquid helium temperatures

ARTHUR S. BRILL, CHARLES P. SCHOLES*
and CHIN I. SHYR

*Department of Materials Science, University of Virginia,
Charlottesville, Virginia 22901, U.S.A.*

Abstract

A small liquid helium double dewar system for routine use in conjunction with the Varian V-4531 multi-purpose x-band room temperature cavity is described. This low cost assembly is compatible with 100 KC modulation, does not adversely affect the cavity Q, and hence permits operation at maximum sensitivity. The use of the system for orientation studies with single crystals and for measurements on frozen solutions and polycrystalline samples is briefly discussed and illustrated.

IT HAS BEEN KNOWN for some time that effective investigation of hemes and hemeproteins by electron paramagnetic resonance (EPR) requires that the measurements be made at cryogenic temperatures (1, 2). Furthermore the resolution of ligand hyperfine structure can only be achieved when the natural line width and the modulation field do not exceed the splittings. The latter have been observed to be as small as 3 oersteds for high spin iron in heme (3, 4), a figure which translates into operation at liquid hydrogen and helium temperatures.

The development of the apparatus described here started with recognition that the reentrant dewar employed by Professor Robert G. Wheeler's group

---

* Present address: The Clarendon Laboratory, Oxford, England.

(Yale University) for low temperature optical spectroscopy could be modified for use in EPR measurements. We have recently learned of the use of a reentrant dewar for liquid helium EPR in a non-biological study (5). The performance of our version, Fig. 1, demonstrates that routine use of liquid helium is economically practical. The dewar consists of a silvered pyrex two-

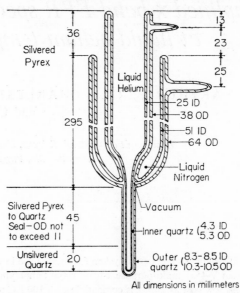

Figure 1   Liquid helium dewar. Dimensions given for the diameters must be maintained. Other dimensions are nominal. Tapers, curvatures, and corners are depicted schematically

refrigerant section graded to an unsilvered clear fused quartz (Amersil, Hillside, New Jersey) single-refrigerant finger. There is but one continuous vitreous surface and a single continuous evacuated volume. (The lower port is for re-evacuation and the extra port at the top serves in the silvering procedure.) Penetration of the liquid nitrogen shield close to the head of the finger is an important aid in keeping down the rate of helium "boil-off", and the skill of Mr. Edward Brosious (glassblower, Yale University) is a major factor in the successful fabrication of this device. Because of the fragility of the pyrex shell which carries liquid nitrogen in toward the cold finger, care must be taken to dry the enclosed volume before reusing the dewar. A small amount of water frozen in this region can crack the glass.

The assembly (designed by Mr. Nathan Mandell, supervisor of the Biology Machine Sh op, Yale University) shown schematically in Fig. 2 permits the dewar (with sample fixed inside) to be rotated about its vertical axis for single crystal orientation studies. The dewar is secured inside a brass cylinder by means of two sets of three nylon-tipped screws. The brass cylinder is supported by and rotates within two aluminum bearings which are ultimately fastened to an aluminum frame fixed to the magnet yoke. The nylon-tipped screws are adjusted to align the dewar so that the eccentricity of the finger does not exceed 0.2 mm as the brass cylinder is rotated. (This adjustment is carried out with the assembly mounted in a jig equipped with an indicating pointer pivoted so as to amplify the eccentricity ten times.) The average spacing of 0.4–0.5 mm between the finger and the inner wall of the Varian V-4531 X-band cavity stack is then sufficient clearance. The nylon-tipped

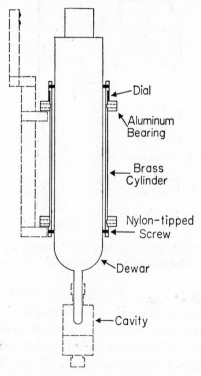

Figure 2 Assembly for holding and rotating dewar. Part of a supporting frame, fixed to the magnet, is indicated by dashed lines, as is the Varian V-4531 multi-purpose cavity

screws also permit vertical adjustment of the dewar to bring the inner tip of the finger, and hence the sample, into the most sensitive region of the cavity. A pointer fixed to the upper aluminum bearing is read against a dial, calibrated in degrees, attached to the top of the brass cylinder.

Shown in Fig. 3 is an assembly consisting of a teflon cap to cover the liquid helium in the inner dewar section and a long tube to introduce the sample into the finger. Triangular teflon "spiders" serve to guide the tube during insertion and then, together with the cap, keep it in place. Into the end of the tube is inserted the rod of a sample mount (which can be of several kinds, Figs. 4A,

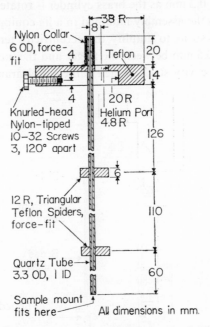

Figure 3 Dewar cap-sample tube assembly. Sample mount rod (Figure 4) is reversibly cemented into the end of the tube with "HISTOWAX"

B, C). Rod of 1 mm D was used (rather than simply an extension of the 3 mm OD tube) in the finger to allow free flow of liquid helium down to the tip. A small amount of Histowax is melted into the tube-rod joint by means of a hot wire. Crystals which do not suffer from exposure to air and refrigerant have been mounted as shown in Figs. 4A and 4B. The crystals were secured to the flat fused quartz surfaces with Apiezon L grease (James G. Biddle Co., Plymouth Meeting, Pennsylvania), and their orientation determined by x-ray diffraction. Protein crystals can be seated in embossed wells as described elsewhere (6). The mount shown in Fig. 4C has been used for polycrystalline and frozen solution samples.

Initially the assembly shown in Fig. 2 is removed from the magnet and both the inner and outer refrigerant spaces are filled with liquid nitrogen.

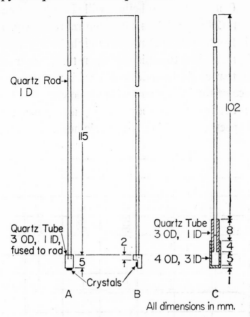

Figure 4   Sample mounts. (A) and (B): for single crystals which can be exposed to air and refrigerant. In A the supporting surface, is ground perpendicular to the rotation axis and in B, parallel to the rotation axis. (C): for polycrystalline (or frozen solution) samples and for single crystals which suffer from drying and/or direct exposure to refrigerant. The quartz parts are reversibly fastened together with "HISTOWAX" or heavy silicone grease, small quartz "stops" determining the depths of insertion automatically

When the dewar has cooled, all refrigerant is poured out, the outer space refilled with liquid nitrogen and the inner one, with liquid helium. The latter operation can be done with the sample (precooled) in place in the finger, or the samples can be introduced directly through the liquid helium at a later time. The dewar-brass cylinder-aluminum bearings assembly is then secured to the frame on the magnet, the finger having been carefully guided into the cavity. Tuning and 100 kc operation are carried out in the usual way. The dewar assembly can be removed from the magnet in order to replenish liquid helium, or refrigerant can be transferred to the dewar in situ by locating the microwave bridge to the rear of the cavity rather than directly over it (and the dewar). The latter method is a convenience in single crystal orientation studies requiring a large number of long sweeps. However it should be emphasized that the dewar assembly can be removed, refilled, and replaced quickly, and that we have done this several hundred times without breaking a dewar finger.

A full charge (160 cm³) of liquid helium will give 60 minutes of operation. The liquid nitrogen (280 cm³) boils off much more slowly. Helium is conserved if the level of outer refrigerant is kept high. We have not attempted to "pump" on the liquid helium to achieve temperatures closer to absolute zero, but the present system could be modified for this purpose.

The performance of the instrument with single crystals of perylene doped with heme can be seen in figures accompanying an earlier publication (3). In Fig. 5 are shown spectra from high spin ferric heme in bacterial catalase in

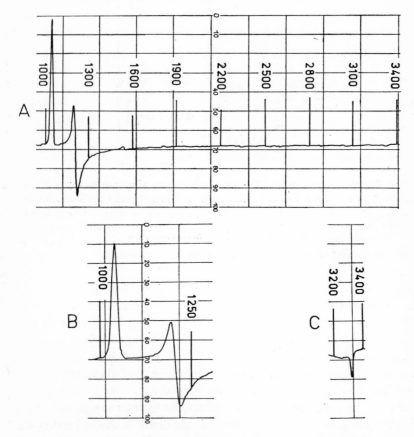

Figure 5   Micrococcus lysodeikticus catalase, aqueous-acetone precipitate, 1.8×10^{16} ferric ions. First derivative presentation. Power at sample: 150 mW. Frequency: 9.245 GHz. (A) Entire spectrum. Modulation amplitude: 0.3 oe. Sweep rate: 250 oe/min. Response: 0.3 sec. (B) Low field spectrum. Modulation amplitude: 0.3 oe. Sweep rate: 100 oe/min. Response: 1.0 sec. (C) High field spectrum. Modulation amplitude: 3 oe. Sweep rate: 250 oe/min. Response: 0.3 sec.

the form of a frozen precipitate at liquid helium temperature. The entire spectrum is shown in 5 A to demonstrate the stability of the system. The high field derivative band at 3300–3350 oersteds is just detectable in this record, this resonance being easily saturated at liquid helium temperature. Indeed all three resonances show saturation at the power level of Fig. 5 and continue to do so until the power is reduced more than two orders of magnitude. An expanded presentation of the low field spectrum is given in 5B to demonstrate the signal-to-noise ratio which can be achieved at a moderate sweep rate. In 5C the high field spectrum is taken under the conditions as in 5A but for a ten-fold increase in modulation amplitude to bring out the high field derivative band. The latter is seen along with a much broader underlying band which arises from the dewar finger. Apparently the silvering process leaves a paramagnetic deposit upon the (inner) quartz surfaces. The weak spurious signal from this film can be accounted for by taking a spectrum from the dewar alone, but at the power levels normally used is of negligible amplitude.

Acknowledgements

The U.S. Public Health Service supported this work through research grants GM-09256 and GM-16504 from the National Institute of General Medical Sciences.

C.P. Scholes was a Predoctoral Research Fellow of the U.S. Public Health Service during a period of this work.

References

1 Ingram, D.J.E., and Bennett, J.E., *J. Chem. Phys.* **22**,1136 (1954); *Disc. Faraday Soc.* **19**, 140 (1955).
2 Bennett, J.E., Ingram, D.J.E., George, P., and Griffith, J.S., *Nature* **176**, 394 (1955).
3 Scholes, C.P., *Proc. Natl. Acad. Sci. (U.S.A.)* **62**, 428 (1969).
4 Scholes, C.P., Ph.D. Thesis, Yale University (1969).
5 Horning, A.W., and Hyde, J.S., *Mol. Physics* **6**, 33 (1963).
6 Brill, A.S., and Venable, J.H., Jr., in *Magnetic Resonance in Biological Systems*, A.Ehrenberg, B.G.Malmström, T.Vänngård, Eds. (Pergamon Press, London, 1967), p. 365.

Some developments in ESR spectroscopy

CAFIERO FRANCONI*

Molecular Spectroscopy Laboratory, University of Cagliari, Italy

---

ALMOST 25 YEARS have now passed since Zavoisky's discovery of the electron paramagnetic resonance in condensed matter[1], however, little significant improvement has occurred in the field of ESR spectrometry after that brought about by the design of the first ESR bridge systems in which magic tees or circulators were employed. It may be reasonably affirmed that there is little room for further significant improvement since the main limitations are imposed by the adoption of the detection technique which uses an hybrid Tee (or circulator) associated with reflex cavities, although an improvement of the individual electronic and microwave components will certainly occur.

The actual ESR spectrometers of customary design do not allow stable, practical monitoring of the ESR line at high power irradiation and even at low microwave power their baseline stability is not good enough for performing a practical integration of the signal. Further sample removal usually causes a transitory unbalance of the system, and critical cavity readjustments are always necessary at any sample removal or substitution.

ESR signals are in fact usually obtained by a sample material containing the unpaired electrons inserted in a microwave resonant cavity placed within the magnet air gap. The commercially available ESR spectrometers consist of an hybrid tee (or circulator) system designed to measure the small changes in the complex magnetic susceptibility of the sample at resonance, detected as

---

* Present address: Dept. of Physical Chemistry, Faculty of Industrial Chemistry, University of Venice, Italy.

change in the complex reflection coefficient of the resonant cavity. On the other hand, changes of the same order of magnitude—or bigger—than the signal itself, also take place following both mechanical deformations of the cavity and displacements of the sample inside it which modify the resonance condition of the cavity itself. Also a deformation of the coupling iris and eventually of the slide screw tuner, modify the standing wave ratio and therefore the microwave power sent back to the detector and superimposed to the signal itself. All these effects add up to form the large spectrum baseline fluctuations actually observed in the commercial ESR spectrometers.

A novel family of ESR spectrometers capable of monitoring the ESR line during high power irradiation, as well as for ESR low power routine work—sophisticated far beyond the possibilities of the reflex cavity ESR spectrometers—has been developed in our laboratory and given fundamental tests. Their principle of operation can be explained by the Bloch phenomenological treatment of spin induction[2].

Although the application of this principle in the NMR field[3] has been quite popular among the NMR spectrometer manufacturers, only a few papers on the development of microwave induction ESR spectrometers have been published[4-7] during the last twenty years, although we know that this problem has also been explored in other laboratories[8]. From these papers one sees that none of the efforts made in this direction seem to have been very successful in producing a practical inductor for routine ESR spectrometry.

As an example, Portis et al.[4,5] described a cylindrical bimodal cavity in which the H_1 microwave field excites two perpendicular degenerate modes of the cylindrical cavity (TE_{111} modes), and in which the signal picked up by the receiver is due to a rotation of the planes of both modes brought about by the resonating sample exhibiting Faraday rotation. The isolation between klystron and detector arms results quite critical also in absence of resonance, since four capacitive tuning plugs and two resistive plugs are employed to adjust the coupling between the input and output waveguides to a minimum. Complete re-tuning of the cavity and decoupling operations are necessary after each sample removal or insertion. As these operations are not independent, the setting of the system is quite annoying and represents in this respect no improvement over the setting of ESR absorption spectrometers using a reflex cavity. Other authors have not improved this apparatus[6].

A different approach has been attempted by Brodwin and Burgess[7] who built a microwave spin induction spectrometer using an H-plane microwave T-junction. Actually, the drawback of this induction system lies in its principle of operation, since to obtain zero e.m. fields at the detector waveguide the superposition of two polarized vectors of high intensity and of opposite phase is made. Therefore this system substantially reproduces the hybrid tee situation of an ordinary ESR spectrometer working with a reflex cavity.

Reflex cavity spectrometers using a circulator instead of an hybrid tee also exhibit similar drawbacks.

The electron spin inductors which we designed are made up of bimodal cavities which allow the oscillation of two uncoupled degenerate modes perpendicular to each other. A cavity of square or rectangular cross section working in TE_{01n} and TE_{10n} degenerate modes proved most suitable for this purpose.

The principle of operation of these cavities may be illustrated with the aid of Fig. 1. One of the two degenerate modes (TE_{012}) is properly excited by the klystron microwave power carried by waveguide **Ia** through iris I' and sets

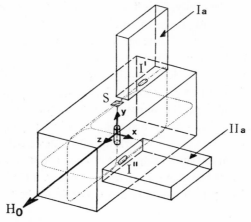

Figure 1 Working principle of a bimodal ESR inductor of square cross section

up the exciting microwave magnetic field, the spin sample being placed along its force lines. Under resonance conditions, the energy irradiated by the spin sample and polarized at right angle, excites the second degenerate mode (TE_{102}). The microwave energy with this polarization is picked up by waveguide **IIa** coupled to the cavity by the iris I'' and conveyed to the detector.

As the two modes are coupled in principle only through the electron spins at resonance, this system might be considered an ideal microwave analog of the crossed coils spin inductor[3]. It is, however, to be observed that a cavity of this design achieves a much higher isolation than a crossed coil inductor, in virtue of the straightness and perpendicularity between magnetic (and electric) field lines of the two modes throughout the bimodal volume.

The simplest cavity consists of a central body of square cross section, working in both the TE_{012} and TE_{102} modes, just as the one shown in Fig. 1. Cavities with cross section of different geometry might be used; however, conditions for a high isolation between modes and proper positioning of the irises are difficult to find.

A different type of microwave inductor may be designed in which two waveguides crossing each other at their narrower side are joined together and the region in which the magnetic force lines of the two modes—each coming from the respective waveguide—are superimposed at right angle, is the area on which the sample should lie. Cylindrical inductors working with two perpendicular TE_{112} (or TE_{11n}) modes could also be used.

In all the bimodal cavities designed the detector receives in principle only the electron spin induction signal polarized as the second mode. Therefore the signal is not affected by a possible fluctuation of the reflection coefficient of the first mode of the cavity. This fluctuation might in fact only modify the value of the exciting field H_1, as in an NMR crossed coil inductor with the transmitter coil fluctuating off tuning. Any change of the tuning conditions and of the Q factor of the second mode—the two modes of such a cavity being tuned separately and independently—brings only a change of the signal level and phase, just as in a crossed coil NMR system in which the receiver coil is undergoing analogous changes of the Q factor and of tuning conditions. Therefore the ESR induction system is intrinsically much more stable than the ESR absorption systems working with reflex cavities.

The main problem in the development of such a system is, therefore, the achievement of a very low microwave leakage between the klystron and the detector waveguides, which would in any case add to the ESR signal at the detector.

Several causes contribute to this leakage. The most important ones are: (a) effects of the input iris; (b) effects of the output iris and (c) accidental couplings between modes brought about by sample positioning, cavity imperfections and deformations.

As to cause (a), there is an accidental excitation of the second mode brought about by the presence of the input iris in the bimodal volume. It would therefore be advisable to include, in the design of a bimodal cavity, one unimodal section working in the TE_{01n} mode for H_1, to substitute for the input iris. With the use of such a unimodal section, the input iris is actually recessed into this section, so that the depolarizing effect on the exciting mode brought about by the input iris is greatly reduced. A detailed description of a bimodal cavity coupled to the klystron waveguide through a unimodal section will be

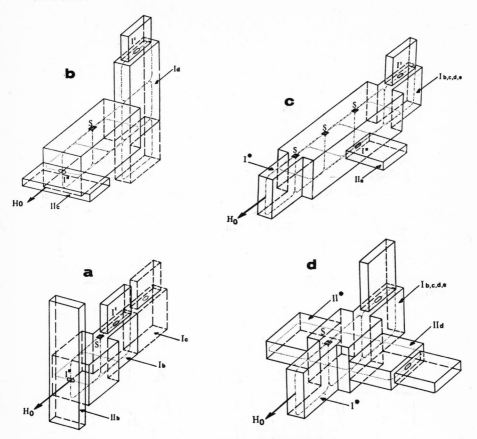

Figure 2 Various configurations of uni/bimodal ESR inductors

given elsewhere[9,10]. In Fig. 2 some possible coupling arrangements are show shown. In Fig. 2a we see resonating unimodal sections **Ib** or **Ic** working in the TE_{011} or TE_{012} mode, coupled to a bimodal section one half wavelength long. The output waveguide **IIb** is coupled through iris I'.

In Fig. 2b we see a resonating unimodal section **Id** working in the TE_{012} or TE_{013} mode coupled to a bimodal section two half wavelengths long.

In Fig. 2c a bimodal section three half wavelengths long, which may accomodate two or more samples, for superimposed resonance experiments, is shown coupled symmetrically to two unimodal sections working in the TE_{011} mode, one of which carries the input iris I', the other being irisless. The presence of two symmetric unimodal sections might actually make the oscillation of the TE_{015} mode more symmetric.

A different cavity configuration, including a few unimodal sections coupled to the degenerate body for mode No 1 and also for mode No 2 is shown
in Fig. 2d. Its principle of operation is self-explanatory.

We have built and tested three bimodal cavities. The results obtained with
the first two are reported in ref. (9) and they are of considerable practical
interest. From these tests it may be said that the addition of a unimodal section to the bimodal body allows for an easier independent tuning of the
first mode, performed outside the bimodal volume. This contributes towards
building a flexible system which is not very critical in its adjustments. In fact
the only plug left in the bimodal section is now the tuning paddle of the second mode, which, although it gives little trouble, could be easily recessed
into a second unimodal section. In Fig. 3 some experimental results obtained with the inductor of Figs. 1a and 2 of ref. (9) are presented showing
the intrinsic baseline stability of these inductors.

In order to obtain a more stable electron spin inductor with an even lower
leakage we have to take particular care of the problems connected with the
output iris. The latter can be positioned—like the input iris—in all the three
perpendicular faces of a rectangular cross section bimodal cavity[8]; however,

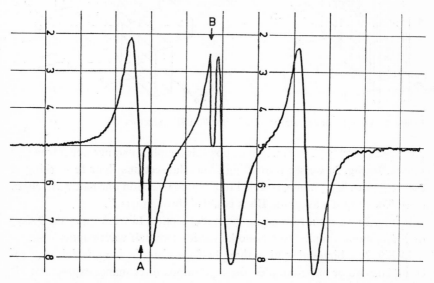

Figure 3 ESR spectrum of a nitroxide radical (2,2,6,6-tetramethyl 1-Aza cyclohexan
1-oxide) at a concentration of 10^{-2} M in Dioxane, taken with the cavity described in
ref. (8). The sample was placed in a 3 mm O.D. cylindrical quartz tube. The receiver was
a Decca Radar Ltd. microwave superhet receiver. The sample tube has been manually
removed from the cavity and put back twice at the A and B points of the spectrum

in all the cases it is difficult to cancel completely a small leakage signal, coming mainly from the geometry of the iris and from a mechanical instability of the unimodal–bimodal system with associated waveguides. In fact it is possible for an output iris placed in the bimodal volume to be excited by the current lines of the first mode if the H_1 field is not perpendicular to the length of the iris itself. An excited output iris actually generates microwaves polarized as the induced spin signal, which are conveyed through the detector arm superimposed to the ESR signal itself.

As a matter of fact mechanical strains do cause H_1 field distortions inside the bimodal cavity so that the H_1 force lines will not be any more perpendicular to the length of the output iris. Thus the output iris itself generates a leakage voltage into the detector arm via the mechanism described above. In order to avoid this kind of leakage a balanced output may be designed in such a way that whilst the induction H_2 field presents itself at the two output irises with the same (or opposite) phase, the leakage signals generated at both irises will have, instead, opposite (or same) phase, so that with the aid of a T-junction on the E or H planes, one can easily add vectorially the induction signals whilst substracting vectorially to zero the leakage signal. We report below some of the results obtained with a bimodal cavity equipped with such an output system.

Figure 4 shows an example of how this can be accomplished in a bimodal cavity, using a couple of slots on opposite faces as output irises and a T-junction on the H plane (which sums vectorially at the output the microwave intensities coming from both arms). It is easily seen that whilst the instantane-

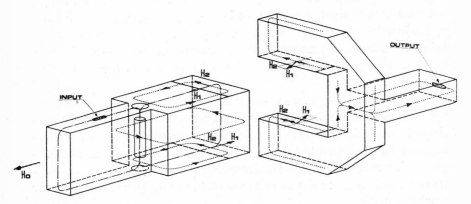

Figure 4 Configuration of a uni/bimodal ESR inductor of square cross section with a balanced output system associated to a T-junction on the H-plane. Instantaneous phase vectors of H_1 and H_2 fields at both output irises are shown

ous phases of the H_2 field at both slots (represented by the vectors $\mathbf{H_2}$) are such that they coherently add themselves in the T-junction, the instantaneous phases of the H_1 field at both slots (represented by vectors $\mathbf{H_1}$) give rise to leakage vectors of the opposite phase when for instance the H_1 plane is distorted about a vertical axis, thus subtracting themselves at the junction output. If a small attenuator and a phase shifter are inserted in one input arm of the T-junction, the output leakage in absence of resonance may be balanced to zero. In this way we obtain a very stable decoupled system which can withstand mechanical – thermal deformations without exhibiting a leakage above the noise level of a superhet receiver under 1 mW of microwave power excitation, equivalent to a stable decoupling of about 90 dB. Pairs of output slots other than these may be selected, according to the H_1 field distortions anticipated, in order to obtain the cancellation of the unwanted leakage.

As to accidental couplings in the bimodal body one may observe that field distortions due to either cavity imperfections or samples of irregular shape may be taken care of by an appropriate choice of a pair of symmetric irises for a balanced output. For extreme cases one can always add a side arm to the cavity to balance out a fixed leakage due to some specific effect.

A bimodal cavity with balanced output has been built and satisfactorily tested in our laboratory also under light mechanical deformations of both sample and cavity. The induced mode of this cavity has been made aperiodic by widening the two output irises to the junction. There are obvious advantages in making nonresonating the oscillation of this mode; the relative limitations are just being analyzed.

A cavity in which the induced mode is resonant throughout the bimodal volume and both separate arms of the junction, together with the third arm of the junction at the end of which the output iris is placed—the third arm being provided with a symmetrical tuning screw for this mode—has also been tested satisfactorily for zero leakage under 1 mW microwave power excitation (Fig. 4).

Thus a balanced output microwave inductor seems to be the right answer for building a highly stable practical ESR induction spectrometer for a variety of uses so far forbidden to reflex cavity spectrometers. In fact the exceptional baseline stability actually makes possible single and double integrations of the spectrum for an easier spectrum presentation and for quantitative analysis work, or else a direct presentation of the absorption signal for lineshape studies.

Further the removal and insertion of a sample does not necessarily bring a readjustment of the cavity tuning and matching, as in the reflex cavity ESR spectrometers. Further, there is no need to lock the microwave oscillator frequency to the cavity resonance frequency by an AFC control, since cavity detuning of the first mode does not affect the leakage and therefore the baseline. However, a frequency stabilized microwave source is needed. Moreover the tuning of both exciting and induced modes are independent to a first approximation; a detuning of the first mode of the cavity would affect only the exciting field intensity; a detuning of the second mode of the cavity would affect—apart from a phase shift—only the signal-to-noise ratio of the receiver.

This type of ESR spin induction spectrometer is also very well suited for high power operation and therefore for monitoring the degree of saturation of ESR line and also for ESR spin echo work. Further there is the possibility of using flat large and thin quartz cells, placed in the nodal *E*-plane of both modes, hosting large water containing samples, making this bimodal cavity, together the above, highly suitable for ESR studies of most biological samples.

The author wishes to thank Mr. P. Galuppi for his contribution to thy design and development of the cavities. This work has been supported be the NATO Scientific Affairs Division Grant n° 294.

Bibliography

1 E. Zavoisky, *J. Phys. USSR*, **9**, 211 (1945).
2 F. Bloch, *Phys. Rev.* **70**, 460 (1946).
3 F. Bloch, W. W. Hansen and M. E. Packard, *Phys. Rev.* **70**, 474 (1946).
4 A. M. Portis and D. Teaney, *J. Appl. Phys.* **29**, 1602 (1958) and *Phys. Rev.* **116**, 838 (1959).
5 A. Teaney, M. P. Klein and A. M. Portis, *The Rev. Sci. Instr.* **32**, 721 (1961).
6 G. Raoult, H. Chandezon, M. T. Chenon, A. M. Duclaud and A. M. Perrin, *Proceed. 12 Coll. Ampere*, Bordeaux, 1963.
7 M. E. Brodwin and T. J. Burgess, *IEE Trans. Instr. Meas.* **IM-12**, 7 (1963).
8 J. S. Hyde, J. C. W. Chien and J. H. Freed, *J. Chem. Phys.* **48**, 4211 (1968).
9 C. Franconi, *The Rev. Sci. Instr.* **41**, 148 (1970).
10 NATO Grant No. 294, Progress Report, 1970.

Subject Index